AF708683

The Passionate Triangle

THE PASSIONATE Triangle

Rebecca Zorach

University of Chicago Press
Chicago and London

REBECCA ZORACH is associate professor in the Department of Art History at the University of Chicago. She is the author of *Blood, Milk, Ink, Gold: Abundance and Excess in the French Renaissance*, also published by the University of Chicago Press, and coeditor of *The Idol in the Age of Art: Objects, Devotions, and the Early Modern World*.

The University of Chicago Press, Chicago 60637
The University of Chicago Press, Ltd., London
© 2011 by The University of Chicago
All rights reserved. Published 2011.
Printed in the United States of America

20 19 18 17 16 15 14 13 12 11   1 2 3 4 5

ISBN-13: 978-0-226-98939-6 (cloth)
ISBN-10: 0-226-98939-9 (cloth)

Library of Congress Cataloging-in-Publication Data

Zorach, Rebecca, 1969–
The passionate triangle / Rebecca Zorach.
p. cm.
Includes bibliographical references and index.
ISBN-13: 978-0-226-98939-6 (cloth : alk. paper)
ISBN-10: 0-226-98939-9 (cloth : alk. paper) 1. Triangle in art. 2. Triangle. I. Title.
N8253.T75Z673 2011
701′.8—dc22

2010052678

∞ This paper meets the requirements of ANSI/NISO Z39.48-1992 (Permanence of Paper).

Contents

Illustrations

PLATES

The color gallery can be found following p. 50.

FIGURES

Acknowledgments

Like any book, this one is a group effort, and I have benefited from the great wisdom of many colleagues and students. Whatever virtues this book possesses are due largely to the many, while its defects derive only from me.

For intellectual companionship, tough questions, and excellent suggestions, I especially want to thank Anne Dunlop, Megan Holmes, and Peter Parshall, who each read a chapter and provided crucial critique and suggestions, and Stephen Campbell, Michael Cole, Michael Gaudio, Patricia Simons, and Claudia Swan, who contributed in many different ways, as did Chicago colleagues Persis Berlekamp and Aden Kumler, and the fabulous group of graduate students in two seminars on Neoplatonism and other classes. I'm grateful for the supportive comments and insightful critiques of Jill Burke and Brian Curran, who were kind enough to reveal themselves to me as reviewers for the press.

For invitations to give a lecture, discuss this research in a seminar, participate in a panel, or contribute to a volume, I am grateful to Lauren Berlant, James Elkins and Robert Williams, Michael

Gaudio, Joseph Koerner, Allison Levy, Maria Loh, Walter Melion, Hans van Miegroet, Rose Marie San Juan, David Sedley, Claudia Swan, Ulrich Pfisterer, Lee Palmer Wandel, and Gerhard Wolf, and to graduate students at Johns Hopkins, Harvard, the Institute of Fine Arts, and Northwestern. Each of these opportunities provided me with crucial feedback and new ideas. Giovanni Careri's invitation to give three seminars at the École des Hautes Études en Sciences Sociales provided wonderful opportunities for intellectual exchange and convinced me that I needed to break from my original three-chapter structure. The volcanic ash cloud that blanketed Europe in April 2010 left me stranded in Germany but presented me with the opportunity for a quick trip to Dresden and hence several late-breaking and useful insights. Many thanks are owed to Ulrich Pfisterer for providing me with a refuge in Munich while I waited out the cloud. Further thanks for hospitality are owed to Sarah Cree, Lia Markey, and Leah Price.

Jeffrey Hamburger, Herb Kessler, Stephanie Leitch, and Luigi Sperti each provided me with a triangle that became key to the book. I wish I could have found a way to incorporate those shared by David Dabbs and Chris Faraone. Comments, questions, and various forms of help and support came from Michael I. Allen, Giovanni Careri, T. J. Clark, Tom Conley, Elizabeth Cropper, Susan Dackerman, Whitney Davis, Charles Dempsey, Heather Dubrow, Colin Eisler, Frank Fehrenbach, Paul Gehl, Christopher Heuer, Suzanne Karr-Schmidt, Joseph Koerner, Maria Loh, Hank Millon, Edward Muir, Alexander Nagel, Richard Neer, Robert Nelson, Will Noel, Christina Normore, Katy Park, Ulrich Pfisterer, Debra Pincus, Verity Platt, Rose Marie San Juan, JB Shank, Larry Silver, Patricia Simons, Irene Small, Carl Strehlke, Claudia Swan, Ben Tilghman, Jonathan Unglaub, Will West, Martha Wolff, Nino Zchomelidse, and Henri Zerner.

I have learned much from participation in two groups. Feel Tank made me think about emotion in ways that may not be immediately obvious in this book, but they're there. The Interdisciplinary Working Group on the Medieval and Early Modern World was a crucial support and provided extensive feedback on early draft chapters: thanks go to Daisy Delogu, Nicole Lassahn, Lucy Pick, Elizabeth Rodini, and Lisa Voigt.

A fellowship at the Center for Advanced Study in the Visual Arts provided the chance to engage in substantial new research and finally to begin to pull the disparate strands of this project together. I thank the deans and staff members for making my research there so pleasant and productive. Earlier fellowships at the Penn Humanities Forum, Houghton Library, and Franke Institute for the Humanities helped advance the research. Two deans of the humanities at the University of Chicago, Danielle Allen and Martha Roth, have generously fostered this project, as have two department chairs, Marty Ward and Joel Snyder.

Particular thanks are due to the staff of the Special Collections Research Center at the University of Chicago Library, the Library at the National Gallery of Art, the Houghton Library, the Walters Art Museum, the Newberry Library, the Minneapolis Institute of Arts, the Staatliche Graphische Sammlung in Munich, the Gabinetto Disegni e Stampe at the Uffizi, and the British Museum's Department of Prints and Drawings. Among individuals at these institutions I am especially indebted to Lamia Doumato, Yuri Long, Mark McDonald, and Alice Schreyer.

Writing that made its way into this book appeared first in a special double issue of *Critical Inquiry* edited by Lauren Berlant, and, in Italian, in the volume *Sesso nel Rinascimento*, edited by Allison Levy. My thanks to the editors and publishers for the opportunity to work out these ideas.

I am indebted to research assistants Kate Aguirre, Irene Backus, Kathleen Mills-Curran, Rainbow Porthé, Erika Tanacs, and Kelli Wood. Alexandra Korey provided indispensable help in tracking down and securing images and permissions in Italy.

At the University of Chicago Press, Susan Bielstein encouraged me to believe in this project and helped me imagine its contours; Anthony Burton shepherded the manuscript with great patience and good humor, and Ruth Goring copyedited it masterfully.

A few special notes of thanks are in order. In the fall of 1993, Linda Seidel showed a slide of Tintoretto's *Vulcan, Venus, and Mars* in a lively class on Ovid, transformation, and violence. Little did she know that she was converting me from a medievalist into a Renaissance scholar. In this book I both return to my medieval roots and explore ideas that first began to percolate in that class. Linda has fostered my work in every possible way as teacher, colleague, and friend, and I am so very grateful.

My triangles are not as belligerent as They Might Be Giants' Triangle Man, antihero of the song "Particle Man," but I'm sure the song's wit and impeccable Neoplatonism lie behind this project somehow. More than that, though, I know this book would not have been written had my parents not given me Clifton Fadiman's *Fantasia Mathematica* and *Mathematical Magpie* at an early age, nurturing an early interest in math that I once thought—wrongly!—I had left behind. I thank them and my whole extended family, including the cats.

This book could not have been written without the keen intelligence and generous, loving companionship of Michael W. Phillips Jr.

1

The Passionate Triangle

As deluged as we may be with digital images, art historians stubbornly maintain that nothing replaces acquaintance with original works of art. For some it may be a matter of an irreducible spiritual experience. It may be the conviction that we can't grasp the labor that goes into the making of an object when we receive it as pixels on a screen. Sometimes this view is simply a utilitarian one: you can't wholly trust reproductions when trying to identify basic visual elements of a large and detailed painting. This is partly, but not only, a matter of scale. Photography is not purely objective, and a trick of light, the texture of painted surface, or imperfect color calibration—all can render a significant detail illegible.

So at the tail end of an all-too-businesslike trip to Europe, with just two days in Munich, I made my way to the Old Masters picture gallery, the Alte Pinakothek, with a bit of sleuthing to do. My goals were modest: I wanted simply to figure out what was going on in the upper right-hand corner of Tintoretto's *Vulcan, Venus, and Mars* (plate 1), a dramatic Renaissance depiction of the most famous love triangle in Greco-Roman mythology.

As it happens, my pilgrimage was rewarded. What in reproductions had been an unidentifiable dark dome turned out to be the cavernous opening of a fireplace; what had seemed like steps leading up to it were actually the bricks of its base. This matters to the painting's meaning: as Vulcan, the blacksmith god, bursts in, seeking signs of his wife's infidelity, he enters from the upper right—direct from his labors. Seen in this way, the room pulsates with parallel heat: heat from the lovers and heat from the fire of the forge. Tintoretto insists on the point that the artisan's place of work is contiguous with the place of his betrayal. A small point, but one of many components that taken together form the texture of art-historical scholarship.

But there was more to this visit. Wandering about the museum's Italian Renaissance wing, I was surprised by a voice: "Typische Dreieckskomposizion."[1] "Typical triangular composition": the words that startled me could have come from the ghost of Heinrich Wölfflin, the turn-of-the-twentieth-century art historian who strongly associated triangular composition with the High Renaissance. Wölfflin trained in Munich and knew this collection well; it provided him with many of his main examples. The voice came, however, from one of the earnest, Birkenstock-shod teenagers standing next to me in front of Raphael's Canigiani *Holy Family* (plate 2) and corroborated my sense that the idea of triangular composition still had currency in textbooks and survey classes. I paused: perhaps my new test subjects would expound further. But they moved on. The phrase was the end of description, not the beginning of it. And indeed, this is the problem with the thing we call "triangular composition." Along with "perspective," it has come to constitute a stereotyped view of Renaissance image conventions. An observation that stands in for argument, it seems more obvious than it is.

Viewing the Italian Renaissance paintings in the museum together, as a group, I could also think about the particular view of the Renaissance that they present. This collection that furnished Wölfflin with so many important examples also happens to include many examples of geometric, and particularly *triangular*, arrangements in its Renaissance paintings. Collected largely in the mid-nineteenth century, the Italian paintings gathered in Munich reflect choices made by German aristocrats in the age of Hegelian idealism and of the Nazarenes—artists who sought to respiritualize painting by taking inspiration from Italian Renaissance art, a process later echoed in art history by Wölfflin and in art by his contemporary Kandinsky.

You may now be asking, "Love triangles and triangular composition: what does one really have to do with the other?" In fact, I continue to ask myself this question: it is one of the central questions of this book. The relationship between abstract mathematical figures and their concrete instantiations was also a pressing question for Renaissance philosophy, and the triangle served philosophers in the Renaissance as *the* example. But this is more than a difference between abstract and concrete. My investigation at the Alte Pinakothek dealt with a love triangle, a metaphor playing out over a pronounced perspectival geometry, full of meaning and pathos. My other topic, triangular composition, seems now a rather sterile visual device, a trick for art history exams. The fact that the students saw fit to declare the composition triangular but had nothing more to say about it was telling. Geometry in art has been either severed from content or connected to it in the most suspect and overdetermined ways. But geometry in the Renaissance, indeed the *triangle* in the Renaissance, was connected to stories, myths, myriad practical uses—and yes, emotions, devotions, desires, and passions.

Pure Form

Sixteenth- and seventeenth-century writers like Lomazzo and Félibien discussed the triangle in painting as a technical precept. In the late nineteenth and early twentieth century Heinrich Wölfflin established it as a pedagogical and hermeneutic principle. In 1971 Ernst Gombrich—rather roughly—characterized Wölfflin's precepts thus: "Where you see a Madonna, I see an equilateral triangle, and that is what you ought to see, or attend to."[2] Gombrich caricatures, but the triangle does haunt Wölfflin's discussion of Leonardo and Raphael. Particularly in their iterations of the Christian Holy Family, Wölfflin finds a decided triangular or pyramidal form. He identifies it with the High Renaissance—"classic art"—and with idealism, seeing it as part of a reaction against and refinement of the (admittedly appealing) naturalism of the Quattrocento.

Art historians ever since have considered triangular composition a hallmark of the Renaissance. A typical and rather judicious example is Harald Keller's account: "Beginning with the first stirrings of the Renaissance, articulation of surfaces, especially when achieved by means of human figures, was effected by applying simple formulas derived from everyday geometry. Among those formulas, none was used more extensively than was the triangle."[3] Textbooks tell us that the triangle reflects High Renaissance ideals and interest in geometry and that it provides stability, hierarchy, coherence, or focus.

As a gimmick of the introductory survey lecture, the triangle is simple and identifiable. It is also, significantly, something one can *say* about visual form, independently of content; without a crutch, even professional art historians can be inarticulate about form. The triangle fits well with the writing exercises we give beginning students—"formal analysis" papers in which we ask them to set aside any ideas they may have about a picture's content and instead to focus on form: visual properties like line, color, and texture. With Renaissance art, it is justly difficult for students to set aside subject matter in describing what they see. To ask them to do so is a deeply, and ideologically, *modernist* exercise. Modernism, particularly though not only in the tradition of Clement Greenberg's criticism, asks artists (and their viewers) to sacrifice content to form—indeed, to rid art of subject matter entirely.[4] Most Renaissance works are not, however, amenable to an argument that completely disregards content. They were made to communicate subject matter, to make arguments, to impress with their lavish materials, to prompt offerings and meditation.

In this last aspect, Renaissance art possesses a surprising parallel with modernism. Formal analysis may be part of a modernist colonization of earlier periods of European art, but the earlier periods have their sly revenge: this ascetic operation of denying oneself "content" is typical of devotional exercises practiced by medieval and Renaissance believers. One might even say that the focus on form is itself a spiritual practice, as Wölfflin's contemporary Wassily Kandinsky, one of the founders of modern art, could tell us. Kandinsky—like Wölfflin resident in Munich in the 1890s—derived the principle of the triangle from past art in his effort to respiritualize form. (He even used some of the same triangular examples from art history as Wölfflin had, and referred to Wölfflin in his own teaching.[5]) In the pages that follow I will argue that Renaissance philosophers and artists indeed imagined the possibility of progressing, by a process of abstraction from the visual, toward pure mathematical form and finally contemplation of the divine—via the triangle. And yet the triangle was not a simple, pure, celestial form, for it was equally grounded in mundane things. The equilateral triangle based on the ground suggests transit between earth and sky. The human condition as understood in the Renaissance was to be in between: aspiring to spirit yet insistently dragged back to the world.

The compositional triangle begets other triangles. At times art historians have been fond of tracing lines—perspectival and compositional—on images as a tool for interpretation. But once we begin looking for triangles and tracing them on images, they seem so ubiquitous as to become meaningless. How to think about them within the context of art? We might start by jettisoning the phrase "triangular composition" and think instead in terms of shapes and arrangements. (We can allow the phrase back into our discussion where it really makes sense.) While it's often possible to identify shapes and arrangements convincingly, the notion of triangular composition relies on questionable assumptions about the way painters "mapped" images. Did complicated arrangements of invisible lines really underlie images? Existing preparatory drawings don't really support this idea. Many have criticized the idea that paintings possess complicated and secret geometries.[6] The fact that geometries can be traced on a picture does not mean they were part of the work's conception; in fact it might be better to consider these geometries as works of art in and of themselves, inspired by the original works they purport to diagram.

But the plotting of a real or symbolic geometry of the picture plane is not an impossibly strange idea, since astronomers constructed constellations connecting points in the sky, surveyors imagined diagonals when making measurements, and navigators plotted points on maps. Artists did plot the geometries of bodies and buildings, dividing them into shapes. Geometric order was imposed on images to depict conceptual relationships, and invisible lines were made visible in diagrammatic drawings and in paintings that use lines of force to map affinity, dependence, and desire.[7]

Significant geometries, simple or more complex, do manifestly appear in works of art. We can "diagnose" a symbolic triangle formed by the wings of two doves in a Ferrarese manuscript (fig. 1.1)[8]—they form a perfectly equilateral "Alpha" adorned with a musical scroll that forms an "Omega"—or a triangular trinity of rosebushes on a Belgian tapestry (fig. 1.2). As early as the ninth century, a crucifixion mural in Trier was composed with (still discernible) snapping lines that determined a triangular composition with Christ's heart as its apex.[9] Describing the composition, Warren Sanderson notes that "this equilateral triangle was located so precisely at the center of the lunette it must have served more than just practical compositional purposes."[10] The triangular arms and torso of Christ in Leonardo's *Last Supper* provide a good example of an "unnecessary" triangle: the shape is not required by the subject matter.[11] Is it, then, a satisfying decorative motif that just happens to hint at the Trinity? The Virgin Mary's body produces an equilateral triangle in Bellini's *Madonna of the Meadow* (plate 3) and in Andrea del Sarto's *Madonna of the Sack* in the Santissima Annunziata in Florence (fig. 1.3); Garofalo's *Saint Jerome in Meditation* is similarly triangular (fig. 1.4)

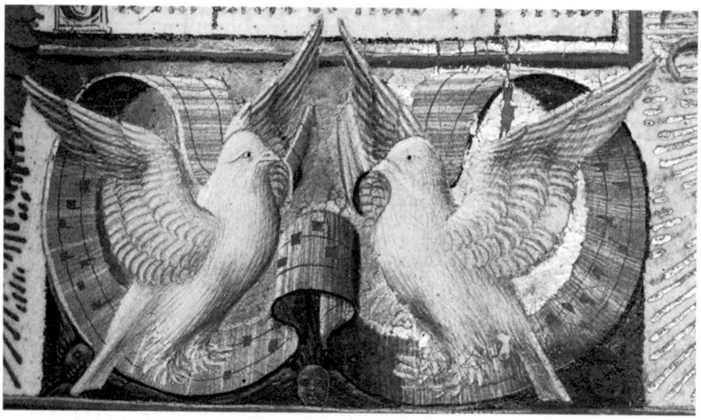

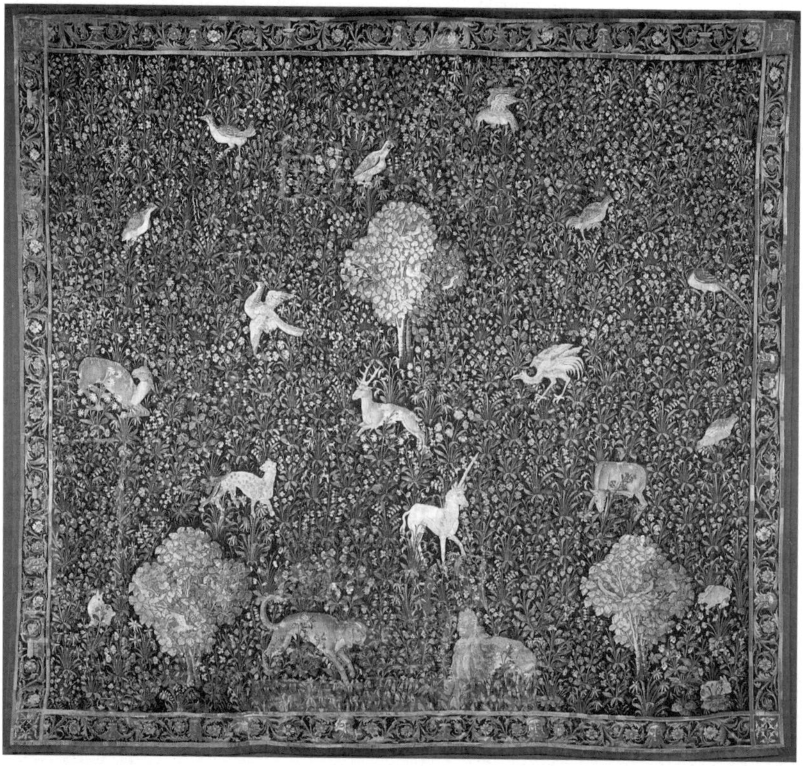

1.1 Taddeo Crivelli, initial *D* with doves, "Domine labia mea aperies . . ." (Matins, Hours of the Cross), *Book of Hours*, ca. 1456–66. Illuminated manuscript. The Pierpont Morgan Library, New York. MS M.227 f. 92r. Photograph: Graham Haber, 2010. By permission.

1.2 Anonymous Bruges tapestry workshop, *Millefleurs with Rosebushes*, c. 1530–45. Wool silk, 352.7 × 400.7 cm (irregular). Minneapolis Institute of Arts. Gift of Mrs. C. J. Martin in memory of Charles Jairus Martin. By permission.

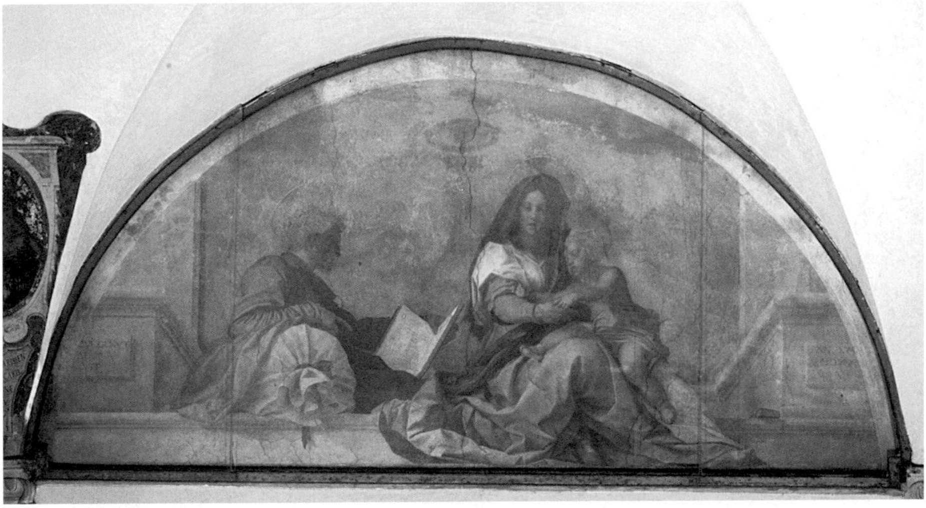

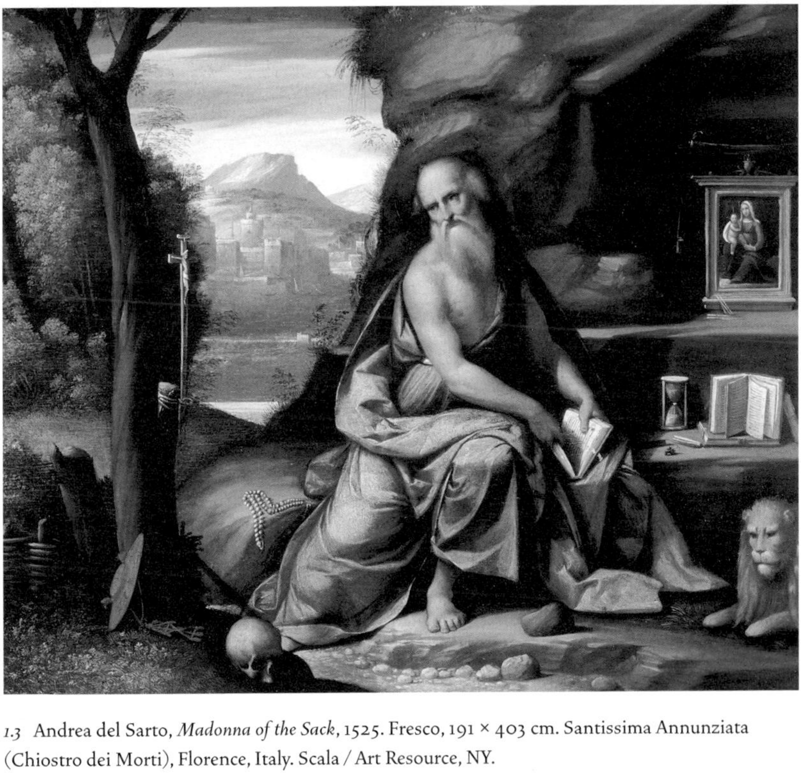

1.3 Andrea del Sarto, *Madonna of the Sack*, 1525. Fresco, 191 × 403 cm. Santissima Annunziata (Chiostro dei Morti), Florence, Italy. Scala / Art Resource, NY.

1.4 Garofalo (Benvenuto Tisi), *Meditation of Saint Jerome*, ca. 1520–25. Tempera on wood, 51.1 × 60.3 cm. New Orleans Museum of Art: The Samuel H. Kress Collection. By permission.

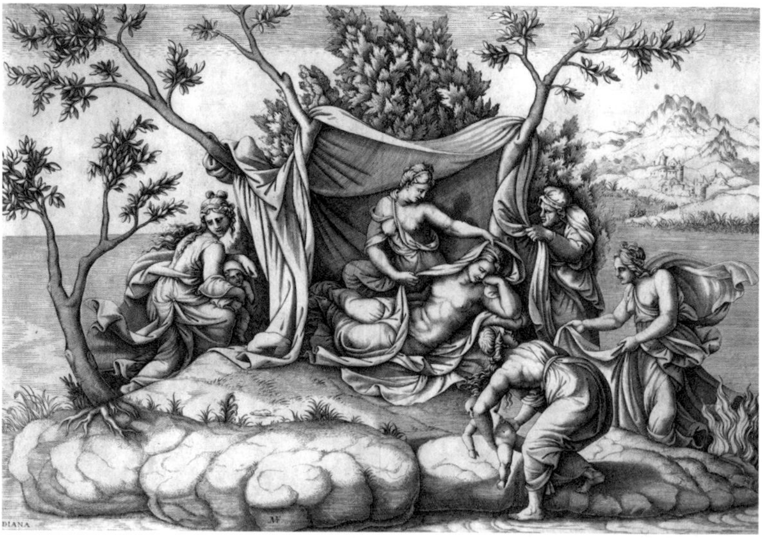
DIANA

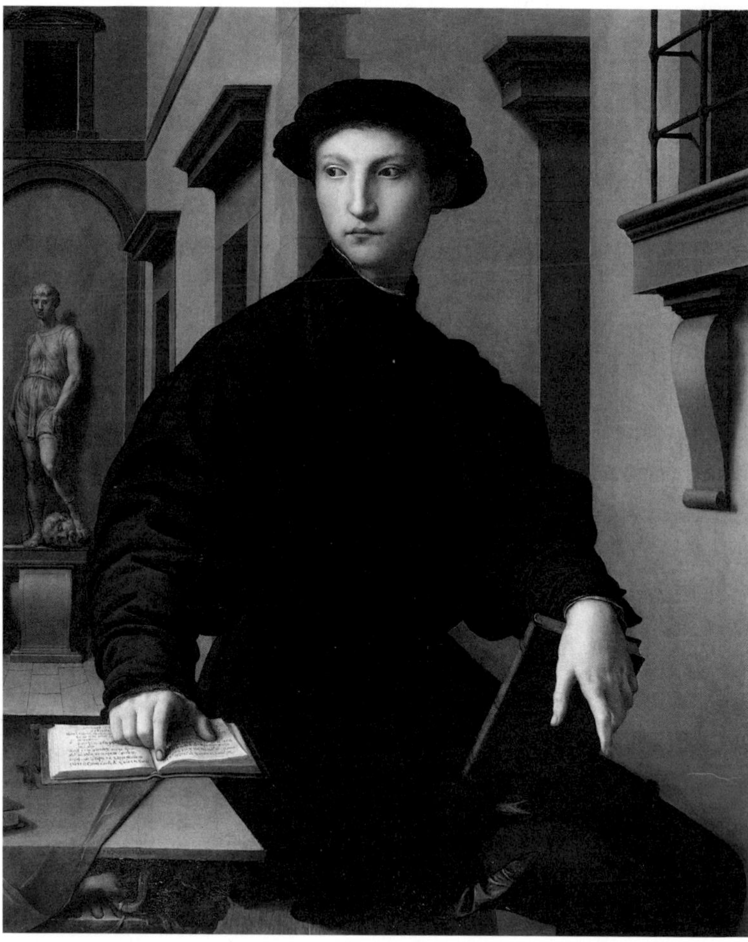

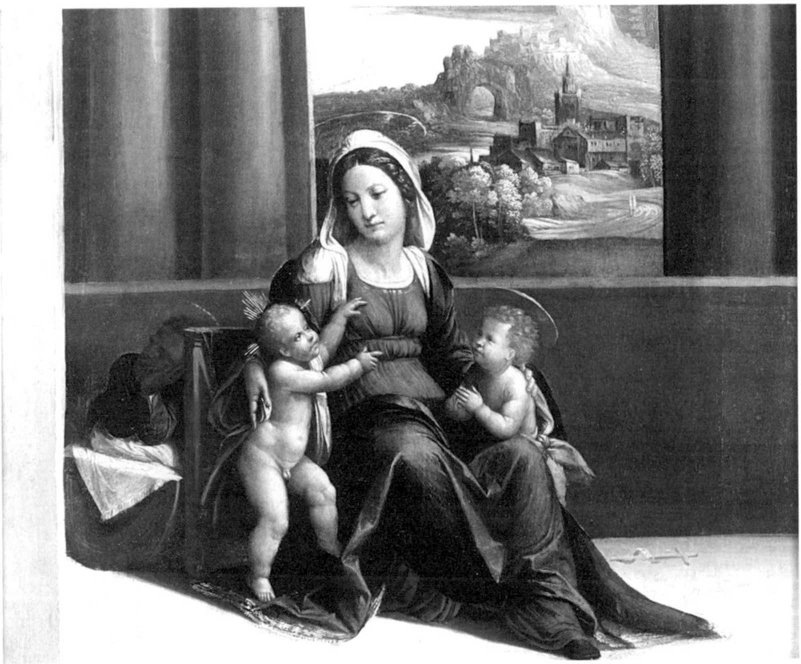

1.5 (opposite top) Diana Mantuana, *Latona Giving Birth to Apollo and Diana on the Island of Delos* (after Giulio Romano), ca. 1580. Engraving, 25.8 × 37.6 cm. © Trustees of the British Museum.

1.6 (opposite bottom) Agnolo Bronzino, *Portrait of Ugolino Martelli*, 1540. Oil on wood, 102 × 85 cm. Gemäldegalerie, Staatliche Museen zu Berlin, Inv. 338 A. Bildarchiv Preussischer Kulturbesitz / Art Resource, NY. Photo: Jörg P. Anders.

1.7 (top) Garofalo (Benvenuto Tisi), workshop. *Madonna with Infant Christ and John the Baptist*, ca. 1525. Oil on wood panel. Pinacoteca Capitolina, Rome. By permission.

to a degree that can hardly be accidental. The outline of a group of figures might follow an unnecessarily triangular shape—as in Raphael's *Holy Family* or Diana Mantuana's print depicting Latona giving birth to Diana and Apollo on the floating island of Delos (fig. 1.5). Supporting figures in a triad group might lean in to suggest the lines of the triangle strongly with their bodies, as in the Quattrocento paintings we will see in chapter 4. Unnecessary triangles also include the nearly equilateral one formed by the corner of the table in Bronzino's portrait of *Ugolino Martelli* (fig. 1.6), the bright triangle of light on the musical score in the foreground of Carpaccio's *Vision of Saint Augustine* (plate 4), or the folded cloth in front of Joseph in a *Holy Family* by Garofalo (fig 1.7). Perhaps we can see such triangles as "ensigns," following a suggestion by Hubert Damisch: "an iconic element which without performing any syntactical or constructive function, broad-

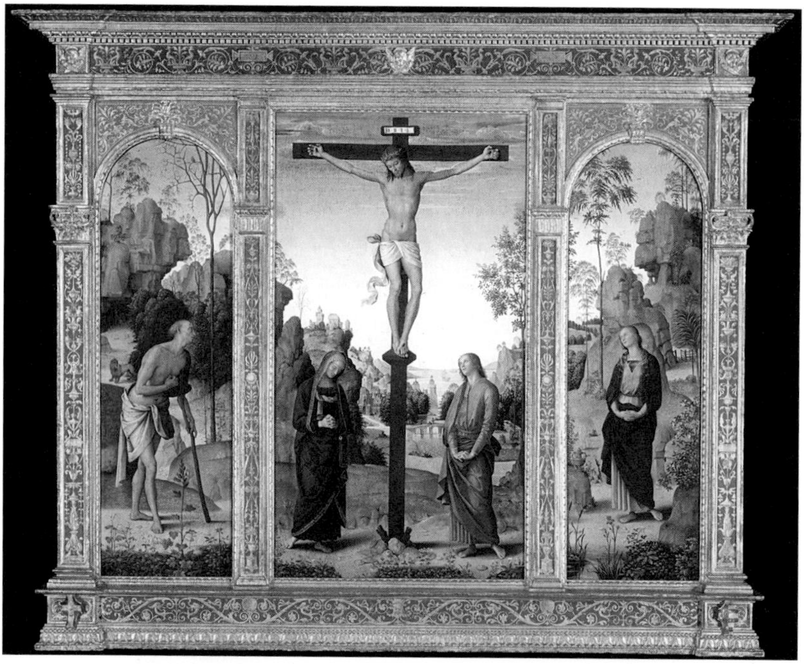

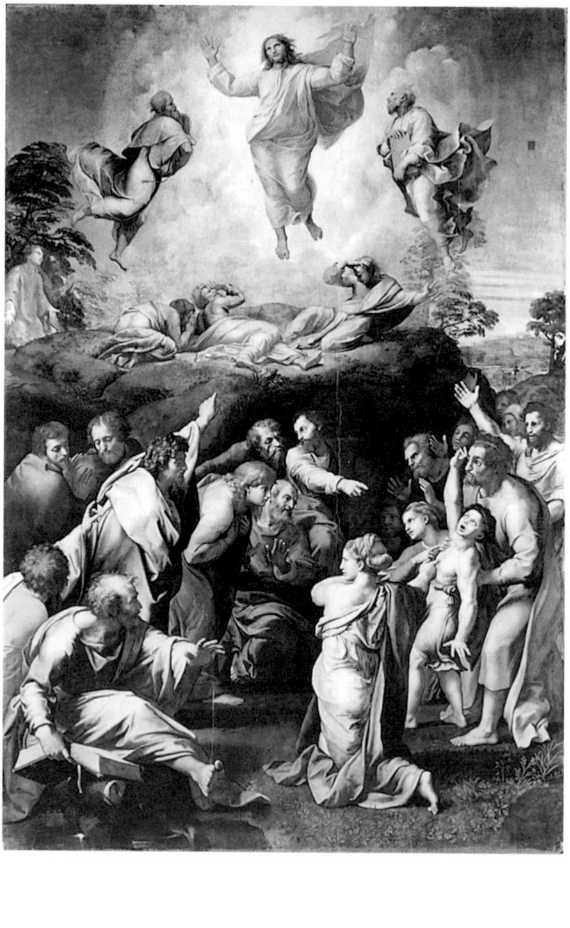

casts, as it were, its operations by miming them" and thus "proclaims and reflects the work of the painting."[12]

A problem arises when in order to reveal a triangle we must select subjective points and plot the shape as if on a graph. How does one choose which points count and which don't? John Canaday argues that Perugino's triptych of the *Crucifixion with Saints* (fig. 1.8) in the National Gallery in Washington is triangularly composed; Canaday plots points and draws lines between them to make two intersecting triangles. Yet Raphael's *Transfiguration* (fig. 1.9), he asserts, can't be mapped the same way: "An attempt to diagram the composition would result in an incomprehensible and irrational web of lines stretching arbitrarily from point to point."[13] This happens because he is trying to diagram *too much* in the composition. Given the angle of the hovering prophets, the *Transfiguration* can easily be mapped (should one wish to do so) as a blank triangle revealing Christ inside a circle—heavenly shapes mounted atop a roiling earthly cube. The *Transfiguration* doesn't fit Canaday's definition for two reasons: the triangle is not comprehensive—it applies only to part of the painting—and the painting itself, taken in its entirety, veers away from the High Renaissance purity and simplicity that Canaday associates with triangles.

To establish ways of talking about triangular arrangement, I ask, "Triangular in relation to what?" and "How does one triangle differ from another?" In this book, I work with constellations of related images, learned and popular, that "think." They define, construct, and express philosophical ideas—not only visual ones. Sometimes this thinking can be documented in textual evidence, and sometimes it can't; images must then constitute their own form of evidence.

To assert that the triangle *matters* in its own right, as opposed to being simply a feature of composition, might seem to promise only limited consequences for how we understand the meaning and significance of images. It lacks the chronological linearity of an artist's or patron's life story, or the material or institutional consistency of an object, the temporal and spatial stability of particular places and periods. As an abstraction, it is neither

1.8 Pietro Perugino, *Crucifixion with the Virgin, Saint John, Saint Jerome, and Saint Mary Magdalene*, ca. 1482–85. Oil on panel transferred to canvas. Center panel 101.3 × 56.5 cm; side panels 95.2 × 30.5 cm. Andrew W. Mellon Collection, National Gallery of Art, Washington. Image courtesy of the Board of Trustees, National Gallery of Art, Washington.

1.9 Raphael (Raffaello Sanzio), *Transfiguration*, 1516. Oil on wood. 405 × 278 cm. Vatican Museum, The Vatican. Photo Direzione dei Musei Vaticani.

topic, nor genre, nor exactly technique. The triangle is a kind of emblem of geometry and related philosophical questions, and thus can be used as an interpretive tool for getting at a certain set of issues. It is also a surprisingly "real" topic. It can also help us escape the frames we traditionally use to study Renaissance art—biography, nation, genre, theme. These frames are useful but they are conditioned by modern institutions; the triangle, as a tool, helps us open up questions from a different point of view.

Medieval and Renaissance Triangles

Once one enters the theoretical world of the Renaissance triangle, things get complicated quickly. For philosophers the triangle is a topic, an example, even a primal site of origin; for theologians it is a conscious and overt trinitarian symbol; for perspectivalists it is the basis of vision; for artists it is a convention for organizing a picture plane. For many of those who thought about it, it was a way of mapping relationships. From the early fifteenth century to the mid-sixteenth century, interests in geometry and in ancient philosophers (Plato, Pythagoras, and various followers) converged with new theological ideas, as in the work of the fifteenth-century German philosopher and cardinal Nicholas of Cusa. Medieval diagrams played a role in medieval scientific culture (constructing diagrams was part of training in geometry), though the results were variable in quality.[14] At the same time geometric diagrams played an exegetical role in theology. Michael Evans argues that diagrams were "the way medieval thinkers approached some of the problems most important to them"—tools for conveying theological ideas through the "logic and exactitude of technical illustrations."[15]

Late in the fifteenth century new printing techniques began to reproduce diagrams with a high degree of accuracy. The first printed edition of Euclid was produced in 1482 in Venice by Erhard Ratdolt, who used shaped thin strips of metal (probably what we call printer's rule) to create the hundreds of diagrams the text required.[16] This was a watershed moment for the history of science and of the visual representation of knowledge, and surely had an effect on the symbolic as well as technical uses of geometry over the decades that followed. Interest in mathematics may also have been inspired by the semi-millennium itself; the work of Dürer and Leonardo particularly exemplifies this. Luca Pacioli wrote in his *Summa* that "Holy Theology cannot reach our intellect well without mathematics."[17] The triangle in particular appeared in early printed diagrams as a tool for understanding the relations between microcosm and macrocosm, human and world. Diagrams schematized knowledge, abstracting linear principles. Alongside illustrations in editions of Euclid and other "strictly" mathematical texts, Renaissance diagrams used triangles to understand and define the "human."

The Renaissance did not, of course, invent the triangle. Triangles appear in the Middle Ages as patterns in textiles and ceramics, in architecture (including both straight-edged triangles and curved triangles delineated by Gothic arches), and in a variety of scientific and devotional diagrams in manuscript culture. Since ancient times, different cultures have employed the triad group, a form of composition that presents frontally an elevated, or larger, central figure and two lower or smaller flanking figures. The configuration can create a centralized, symmetrical, and hierarchical composition, but it is not entirely obvious that it should be "plotted," as if on a graph, as a geometric triangle.

Ancient, medieval, and Renaissance authors had much to say about triangles.[18] In a frequently cited passage of the *Timaeus*, Plato describes the triangle as a basic building block of all matter: the four elements are the products, in their basic structure, of different kinds of triangles.[19] Plato's view rested on the idea that, mathematically, the triangle is the first plane figure (that is, at least three sides are necessary to form a figure that contains a non-zero area) and all other polygons are resolvable into triangles. A basic statement of this idea comes in the ancient author Nicomachus of Gerasa: "The triangle . . . is found to be the most original and elementary form of the plane number. This we can see from the fact that, among plane figures, graphically represented, if lines are drawn from the angles to the centers each rectilinear figure will be . . . resolved into as many triangles as it has sides; but the triangle itself, if treated like the rest, will not change into anything else but itself. . . . It is therefore the element of the others, and has itself no element."[20] Nicholas of Cusa follows up on the philosophical implications of this geometric axiom. In his dialogue *De possest*, he observes that "the first figure [composed] of a combination of quantities is a triangle, to which the other figures are reducible—a fact which shows that it is the first."[21] He then goes on to use this fact about triangles as the basis for an analogy with the triune God, who is three (though in a way that's beyond mathematics) yet indivisible.

This primacy of the triangle had metaphysical implications. Though today we may, rightly or wrongly, think of mathematics as the foundation of science and rational thought, the pursuit of mathematics in the premodern world was often closely connected to mysticism. As more Greek texts began circulating in Western Europe in the fifteenth century, this synthesis of mathematics and mysticism—supported by fragments from Pythagoras and the writings of late-antique Neoplatonists—gained ground. In the Christian context, ideas about the Trinity enriched the shape's connotations.[22] Nicholas of Cusa points to particular properties: all triangles have three angles, and all have the same, single sum of their three angles, whatever their individual magnitudes. The triangle thus possesses both threeness and oneness. It is a "beginning" that is "both one and three," hence likenable to the Trinity.[23] Charles de Bovelles, presenting the human senses as a triangle (a diagram we will see in chapter 5), describes the triangle as "the first and most mystical of all the angular figures."[24] He compared the triangle—the "first, simplest, and most excellent figure"—to the sun, for as the sun gives light to all other bodies, so the triangle gives shape to all other figures.[25]

Such ideas took shape within the context of Christian ideas about the theological implications of geometry.[26] Mathematical theology was a minor but insistent thread in medieval and early modern writings. It drew on those ancient authors whose works were available and some (like Pythagoras's) that were not. Throughout the Middle Ages, the mathematical primer attributed to Boethius provided inspiration for mystical or "speculative" geometry. Boethius's *Geometry* is lost, but his *Introduction to Arithmetic*, a Latin adaptation of Nicomachus's *Introduction to Arithmetic*, was widely read as a basic textbook. The *Arithmetic* also contains "geometric" ideas: numbers are understood as shapes (e.g., triangular numbers that can be diagrammed as an equilateral triangle of their units, such as 1, 3, 6, and 10). The number 6, in particular, was not only a triangular number but a perfect number: the numbers into which it can be evenly divided (other than itself, i.e., 1, 2, and 3) also add up to it. As read in the Renaissance, this significance extended to spiritual matters. One commentary, for instance, described perfect numbers as signs of the Trinity: "Each perfect number is a triangle, maintaining equality of three sides; which is a particular sign of the Trinity. . . . [The triad] is an analogy for the highest unity, equality, and connection."[27]

Nicholas of Cusa (in Latin Cusanus, from his birthplace of Kues/Cusa) summed up centuries of conviction in his 1453 *De theologicis complementis*, an effort at elucidating the theological significance of mathematical speculations he had already discussed in a text written for Pope Nicholas V. Mathematics, he asserts, assists us with "theological befigurings": "in the mirror-of-mathematics there shines forth that truth which is sought in and through everything knowable—shines forth not only in a dimly remote likeness but also with a certain bright-shining nearness."[28]

This is heady stuff. But geometry also had its practical uses: it was part of the basic university curriculum as one of the four arts of the *quadrivium*, and there it was divided into "speculative" and "practical" aspects. In the more practical vein, surveying—geometry as the measuring of the earth, *geo-metry*—made ample use of the triangle. In the early twelfth-century *Geometriae practica*, attributed to Hugh of St. Victor and composed almost entirely of different uses of the triangle, the author writes that in the "figure of the triangle consists the way [*ratio*] of all measurements [*omnium dimensionum*]" and adds that "it behooves us to adapt the whole world to this artifice."[29] This is one of many medieval accounts of practices that date back to Roman times. The triangle was understood as the first two-dimensional shape. All other polygons can be divided up into triangles, whose area can then be readily found through simple measurements. In practical terms, the triangle helped measure the surface areas of plots of land. With the Pythagorean theorem it also provided ways to measure heights and distances, for purposes of land management, navigation and cartography, astronomy, and military strategy.

Architecture likewise could not do without triangles. As structural components, they were (and are) especially stable: once the length of its sides is fixed, a triangle's angles also remain fixed, unlike those of a square or rectangle. In ecclesiastical architecture, architects and masons necessarily made use of geometry, including methods known as *ad triangulum*, which based building proportions on an equilateral triangle. Perhaps the most significant instance of the triangle in Renaissance building is Milan's cathedral, begun in the late fourteenth century. There the symbolic importance of the equilateral triangle trumped the practicalities of building. Given an established ground plan—such that the length of the base, hence of each side, was a measurable quantity—the height would be an irrational number, and difficult to calculate, so the builders hired a mathematician to provide an approximation. Even so, the plan provoked controversy. Though the cathedral's style is what we now call flamboyant Gothic, its deployment of proportion prompted Cesare Cesariano, in his 1521 translation of Vitruvius, to present it as an instance of classical Vitruvian geometry.[30]

Theories of vision also had their basis in the triangle. Matter itself, and vision, and land—object of the surveyor's art—were all considered to be built up out of tiny triangles. This gave the impression of a fundamental coherence in the natural world. The use of the triangle in surveying is especially relevant to artists because they were trained in geometry and often acted as surveyors themselves, among other employments. The professions of geometer and artist were linked. For at least one scholar they were linked specifically through the triangle: in his unprinted work on astrology, the *De magia naturali*, Jacques Lefèvre d'Étaples describes the extraordinary influence of the now largely forgotten "Triangle" or Deltoton constellation that represents justice and equality and governs geometers, on the one hand, and artists, on the other: "It creates admirable geometers, astronomers, and experts of the abacus . . . the benefits to painters, sculptors, makers of images and architects are also very great."[31]

The triangle was thus a habit of thinking about the natural world. In conjunction with efforts at mathematizing knowledge, it promised insights into the stars and the earth, as well as questions of perception, logic, memory, and physical change. Renaissance philosophers believed that Plato had had written above the doorway of his Academy "Let no one ignorant of geometry enter."[32] Certain Greek thinkers following Plato had argued that his Forms were in fact numbers, though Plato seems not to have said this specifically. Gérard Roussel notes this absence, stating in his commentary on Boethius that the "numbers immanent in the divine mind are the exemplars of things and these are the ideas of the divine mind, which the Platonists maintain but not appositely enough."[33] Renaissance authors cited the Pythagoreans—probably filtered through Aristotle—in the belief that *number*, understood as a quality far beyond the human experience of arithmetic, was the essence of the identity of things in the world: to every thing its number.[34] (According to Ficino, it was specifically the number three that for the Pythagoreans was the measure of all things.)[35] In the fourteenth century, Ramon Llull conceived a system that sought to mathematize the knowledge of the natural world, assigning numerical properties to natural substances as a way to standardize knowledge of them and graphing them geometrically; his charts were often composed of multiple triangles.[36] Also in the fourteenth century, in France, Nicole Oresme used the triangle as a way of graphing continuous and discontinuous physical properties, creating a form of coordinate geometry where the *X* and *Y* axes specific to each graphed quantity often form a triangle (as in the case of uniform acceleration or the way a material's retention of heat is related to its thickness).[37] As the Middle Ages progressed, so too did the ambitions of mathematical knowledge.

But was mathematics inherent in the world—or quite distinct from it? Different philosophers held different views on this topic, but the fifteenth century was characterized by a certain ambition for mathematics as the link between the two: the material and immaterial, the human and the divine.

Cardinal Bessarion cited the Pythagoreans on the theory that mathematics offers a midpoint between the material and the immaterial.[38] Many others expressed similar views. The French scholar Charles de Bovelles echoed this idea in the early sixteenth century in his book *On Proper Reason*, writing that "mathematical entities hold a middle place between physical and metaphysical substances."[39] Nicholas of Cusa, as we will see at greater length, viewed mathematics as the specific language of the mind, placed midway between world and God.

Despite its abstraction, mathematics could not entirely escape the material—the earthly, the earthy, even the erotic. Even the abstract thinker Cusanus writes that "those who taste of geometrical learning cling to it with a marvelous love, as if a certain nourishment for the intellectual life were very purely and very simply contained therein."[40] Proclus also writes of mathematics that its practitioners "are fond of it and occupy themselves with it to the neglect of other concerns."[41] When perspective appears in Giorgio Vasari's life of the Florentine painter Paolo Uccello, it risks being better than sex: the artist exclaims to his wife, when she asks him to come to bed, "How sweet perspective is!"[42]

In the fifteenth and sixteenth centuries many philosophers hoped mathematics could forge a link between the material and immaterial, the human and the divine. In the associations they made between mathematics and divine truths, geometric figures acquired meaning. The triangle was commonly used in philosophy as an example of a geometric figure (as "white" is for color and "man" is for animal).[43] The fact that it could be defined a priori (i.e., in the absence of any particular triangle) made it available as a popular example in philosophical texts from ancient times as an example of an idea that could be known purely theoretically, without reference to the observable world. The triangle was a privileged form, an object of admiration—in some cases obsession—crystallizing philosophers' interest in the relationship between form and knowledge.[44]

The Triangle as Origin of Form

The triangle took on a special role in artistic creation because it promised to join together philosophy and theology. For Cusanus, the triangle was not merely one example among many. Like Aristotle, he used it to exemplify "figure." But in his *De beryllo* it was more than just an example; it was a privileged figure, "likenable unto incorruptible specific form and unto incorruptible first substance."[45] The triangle is the basis of all shape, because while all polygons can be divided up into triangles, triangles are not divisible into any simpler form. All triangles possess the same abstract properties—they have three angles that add up to 180 degrees—and yet there are many different triangles. Thus they both exemplify and symbolize *form itself*, the basic principle, in Western philosophy, of all creation.[46]

With such ideas as a backdrop, it is possible to identify particular philosophical triangles that emerge—whether in texts or in diagrams—as especially significant, and also to look at triangles in art armed with new interpretive tools. Thinking about perspective in terms of triangles allows us to see how often it is used as a compositional tool to structure surface geometry, not only to create three-dimensional space. If God was understood as trinitarian and the earth was stamped with this image, it made sense that things in the world would be made up of triads and triangles. For Charles de Bovelles, the "substance of the physical world is perfected in the triangle": the triadic figure contains the two positions that correspond to binary entities (earth and sky, form and matter) plus the third position that is the relationship between them.[47] Many authors understood the triangle in a dynamic way, as a figure-in-process, movement from point to line to plane. In this way, it concretized relationships. It allowed for an understanding of transit between material and spiritual, concrete and abstract. It served as a symbol for the soul, the mind, the senses, the intellect, the memory, and causality.

The problem of the triangle, and the problem of geometry, is on one level the problem of form, design, and abstraction itself. As the first geometric figure, the triangle also comes to denote the general idea of "figure": shape itself. The basic figure, the first figure, it represents form, composition, multiplicity-in-unity. In *The Theology of Arithmetic*, attributed to Iamblichus, the author writes that the dyad (that is, the number two in a philosophical sense) "is clearly formless, because the infinite sequence of polygons arise in actuality from triangularity and the triad, while as a result of the monad everything is together in potential, and no rectilinear figure consists of two straight lines or two angles. So what is indefinite and formless falls under the dyad alone."[48] Combined with the monad, the dyad forms a triangle or triad and accedes to form.[49] The triad "has a special beauty and fairness beyond all numbers, primarily because it is the very first to make actual the potentialities of the monad—oddness, perfection, proportionality, unification, limit."[50] For Plutarch this implied a masculine generative principle. The number three has a phallic middle: "The even number . . . has at its middle an empty space, capable of reception, whereas the odd number has a middle member with procreative power."[51] Extending these ideas drawn from ancient authors, Giovanni Pico della Mirandola saw a triangle in the human soul and called it "prima forma formans": a kind of cosmic triangle, the first form, the generative principle of creation.[52] As three is the number that opens the way to multiplicity, the triangle is the first shape, hence the beginning of order, structure, system, even existence.

To think of the human subject as geometric—a triangle in the soul!—is also to challenge our notions about the historical development of individual subjectivity, in which the Renaissance has been viewed as a key moment. One central question in the historiography of the Renaissance is its putatively central place in the constitution of a modern philosophical notion of "subjects" and "objects." Renaissance visual representation, in particular perspective, has often been closely connected to theories of modern ideology. In an understanding of perspective that has become standard, we imagine a solitary observer, gazing as if through a window onto a scene from which he, or she, is cut off. Modern ideas about perspective have associated it with a distancing effect—a strict distinction between subject and object. And because the development of geometric perspective has been understood as the assertion of an objective (sometimes "hegemonic") viewing position, the Renaissance has been considered central to the development of a modern notion of the individual subject. Jacob Burckhardt's foundational work *The Civilization of the Renaissance in Italy* (1860) argued that a new consciousness developed in the Renaissance, in which an objective world stood, like a picture plane, before a unitary subject. Following Burckhardt, Ernst Cassirer asserted in his *Individual and Cosmos in the Renaissance* that the Renaissance, and its art in particular, ushered in "the new, specifically modern view of the relationship of 'subject' and 'object.'"[53]

Cassirer gave art a fundamental place in this perceived shift. He argues that art created an independent concept of nature and secured for it "a strong, strictly 'objective' character"[54] that he associates with artistic naturalism. In this, he sees mathematics— specifically the Renaissance mathematization of art—as key:

> This separation of the "necessary" from the "accidental," this distinction between that which obeys laws and that which is fantastic and arbitrary, was brought about not by the empiricism and sensualism of the philosophy of nature but by the intellectualism of mathematics. . . . [T]he logic of mathematics goes hand in hand with the *theory of art.* Only out of this union, out of this alliance, does the new concept of the "necessity" of nature emerge. Mathematics and art now agree upon the same fundamental requirement: the requirement of "form."[55]

For Cassirer, the freedom of the human subject is defined in relation to the objectivity of nature—its rule-boundness, but also its independent existence. This objectivity is, he suggests, established by Renaissance art as much as by science.

But Renaissance views on mathematics, perception, and the mind suggest a continuing permeability of subject and object. In Renaissance aesthetic theories, objects influenced people in ways that suggest the circulation of mutual imitation rather than the calcified world of subject-object distinctions that modern European philosophy has encouraged us to see in the Renaissance. Theories of perception stressed similitude between the organs of perception and their objects. In the process of collecting sense perceptions, the senses and then the imagination were impressed with and took on the shapes of the objects of perception.[56] As we will see in chapter 3, Augustine's influential text *De Trinitate* (*On the Trinity*) presents a relational view of the world. It can thus help us explore relationality in the context and operations of images and ideas about form and design in Renaissance art. A focus on such relationships can help us get some distance from Renaissance subjectivity in order to understand its development.

The operations of vision did serve, in the Middle Ages and Renaissance, as a metaphor for knowledge. But how did people imagine vision? In the fifteenth century visual theory was, as it had been for centuries, dominated by triangles. The science of vision known as *perspectiva* in Latin drew heavily on Arabic precedents. According to the theories of the Baghdadi scholar known in the West as Alhazen (Abu 'Ali al-Hasan ibn al-Hasan), whose work on optics was especially influential, it also worked geometrically, through a pyramid. Alhazen argued that rays/species are emitted in all directions from all objects, but the rays that land perpendicularly on the eye are the most powerful and hence produce the strongest visual effect. This allowed Alhazen to posit the centric cone of vision that later became so influential in the theory of perspective. Alhazen's innovation was a major component of what David Lindberg calls the "Baconian synthesis," the dominant approach in fourteenth- and fifteenth-century Europe, based in the work of the thirteenth-century English friar Roger Bacon. Drawing on many ancient and medieval authorities, Bacon brought together existing theories of intromission (in which things in the world emit images, or "species," that enter the eye) and theories of extramission (in which the eye sends out rays that gather visual information and bring it back). According to Bacon's synthesis, vision operates by intromission, but extramission plays a role. Visual objects emit species, but they need the ennobling power of the visual rays emitted by the eye in order to make an impression on the eye: thus vision is a meeting and joining of two forces—a collaboration.[57]

As the visible world *acted upon* the organs of perception (for Witelo, theorizing vision in the wake of Bacon, vision was painful by definition), so too was the mind shaped by the objects it contemplated. In Nicholas of Cusa's dialogue *Idiota de mente* ("The Layman on Mind"), written in 1450, the mind assimilates itself to its objects like a ball of wax, blurring any possible distinction between subject and object. In this dialogue, an orator and a philosopher come upon a layman—an artisan in the process of carving a wooden spoon—and ask him to discuss the concept of "mind" with them. The layman turns out to be suspiciously well versed in philosophical concepts and proceeds to lecture them about the mind, its powers, and its relationship to the world and the divine. The mind is rigorously placed between world and God and able to assimilate itself to either: it *becomes like* its objects. Where Alhazen had suggested a way to mathematize vision, Cusanus mathematizes the mind. As it had for Augustine and Aquinas, the mind operates by means of its similitude to the Trinity. Most significant for our purposes, it operates in the language of mathematics. When it takes on the shapes of things in the world, it imitates only their appearance. But when it moves to contemplate mathematical truths, as though seeing shadows of the divine ideas, it can leave the material world behind to exist in its proper sphere. The fact that the human mind can assimilate itself to both material and spiritual things makes it a special hinge in the cosmos, and for Cusanus, this assimilation happens through mathematics.

In the early sixteenth century, French scholars built on Cusanus's ideas, drawing out the triangularity of the mathematical mind even more overtly. In his 1521 commentary on Boethius, Gérard Roussel extends the Augustinian triads of mind and God. They are not only triadic but also triangular: following Boethius's notion of figured (geometric) number, Roussel sees the "number" of the human mind as a triangle that echoes the "superimmense" triangle of the divine mind:

> This triangle is not so much a number as the principle of number, not so much a figure as its principle, equally prior to every number and figure. But what indeed is this triangle forming and emitting every shape, anticipating every number of our mind, unless the immanent triangle of our mind and likewise substantial unity of the soul? . . . Nor are the numbers of our soul anything but its images and contractions. . . . Therefore in the same way that the triangle of the human mind, in its unity, and this unity in the triangle, creates, forms, and

> terminates every number, so the superimmense triangle in the divine mind in its highest unity, and this highest unity in the superimmense triangle, creates, forms, and defines every creature.[58]

For Charles de Bovelles, the wise man in particular is trinitarian: "It is clear that the soul of the wise man is happily completed in the number of its foundation, that is of the Trinity."[59] Contemplation is a triangular process, and knowledge is a trinity: the looker, the object looked at, and the action of the two, just as in the soul a triad is formed by the intellect, the memory, and contemplation, the function of the two.[60]

The idea of the mind assimilating itself to God or embracing the world is suggestive of an intimate tactility, and indeed, often these discussions were infused with eroticism. Neoplatonic writers, whether late-antique or Renaissance, were generally more optimistic about the material world than Plato is usually understood to have been: they chose particular elements of Plato's writings and synthesized them with the ideas of other philosophers and Christian theology. In the Neoplatonic doctrine of emanation, God's spiritual force was held to emanate toward matter through a hierarchical chain of beings, becoming more attenuated, but still present. Thus the world contains traces of divine light that can spark a spiritual flame. For some Neoplatonic thinkers, these traces were erotic: the emanations of divine love. In his commentary on Plato's *Symposium*, Ficino suggested that eros was attached to everything—plant and animal, natural and artificial things.[61] He and other Florentine Neoplatonists suggested notions of cosmic sympathy in which desire and looking produce likeness. For some authors knowledge was an experience of "coitus" with the object.[62] Bovelles himself asserts that learning and knowledge can be understood as a type of procreation. The triadic combination of teacher, student, and object of knowledge provided a direct analogy with father, mother, and child. He calls this an analogy of natural and spiritual action (*analogia naturalis actionis & spiritalis*): "spiritual action is indeed knowledge . . . transmitted from one mind to another. . . . What begetting is in natural action, teaching is in spiritual action."[63] This means that the learning student occupies the female position, penetrated by the teacher's knowledge.

Can and should we carry through such theories of the mind to our understanding of Renaissance art? We already have theories of how Renaissance minds work; art historians still often take for granted a Burckhardtian view of the Renaissance individual. Though we know that the ideas and institutions we associate with modernity were only "under construction" in the Renaissance, we proceed as if they were already in place. Many scholars view it as anachronistic to read Renaissance paintings through the lens of psychoanalysis—which, whatever its faults, provides a critique of the unitary individual—while projecting other normative, "commonsense" ideas about modern psychology onto our objects of study. A secondary goal of this study is to propose alternative models and metaphors to frame our approach to Renaissance art. What would it mean to think of the relationship between persons and things as triangular, polygonal, rather than one to one?

I find a precedent in Hubert Damisch's *Origin of Perspective*, in which he argues that perspective requires a third party—not just viewer and artwork, but a third position outside this absorptive dyad. Perspective diagrams often pose a point (the "distance point") outside the frame of the picture, or at its edge, that represents a two-dimensional projection of the position of the viewer standing in front of the picture. Technically, to read such images properly, we are to understand that this is not a doubling at all, simply a different representation of the viewer's position in front of the picture. But the point creates the symbolic possibility of a third place, a threeness. To think in terms of triangles is to begin to open up the dyad of subject and object, viewer and work, or even artist and work, into multiple relationships—to something that accounts for the plurality of the single mind (in the Renaissance the mind was understood to be composed of many faculties), the complexity of social relations, and the web of references of one work to another.

The triangle is, thus, not just a topic. It offers a different way of dividing "figure" and "ground" for Renaissance art. As I move from one significant triangle to another, I hope to illuminate a different set of issues from those I would when working iconographically or biographically. This might mean to think triangularly about viewer, artist, and artwork. But it might mean simply to consider each as multiple and multiply connected to networks of other persons and things. Compared to the individualistic models of art viewing prevalent in the twentieth century, it's worth emphasizing that in early modern Europe the viewer was *not alone*—with all the comforting and potentially menacing implications of that phrase. And objects possessed agency—sometimes mysteriously hidden, sometimes quite evident. Multiple viewers together created social performances of looking. Viewers lived with their artworks, saw them daily, and had the potential to invest them with rich individual and social meaning—as well as sometimes letting them fade into the background. Where people were not able to circulate, tokens that represented them did. The topos of the portrait that lacked nothing but breath or voice, that of the animated statue, was widespread, and animation also operated in more routine ways, via the imagination. In this context to see was to triangulate—to enter into networks created by multiple objects and viewers.

For the Renaissance, the triangle's imaginative capacity to bring together theology, philosophy, and visual expression, earth and sky, body and soul, gave it a special appeal, even if these aspirations were often disappointed. But my story is also inflected by the preoccupations of mid-twentieth-century art history. Developing in tandem with modernism, art history's methods separated form from content and pitted them against one another, as Gombrich's skewering of Wölfflin on the triangle makes clear. In the Renaissance, the triangle—the very origin of form—is the basis for theological, political, philosophical, and anthropological explorations that take place in and through images. At the same time, the notion of abstraction itself, so central to modern art, is prefigured by the aspirations for the triangle—and the mind—to transcend the material.

The thread of references that can constitute a partial history of this privileged geometric form also illuminates many different ideas and art objects. Does the triangle have historical agency? How might we use it to explore the purchase of abstract theory on concrete material cases? Most importantly: *how can a shape have a history?* This is not a trivial question. As I write about the triangle, historical tenses get complicated, because the geometric shape appears to exist in an eternal present whose rules do not change. This is itself a version of the questions Renaissance thinkers faced. The functions of the triangle, ideas and emotions about it, of course, do change. The triangle at its most abstract has no material reality—yet it is persistently taken to have reality. This is thus a history of the reception, and the concretization, of one of the most abstract of abstract ideas.

Following this introduction, the book is divided into six chapters. Three deal with the period before 1450, studying the era that set the stage for Renaissance image-making conventions, with the development of theoretical perspective and the independent altarpiece. I address issues of equality and hierarchy as expressed through the triangle, along with the Trinity's fraught relationship to visual representation (including through the triangle) and the special, and especially complex, place of Mary in relation to the Trinity and to visual forms. Chapters 5 and 6 address the deidealizing potential of geometry, the melancholic position of the mathematical mind, and the oblique triangles of perspective for artists and philosophers working in the late fifteenth and early sixteenth centuries. Here, the position of the human poses problems for geometry. Chapter 7 is an extended chapter on love triangles, focused on the mid-sixteenth century, a time when overt reflection on the artist's craft had taken shape in expanding literary genres. In that concluding chapter I build on the affective aspects of the triangle (melancholy, in chapter 5) and relational, conceptual triangles (studied in chapter 6), to examine how artists created a meditation on creative identity through the figure of the blacksmith god Vulcan, chief player in mythology's most famous love triangle. Mathematical abstraction, I suggest, opens new possibilities for defining viewers' relation to images, their subjects, and their makers. Chapter 7 addresses the erotics of these relationships; in an epilogue I address their gender. In its triangulations, the female eye—receptive and active, troubled and troubling—poses a final riposte to the masculine myths of perspective.

2

The Triangles of Perspective

A popular textbook describes the High Renaissance as a culmination of "composition, ideal proportions, and perspective."[1] While triangular composition and perspective are both often held up as characteristic "achievements" of the Renaissance, they are rarely examined together. Both perspectival geometry and compositional conventions produce triangular forms. In one case, the triangle extends into imaginary depth; in the other, the triangle is read from the flat surface of the picture plane. In a sense they are incommensurable echoes of one another, and the fact that they are rarely discussed together reflects a divide within the twentieth-century study and criticism of art, a divide between the flatness of the surface—prized by modernism—and perspective's illusionistic space projected beyond the picture plane. Modernism rejected as false the inherited conventions that it labeled "Renaissance space." Modernist painting continued, as Thomas Puttfarken has pointed out, to work with the basic inherited convention of putting color on a surface, often a rectangular one.[2] But this shape, understood as a window *through* which to look for centuries after Leon Battista

Alberti's *De pictura*, was read as a surface *on* which to paint by twentieth-century modernists. Understanding a painting as a surface with pigments on it was seen as true to materials and to the two-dimensional field.

Modernism's focus on formal properties, flatness, the suppression of narrative, and the "honesty" of the painted surface (contrasted with the fiction of three-dimensional space) forms the backdrop for the mid-twentieth-century interest in triangular composition. In survey texts it provided a simple, identifiable paradigm that could be understood (vaguely) as expressive of the supposed values of the Renaissance—order, stability, ideal beauty. Modernist painting generally inspired critical writing that shared its set of values, or at least existed in the same conceptual universe. Scholarship on Renaissance art, however, has reflected a variety of approaches, some of them more closely linked to modernism than to Renaissance concerns. The history of Renaissance art was often told in the twentieth century as a story of secularization, as the liberation of form from content—a view inspired directly by modernist preoccupations.[3]

Yet the dominant method in Renaissance art history for much of the twentieth century was iconology, or in more practical terms iconography, which emphasized the deciphering of narrative content. It allowed for the disciplining of visual material by erudition, a rationalist response to irrational times. Iconology incorporated the study of perspective—most prominently in Erwin Panofsky's *Perspective as Symbolic Form*—because perspective provided the fictional space within which narrative could play out. For Panofsky and scholars who followed his lead, perspective did more than represent the world; it constituted a unique contribution made by art to modernity, foreshadowing Cartesian thought, the subject-object divide, and modern rational mappings of space, time, and matter.

The emphasis on perspective is such that it is sometimes taken to be the only form of mathematical knowledge for Renaissance artists. A quick encapsulation of this bias toward perspective can be seen in Elizabeth Holt's *Documentary History of Art.* Bartolomeo Fazio wrote in 1456 that Jan van Eyck was "quite learned in letters, principally in geometry and her arts that relate to the adorning of pictures." Holt footnoted "geometry" with the two words "Meaning: perspective."[4] That simple equation contains one of the longest-running themes of art history and cultural studies: the assumption that perspective—and with it the representation of "real space"—was the only kind of mathematics that artists practiced.

But surface geometry—the mapping of geometric lines and shapes on the surface of a painting—has also had its partisans. When scholars from the early twentieth century onward mapped complicated geometries onto Renaissance paintings, some argued for a symbolic role for geometry, others for an empirical or formal one. It has often been difficult to establish that any of these ideas were part of the artist's conscious intention. Hence the attribution of complicated surface geometries to Renaissance paintings has come in for much critique.[5] I too am skeptical of any highly complex, secretly embedded geometric tracings believed to hold the indispensable key to meaning—but simpler geometric arrangements are no secret at all. Instead the question is what we can or cannot illuminate by identifying them.

How did triangular composition become a convention? How did this particular mode come to *feel right* to artists and viewers? As with symmetry, the triangle, as a basis for pictorial arrangement, may be more or less conscious. A basic use of geometry (as indeed Michael Baxandall argued in *Painting and Experience in Fifteenth-Century Italy*) was integral to both the training and the ambitions of artists (and their viewers).[6] Painters were thoroughly trained in the practical geometry of their time; simple plane and solid geometry were their bread and butter, and the task of creating patterns on surfaces was part of the artist's repertoire. Sometimes an artist makes compositional structure in a figurative work especially obvious, and we can trace geometric shapes—often triangles—that enclose a figure or arrangement of figures.

At the start, in fact, perspective was linked to surface geometry. Leon Battista Alberti's *De pictura*, written in 1434 (and finally printed in Basel in 1540), is a foundational text for our understanding of Renaissance art, as the first textual manifestation of "single-point" perspective. For Alberti, perspective serves the purpose of properly mapping the surface of the painted panel: there is no contradiction, here, between surface and depth. What is more surprising is that his description of the purposes of painting is at odds with much of what painters were producing—both as he wrote and in the wake of his text.[7] What other purposes did his text serve? What can Alberti and others tell us about how to understand geometry, more broadly, on the surface of Renaissance paintings?[8]

Alberti's Perspective and the Geometry of the Intercisio

Alberti has a place in the history of the triangle as the first writer systematically to apply the triangles of vision to the making of an artwork. To do this, he drew on the perspectival experiments of Filippo Brunelleschi, medieval optical treatises, and his own training in surveying. While modern scholars have often split perspective and composition apart, for Alberti they are tightly joined in the practice of measurement and integrated in Alberti's ingenious use of geometry.

In her book *Painting and Politics in Northern Europe*, Margaret Carroll relates a prevailing view of Italian single-point perspective as follows: "The perspective system presupposes a single, centrally placed viewer.... The painting privileges a hegemonic viewer for whom, one might say, the entire composition is planned.... The artist assumes the role of the rational inventor of a composition, not the recorder of what he beholds."[9] Though Carroll cites Alberti as the exemplar of this view, his own writings reveal a more complex picture in which the artist is precisely the recorder, the measurer, of what he beholds, and no more. Alberti's emphasis on measurement actually leaves little room for fiction (an absence that artists remedied in both theory and practice) or even for religious pictures. His understanding of geometry, grounded in earth-measuring, is at odds with the art's more mystical uses that we will see in future chapters. The one place where Alberti departs from the empirical and into the realm of fiction or representational convention, it is to gesture at social and political meaning for picture-making and viewing—one that, I will argue, defies the way Albertian perspective has often been understood. Rather than being hegemonic, the viewer is the *equal* of the picture. The triangles of vision that underlie perspective suggest the participation of the viewer in the work.

A member of a family that had been exiled from Florence in 1387 and was allowed to return only in 1428, Alberti worked in the papal chancery, designed buildings, and wrote fictional dialogues and numerous treatises of interest to the history of art and architecture. Inspired by the artistic activity he found on his arrival in Florence following the end of his family's exile, Alberti decided to inject his knowledge of liberal arts—mathematics and rhetoric—into artistic practice.[10] Along with *De pictura* (*Della pittura* in the vernacular version)—written soon after his arrival in Florence from exile—his *De re aedificatoria*, on building, is also frequently cited.

Alberti presents the first book of *De pictura* as a work of mathematics that will show the "roots in nature" of painting. He says little about the triangles formed on the surface of the painted panel through the use of converging orthogonals—lines that might be echoed or emphasized by triangular compositional strategies. But he has much to say about the triangles and pyramids of vision. If we read Alberti's account of perspective with an eye to triangles, they are everywhere. "It is usually said," Alberti begins, "the sight operates [*fieri*] by means of a triangle whose base is the quantity seen,

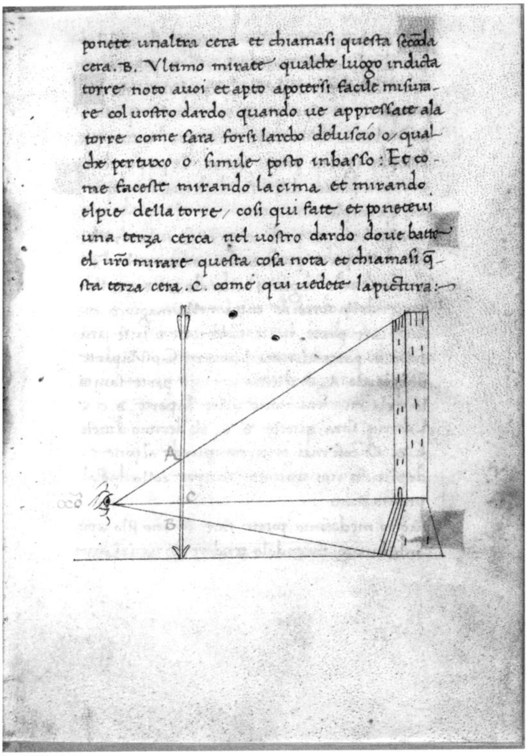

2.1 Leon Battista Alberti (?), technique for finding the height of a building, diagram in *Ex ludis rerum mathematicarum*, ca. 1450. Ink drawing. Houghton Library, Harvard University, MS Typ 316, f. 2r. By permission.

and whose sides are those same rays which extend to the eye from the extreme points of that quantity."[11] For Alberti, human perception is the product of nearly infinite numbers of labile triangles composed of elusive "visual rays." This notion of vision dependent upon triangles coincides with Plato's view of triangles as the building blocks of the world.[12]

If artists allied themselves with mathematics in the later fifteenth and early sixteenth centuries as a way of elevating their profession, this was in keeping not only with philosophical speculations but also with the longstanding practices of builders and surveyors. Indeed, Alberti's description of the practice of perspective has much in common with the surveying exercises he employed in his *Description of Rome* (written before *De pictura*) and later wrote about in his profusely illustrated vernacular text on the apparently mundane practice of surveying, the *Ludi matematici* (Mathematical Games; fig. 2.1).[13] It also corresponds to Filippo Brunelleschi's training as an architect (Alberti probably learned much of this particular application of geometry from Brunelleschi). As Jehane Kuhn writes, "Perspective was not only inspired by the geometry of surveying, but arose within surveying practice": that is, as a direct consequence of surveying rather than of painting practice.[14]

In the *Ludi* as well as in the *De pictura*, Alberti drew on the existing body of knowledge already known as *perspectiva* (a science of seeing with roots in Greek and Arab studies in optics that did not necessarily assume an application to the work of painters). The *Ludi* demonstrate the use of right triangles and the Pythagorean theorem as tools for measurement, and can enlighten us about the techniques that underlie Alberti's painter's perspective. Alberti's mathematical games were not especially novel; the techniques he presented had been in use for centuries. Right triangles, in particular, had long been a crucial part of measurement technique, whether for determining the surface area of fields or measuring the heights of buildings. Triangles of all types were employed (and still are) in the practice of triangulation. These forms of knowledge were part of an artist's basic education and, often, professional repertoire. Measurements enabled mapping and the adjudication of land disputes, and artists were often employed as mapmakers and surveyors, whether for peaceful purposes—boundary drawing—or as part of military reconnaissance teams. Such practices must have structured their habits of seeing.

Geometry—as measurement—relied heavily on the triangle, as Sebastian Münster points out in his *Rudimenta mathematica* (1551). "With the triangle," he writes, "the distances between places are discerned."[15] The right triangle is the main protagonist in the many illustrations of such texts as Münster's, Peter Apian's *Cosmographia*, and Oronce Fine's *Protomathesis*—demonstrating uses that include military engineering, architecture, surveying, and astronomy (fig. 2.2). These cosmographies, surveying handbooks, and books on practical geometry trained an ever-broader audience in surveying techniques and the use of tools such as the quadrant, compass, and Jacob's staff.

Indeed, etymologically, *geo-metry* was the "measurement of the earth." Just how straightforward this equation was is demonstrated by Faustino Perisauli's late fifteenth-century *Triumph of Folly*, in which the author castigates, one by one, all the arts and sciences. When Perisauli comes to geometry, his complaints have to do with counting and measurement: geometers waste their time counting the horses of the sun or measuring how distant the "violet beds of Venus" are from the gardens of Adonis.[16] Though the objects being measured are frivolous, the practices he describes typify the contemporary understanding of geometry. It is not theoretical or abstract; nor is it a series of rules for constructing figures, as in Euclidian theorems.

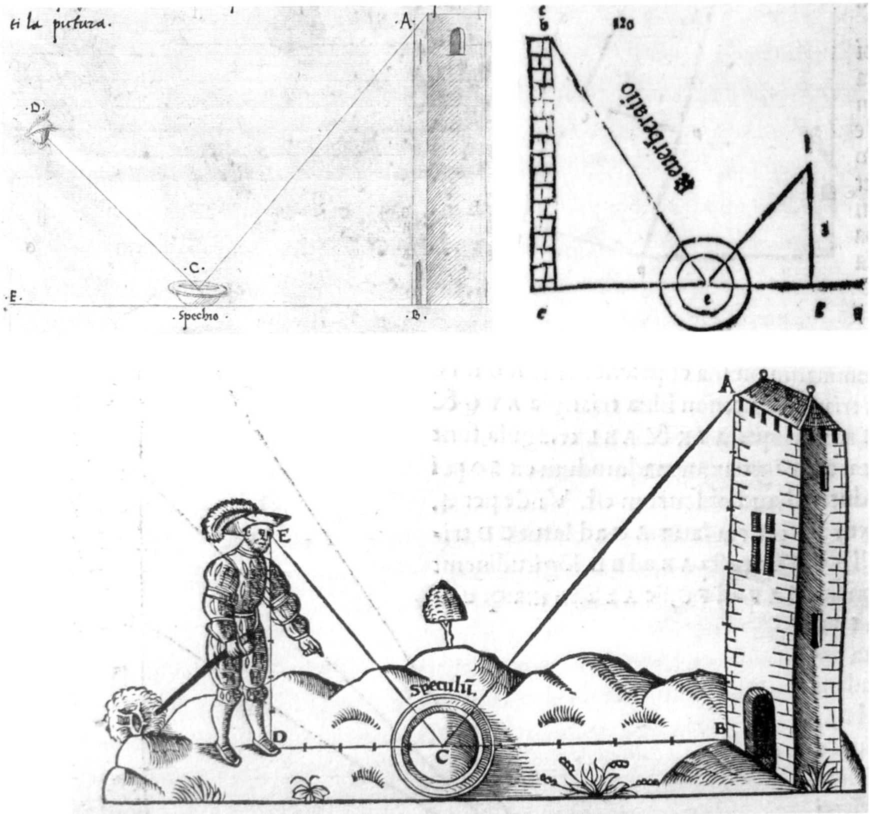

2.2 Three illustrations of the technique for measuring the height of a building with a mirror:

- Leon Battista Alberti (?), diagram in *Ex ludis rerum mathematicarum*, ca. 1450. Ink drawing. Houghton Library, Harvard University, MS Typ 422.2, f. 2r.
- Anonymous woodcut diagram in Luca Pacioli, *Summa de arithmetica geometria* . . . (1523), book II, f. 50v. Houghton Library, Harvard University, Typ 525 23.669F.
- Oronce Fine (?). Woodcut diagram in Oronce Fine, *Protomathesis* . . . (1532), f. 72v. University of Chicago Library Special Collections Research Center. By permission.

Rather, Perisauli creates humor by applying geometry in its etymological sense of "earth-measuring" to a series of ridiculous objects.

The place of practical geometry is explained by Charles de Bovelles in his 1542 *Livre singulier & utile touchant l'art et practique de Geometrie* (later titled simply *Geometrie practique*), where he argues that arithmetic is a higher science than geometry: "Between these two sisters is a similar difference to that between the soul and the body. Arithmetic is dedicated to numbers, which lie situated in the soul. Geometry considers corporeal measures, quantities, and dimensions, that exist in the body, and in any solid and material thing. Thus Arithmetic greatly exceeds Geometry in excellence of dignity and natural perfection."[17] In this vernacular text written for artisans, Bovelles strongly emphasizes the practical and earthly. Such ideas run counter to the modern sense that geometry is the *more* abstract branch of mathematics. Evgeny Zaitsev argues that Greek philosophers considered geometry a middle realm that, as he puts it, "had to be separated from metaphysics on the one hand and from practical surveying and calculation on the other."[18] But even Plato, whom Zaitsev cites as an adherent of this view, addressed its military uses, along with metaphysical ones.[19] Throughout the Middle Ages and well into the Renaissance, geometry had a double life as both a mundane, functional tool and a highly abstract and mystical discipline, serving theological speculation.

Circumscription: Painting as Surveying

Alberti, aware of both components of the art of geometry, begins his *De pictura* by distancing his technique from "pure" mathematics: he will pursue a very practical goal. In geometry, a point—according to Euclid's famous line, to which Alberti alludes—is "that which has no part." In the practice of painting, Alberti suggests, a point is only the smallest dot that it is possible for a pencil or paintbrush to make. Similarly a geometric line has length but no width, and must be held to be invisible or imaginary. Lines in painting may—should—aspire to geometric thinness, but they exist in the realm of the visible and hence material.

According to Alberti, artists must convert their geometrical knowledge to what he calls a "pinguior Minerva," a "fatter" (or "oilier"), more material form of knowledge (colorlessly translated by John Spencer as "more sensate wisdom" and by Cecil Grayson as "cruder terms").[20] When Alberti stresses the difference of the painter's mathematics from "pure" mathematics, by which mathematicians "measure with their minds alone the forms of things separated from all matter," he is making a large claim for both mathematics and the mind, even as he situates his own endeavor in a more concrete realm.[21]

The painted panel as "window" on the world is the perspectival metaphor most closely associated with Alberti's name.[22] This notion necessarily downplays the painted surface in favor of the illusion of space. It has had a long afterlife, but it is just one among many material metaphors Alberti uses. Engaged in the task of building an entirely new vocabulary, he creates concrete metaphors for technical concepts, reaching into everyday life and ordinary objects to find language for the invisible structures of vision and techniques of art. A circle is like a crown; folds of cloth are like the branches of a tree; a geometric surface is like a woven basket; visual rays are like chameleons. Other metaphors include skins, reeds, eggshells, a stalk, and the hairs of the head. The invisible "triangles" and "pyramids" of vision may reflect the immaterial points, lines, and surfaces of pure geometry, suggesting a kind of longing for the higher realms of mathematics, but Alberti's language relentlessly pulls us back down into the material world.[23] In contrast to the window, these metaphors have attracted little scholarly scrutiny. I would suggest that they provide a different and more material Albertian context for thinking about the perspectival image.

Let's consider Alberti's idea of the image more closely, in light of surveying practice. To construct a perspectival image, Alberti extends the property of similar triangles to the pyramid of vision formed with any object as its base. If I look, for instance, at a building, the straight line of its bottom edge forms a triangle with my eye. The building's dimensions present a series of such lines; its "pyramid" is thus produced by the layering of multiple visual triangles. If I imagine a picture representing that building, I see its pyramid knitted together with others to form a picture plane whose accuracy is based on the mathematical principle of similar triangles. (Similar triangles are equal in their angular measurements but do not necessarily have equal lateral lengths: they are proportional but not necessarily equal in size.) The picture plane forms, with my eye, a smaller triangle whose angles are equal to the angle measurements created by the *whole view* that it represents.

The panel to be painted takes the place of an intersection of the visual pyramid: Alberti calls it an "intercisio," from the verb *intercidere*, which means to intersect or cut through but also to perish or fall from memory. This word is worth keeping in mind for its metaphoric richness: the *intercisio* severs the world, substitutes the painting for a world of vision—the obverse of the way in which painting for Alberti "makes the absent present."[24] It does not neatly cohere with the metaphor of the window. The intercision is as much an obstacle to vision, even a forgetting, as a clear view through to the world. Already present in Alberti's own language, then, is a tension between the presence and the absence produced by the image.[25]

Part of the importance of this model for Alberti is that it produces relative proportionality: the painting is proportional to the "view"—already an approximation when understood as a simple plane with no relief of its own. Thus perspective is not only, even perhaps not primarily, a way of creating the illusion of space; it is, significantly, a method of *proportional* composition based on similar triangles, a way of guaranteeing proportionality.[26] (As we will see, proportionality extends to a notion of equality between the viewer and the depicted figures, with the vanishing point at human height.) As an account of the relationship of mental images to the objects from which they spring, this might have helped Alberti develop his notion of the workings of the visual triangles, the *intercisio*, and the formation of a proportional picture plane. But more directly, it is the types of triangles illustrated in surveying manuals that suggest this understanding of perspective.[27] The construction of the perspective panel itself is made possible by analogy to the surveyor's measuring staff. The intersection of the staff—allowing measurement by means of similar triangles—is the same thing that Alberti describes when he addresses the picture plane's relationship to the view represented on it.

Measurement is also essential to what Alberti calls composition, which requires marking off divisions on the surface to be painted.[28] In his practical instructions on drawing a perspectival image, Alberti begins by directing artists to divide up the surface of the picture plane, proportionally, into *braccia* (a unit of measure associated with an arm's length). If we think of the surface of the painting as the representation of a field to be surveyed, a further assumption made by fifteenth-century theoreticians and artists makes more sense: the assumption that the aim of drawing is to *circumscribe* objects, giving them their proper boundaries. Alberti and his contemporaries equated drawing with this practice of circumscription. Drawing does not establish the internal proportions of a body, nor what we would call the compositional structure of a scene. Rather, it serves to outline objects and figures properly, giving them their appropriate size and, significantly, place. Circumscription begins, indeed, with the notion that all things are to be identified with their position. This view accords better with an Aristotelian, that is ancient, conception of space—described by Panofsky as an aggregate of solids and voids, all finite—than with continuous and systematically measurable Cartesian space.[29]

Thus concerned with the proper placement and boundaries of things in the world, composition shares much with the surveyor's art. As Joan Kelly Gadol puts it, Alberti thought in terms of a "geometric copy which the map and the perspectival picture embody."[30] Piero della Francesca echoes this view in his own later treatise: measurement (*commensuratione*), the second part of painting and the one he is studying in his work on perspective, is about objects' "profiles and contours proportionally positioned in their place."[31]

More than a century later the art theorist Giovanni Paolo Lomazzo echoes this language of circumscription—tinged with geometry—in describing how the artist gives form to things just as nature does. Lomazzo both sums up the assumption of the preceding two centuries and stakes out more ambitious claims for art in its capacity to do the work of nature; he actualizes what for earlier writers and artists was only a simile. Nature, Lomazzo writes,

> takes matter, which is a thing without form without beauty and without end [*termine*], and then introduces, into matter, form, which is a beautiful, and finite [*terminata*] thing. Thus does the painter who takes a panel [*tavola*], which on its face has only a surface or plane void of beauty, whose parts have no limits or end [*fini ne termine*], and he embellishes it, and delineates limits [*termina*], and draws on it a man, a horse, a column; and forming it and polishing all its contours, and finally imitating with lines the nature of whatever is being depicted, in its length, height, and width.[32]

Lomazzo's equation of artistic and natural rule hints at how *disegno* was related to mathematics in a deeper sense. The things in the world were fundamentally ordered by mathematics. Though today we tend to associate this idea with Galileo's famous statement that nature is a book written in the language of mathematics, it comes to the Renaissance, as we saw in chapter 1, through interests in ancient Neoplatonic and neo-Pythagorean thought. In Nicholas of Cusa's understanding of mind, not only was number the essence of things but the task of the mind was to circumscribe things properly, according to their numerical values.[33] God's work, too, involved "delimiting": in Cusanus's words, "Properly speaking, God does not understand; rather, He imparts being. And [for Him to do] this is [for Him] to be the Delimitation of all things."[34]

The idea of *disegno* as delimitation thus has philosophical implications.[35] It relates to a notion of quantity, or "number," as the defining characteristic of created things. In the early sixteenth century, Luca Pacioli writes in his dedication to *De divina proportione* that "quantity . . . is such a noble and excellent thing, that many philosophers have judged it the peer and coeternal with substance itself."[36] When he describes what perspective does, he writes that it pertains to "the natural number according to the definition of each thing and to the measure of the visual line."[37]

Circumscription involves both the correct placement and the correct mathematical measurement of each object, and it was key to Alberti's understanding of the purpose of both perspective and composition. This emphasis on circumscription may help us think of "surface geometry" not as a network of intersecting lines traced through objects and bodies—like a wire skeleton that underlies the flesh of the work—but rather as the geometric *contours* of depicted figures and objects.

The Pilgrim and Perspective

But what does circumscription produce on the painted surface? If drawing and perspective consist in dividing the picture surface into circumscribed objects, the product might be understood (as it is by Alberti) as a precise, geometrically defined map of the visible world. But the vast majority of perspectival paintings in fifteenth-century Europe were *not* views of things seen. Not views of things seen by the body, at any rate: they are views of the invisible. One wouldn't know it from Alberti's text, but most paintings in his time, and for many years thereafter, were devotional. Some had a basis in existing narratives, like the Bible and saints' lives. But many did not. The iconic devotional arrangements (now often called *sacre conversazioni*) that became a central motif of Renaissance altarpieces had no such textual basis; the new arrangements they created might be viewed as the conceptual art of their time. Among devotional works, Masaccio's *Trinity* (fig. 2.3), with its striking use of geometric perspective, was the major event of the Florentine context within which Alberti wrote. Alberti mentioned it not at all.

To understand how pronounced Alberti's secularism is—at least where the visual arts are concerned—we can take a detour through a somewhat later text, the first printed and illustrated manual on perspective, written by Jean Pèlerin and first published in 1505.[38] Pèlerin's *De perspectiva artificiali* focuses on the spiritual function of perspective. Like Alberti, he was a secretary, in his case for the French king. He used the name Viator (wayfarer), a Latin version of his French name, which means "pilgrim." Pèlerin's text was first published in 1505 in Toul (now part of France, then an independent bishopric in Lorraine) and seems to have been widely read: it was published in multiple editions including a pirated one, and some of its illustrations were copied and inserted into Gregor Reisch's *Margarita philosophica*.

Pèlerin's text is profusely illustrated with perspectival renderings of

2.3 *(opposite)* Masaccio (Maso di San Giovanni). *The Trinity*, 1427. Fresco, 667 × 317 cm. Santa Maria Novella, Florence. Erich Lessing/Art Resource, NY.

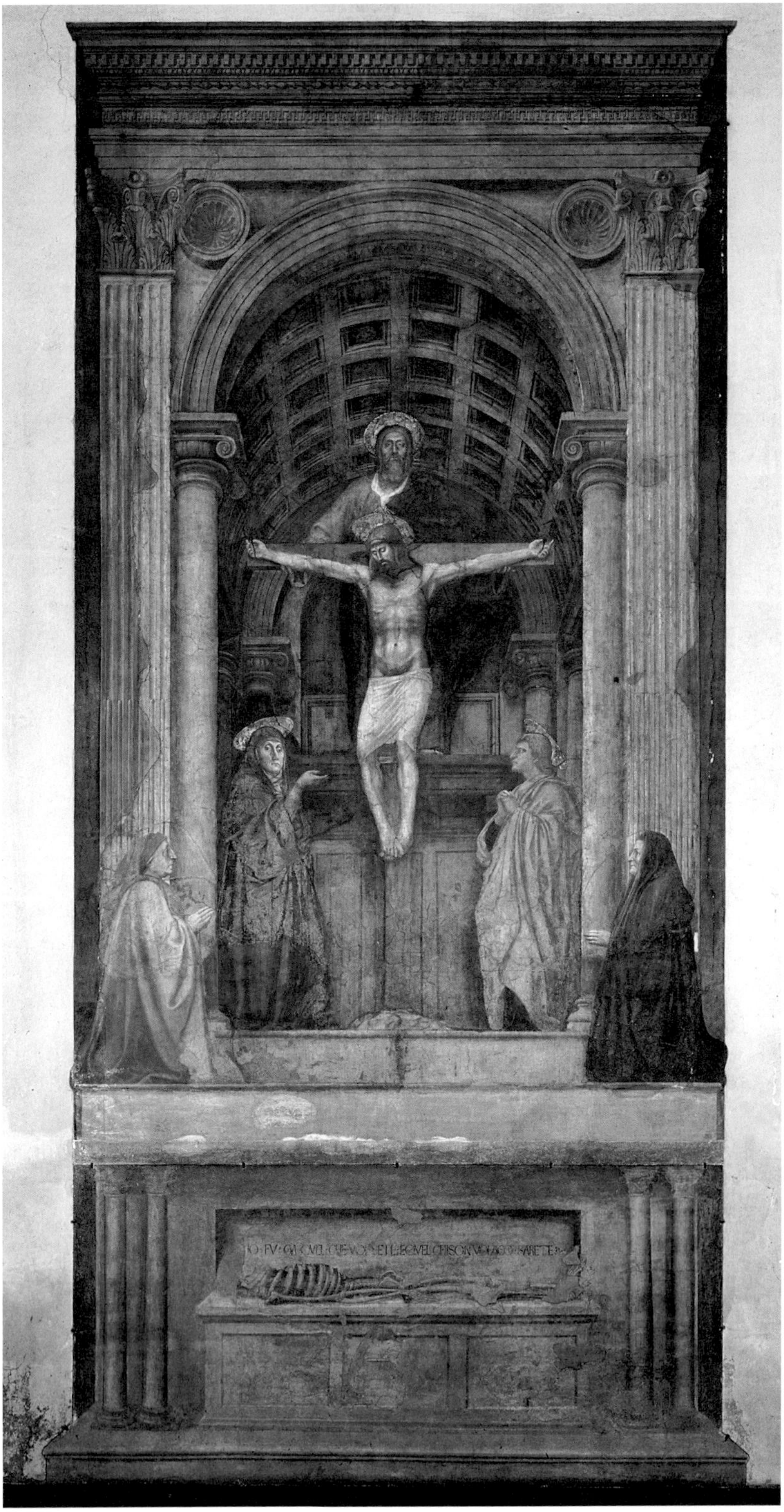

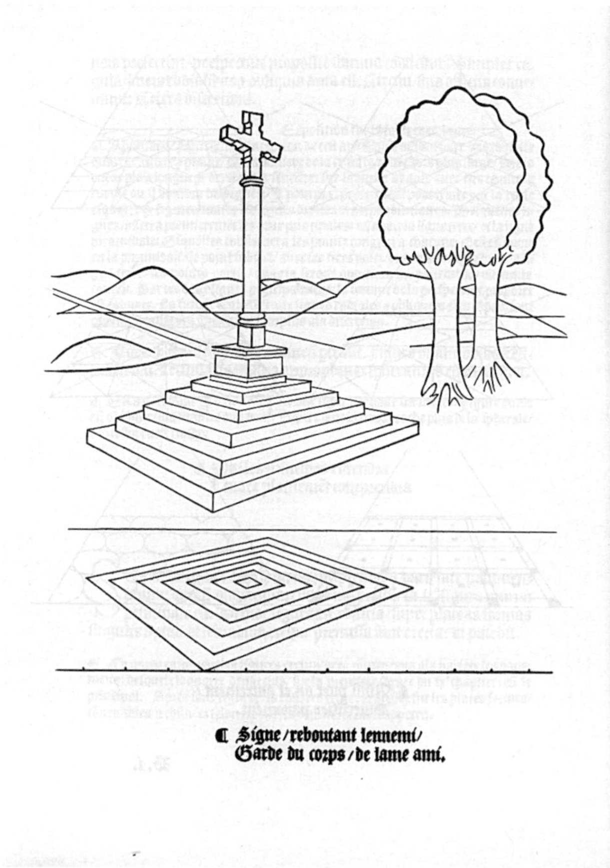

2.4 Anonymous. Cross in Jean Pèlerin, *De artificiali perspectiva*, 1509. Woodcut. Facsimile (Paris: Tross, 1861).

Gothic buildings. Most are religious structures, and some relate to pilgrimage, picking up thematically and punningly on his name. Pèlerin begins "in the name of the sovereign artisan" and goes on to list among his goals "promoting benedictions and prayers" and "virtue and divine action."[39] At the end of his expository text, he states that it has been composed "in praise of the sovereign artificer God the Father the Son and the Holy Spirit." He continues: "May it please [God] to conduct, *to the perspective* and contemplation of [his] royal city, all peaceful wayfarers [*viateurs*] from the earth, in which we, like our fathers, are strangers and pilgrims [*pellerins*]."[40] Pèlerin is playing on his own name, yet the use of perspective to describe the apprehension of a heavenly sight is striking. The sense of a precarious human hold

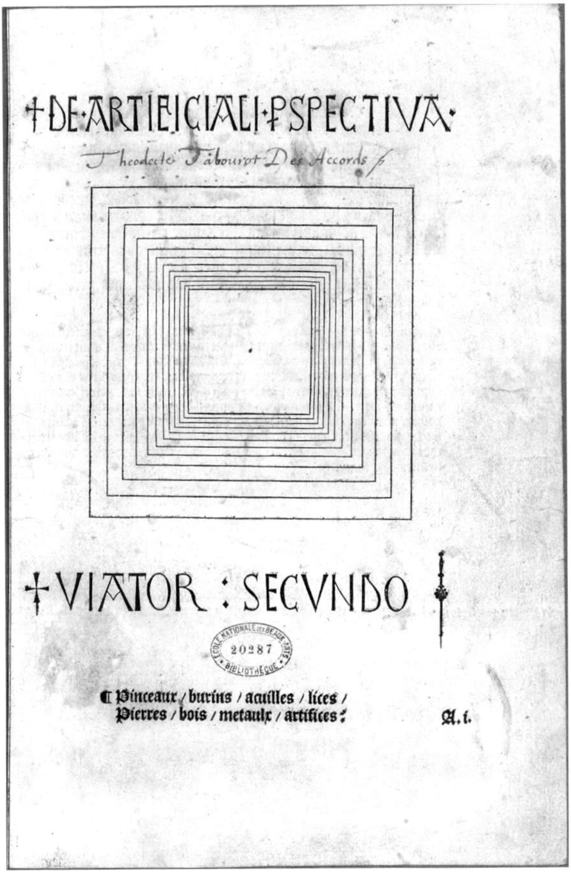

2.5 Anonymous. Title page with concentric squares, in Jean Pèlerin, *De artificiali perspectiva*, 1509. Woodcut. École nationale supérieure des beaux-arts, Paris. Photo: Jean-Michel Lapelerie.

on the earth contrasts strikingly with modern assumptions about perspective as a form of control over space.

Pèlerin begins and ends with religious sites. The first of the series of full-page images is a cross (fig. 2.4) ("Sign repelling the enemy, guard of the body, friend of the soul"[41]). As with other images in the book, Pèlerin displays it in elevation and plan: since it stands on a square, stepped pedestal, the plan is simply a series of concentric squares. In the second and third editions these concentric squares are inserted into the frontispiece (fig. 2.5), suggesting both the recession of perspectival viewing and a particular emphasis or symbolic precedence for the image of the cross—indicated, however, in its absence. The penultimate image is a view through a window of

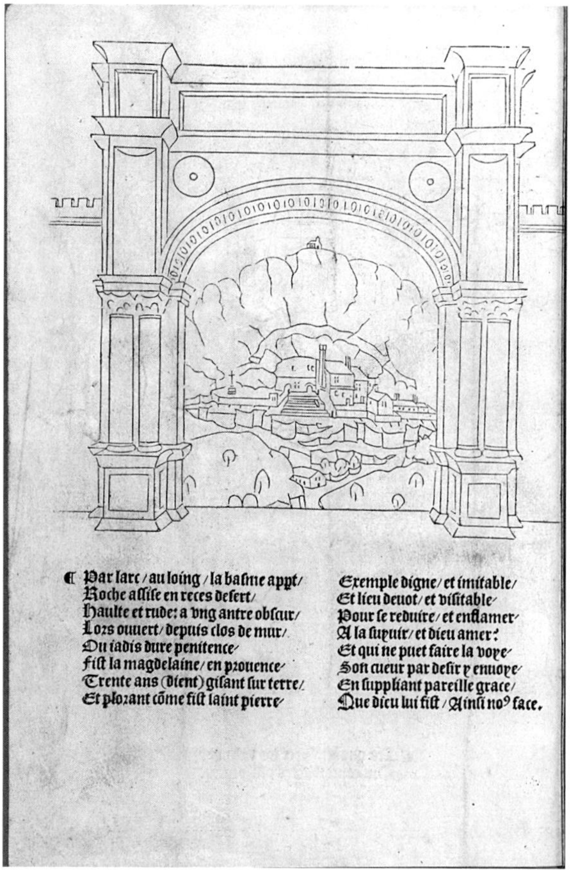

2.6 Anonymous. Pilgrimage site of Mary Magdalen, in Jean Pèlerin, *De artificiali perspectiva*, 1509. Woodcut. École nationale supérieure des beaux-arts, Paris. Photo: Jean-Michel Lapelerie.

the pilgrimage site of Ste.-Marie-Madeleine in Provence (fig. 2.6). Its poetic caption suggests that the window might be not a physical window but one of the heart: "Whoever cannot travel there can send his/her heart by desire."[42] The final image appears to represent Pélerin's own private chapel. An inscription is addressed to God: "Greatest craftsman [*faber*] of things who alone see all things: May the Pilgrim go to you by the direct path. Amen."[43]

Like Alberti, Pèlerin presents perspective as a series of techniques built up from points to lines to triangles and pyramids. In diagrams from the *De artificiali perspectiva* (fig. 2.7), more clearly than in Alberti's text, we can see how emphatically the actual construction of the perspectival space is conceived in terms of triangles. If we think of perspective as made up of such

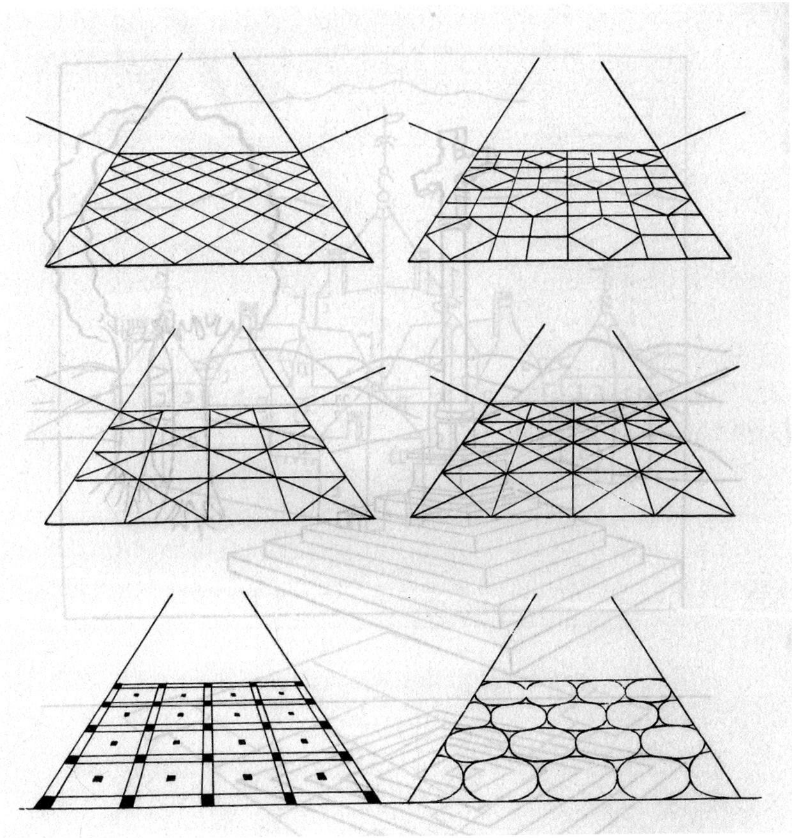

2.7 Anonymous. Triangles of perspective in Jean Pèlerin, *De artificiali perspectiva*, 1509. Woodcut. Facsimile (Paris: Tross, 1861).

triangles, we can also think of it as a compositional tool, not only a way of describing three-dimensional space. It could thus be put to devotional ends—making distinctions between human and divine measure (the eyes of the body and the eyes of the mind) or simply alluding to the Trinity.

In an introductory statement on vision, Pèlerin writes, "As for the principles, it is seen clearly [*perspect*] and deduced [*deduit*] by the speculations of philosophers that all things are seen as if by lines proceeding from the eye: that is, by the triangle: whose base is the thing seen, and [whose] diameter runs [*discourt*] by the motion of the eye across the parts of this thing seen." Perspective appears not only as a mathematical method of composition and visual construction of a space but also as the clear sight of philosophical knowledge. While Alberti stresses the scale of the human and the urban built environment, Viator focuses on the sacred and on images of traveled spaces retained in the memory. He also provides a means to order uncivilized spaces through an "underlying fictive pavement," a device for the representation of a country landscape.[44]

Perspective, therefore, could also represent conceptual space. It could suggest the space of immaterial vision or virtual pilgrimage, explore the psyche's population of allegorical figures, or create memory spaces in which symbolic objects could be placed for purposes of remembering. These functions make it more a matter of *introspection* than a window on the world—the creation of an internal space, as opposed to the depiction of external space. Pèlerin avoids elaborating on the human figure: men and women appear simply as contours, outlines without individual characteristics, purged of their material specificity, which he leaves to the "pleasure" of the artisan.

Pèlerin's focus on the constructive activity of viewing and the spiritual function of perspective suggests by contrast how strong and even idiosyncratic Alberti's predilections are for the measurement of what is seen to the exclusion of what is imagined. This extends to reticence about religion. Alberti's cultural references are found in pagan literature: he mentions the Christian God only once, at the end of his text, thanking God that he is done writing. He even quotes Hermes Trismegistus to the effect that humans created gods in their own image; while one may assume he refers to pagan "idols" and not the Christian God, the possible subversiveness of this view is indicated by the fact that the line exists in the Latin but not the Italian version of his text.[45] In his *Art of Building* he goes a step further, citing the view of some that "a temple should contain no statues" and explicating it approvingly: "Instructed by our elders and appealing to reason, we would argue that no one could be so misguided as to fail to realize that the gods should be visualized in the mind, and not with the eyes. Clearly no form can ever succeed in imitating or representing, in even the slightest degree, such greatness."[46]

What Alberti and the much more overtly spiritual Viator share is uncertainty about how to talk about the artist's creative composition of a work of religious art. Both hesitate to describe or depict a view that could be considered "fictive" or imagined. Scholars have tended to see Alberti in particular as a harbinger of naturalism, but what if these efforts at theorizing the naturalistic construction of the image actually translate anxieties about the status of the devotional image as made object? Alberti works at cross-purposes with what will remain the dominant modes of painting for many years to come. His perspective is about the measurement of things that are or could be; it is vague on the topic of things that are not, or not straightforwardly, the subject of an actual view.

Fiction, Convention, Composition

As Anne Dunlop has shown, fourteenth-century Italians developed an extended reflection—in both texts and images—on the status of the fictive image. How indeed could an image be fictive but not false?[47] As she notes, this is a problem for the entire history of Western art. But in the early fifteenth century, geometry posed, with Alberti and other theorists of perspective, a new set of questions. Alberti suggests tools for the precise measurement of things seen. He sidesteps the problem of the iconic image by focusing on the *historia* or narrative and avoiding any discussion of the particular needs of devotional (nonnarrative) religious art. But what happens, indeed, when a human artisan dares to take on divine subjects using such tools?

From a theological point of view, the idea that God was as immeasurable as the world was finite was a crucial one. Contemplation through mathematics, as we shall see, enabled Christian viewers to make distinctions as well as connections between the human and the divine, the eyes of the body and the eyes of the mind. Antonio Pierozzi (St. Antoninus, the archbishop of Florence) wrote in his *Summa theologica* that "spiritual geometry causes [one] to measure temporal things."[48] It is by comparison with the immeasurable character of God that we recognize the humble limitations placed on our worldly existence. Alberti's measuring of temporal things seems to exclude the divine from human conceptualization in the same way that his insistence on the "pinguior Minerva" excludes pure geometry. What might it mean to "measure" some theological truth visually for devotional purposes? If an image possesses measured pictorial order, but that order is not imposed by the subject matter, what is the ontological status of this "figure" fabricated by the artist? Would it imply a godlike act on the part of the artist, who creates and orders a fictive world?[49]

A way of considering the fictive world of painting as a distinct type of truth appears in another writer's account of a famous anecdote from the history of perspective, that of Brunelleschi's mirror. Brunelleschi had placed a mirror opposite a panel with a depiction of the Florentine Baptistery in perspective. A viewer could look through the back of this panel, through a hole in the middle of it, at the mirror image—while standing in front of the actual baptistery to verify the depiction.[50] In the architectural treatise by Antonio Averlino, called Filarete, the author anticipates an objection: his interlocutor in the dialogue may say that the reflection in the mirror is not true. "This is correct," Filarete says, "but nonetheless in *drawing* it is real. For drawing itself is not a true thing, but is a demonstration of what you are portraying, or what you wish to demonstrate."[51] An image is not real—but fictive lines can be real *in an image.* The distinction Filarete makes between portraying and demonstrating seems important, for it suggests that an image may demonstrate an idea that it is not directly able to depict. Yet the idea of demonstration itself remains slightly murky, thanks to the repetition of its nominal and verbal forms. As an architect, Filarete may have in mind a plan for a building that does not yet exist. But this, like the "conceptual" devotional image, opens the door to fiction.

During Alberti's time, devotional arrangements that had been more or less "diagrammatic" came to be consolidated into a naturalistic, perspectival space—even if the scenes depicted were not "natural" at all. Perspectival geometry, then, was often put to symbolic rather than empirical ends. The visible perspectival lines of Masaccio's *Trinity* are all about the architecture. Perspectival lines constituted by the barrel vault above the head of God the Father meet at a vanishing point that may correspond to the viewer's height but is far below the depicted figures' heads—contrary to Alberti's precepts. Perspective thus seems to be heavenly, descending from above. Meanwhile, using a striking alternation of the color of the figures' garments, Masaccio links the depicted votaries to the figures of God the Father, Mary, and John, in a compositional triangle.[52] Another example is Fra Angelico's San Marco altarpiece (plate 5), a painting that could be described as diagrammatic or decorative in its repeated motifs, symmetries, and highly geometric arrangement of figures. Like the *Trinity*, it uses perspective along with a triangular compositional arrangement but violates Alberti's rule for the placement of the vanishing point. We will see more of this painting in chapter 4.

In the covering of the painted surface, Alberti stresses boundedness: the fact that composition delineates the proper place of each depicted object. The painted surface is like a map. What's represented on the surface is determined by the painter's view, and the most beautiful proportions are those existing already in nature. The *intercisio* depends on a notion of the painting as surface to be covered. Thinking of the painting as a surface might also be conducive to decoration: geometry used with an eye to a pleasing arrangement, rather than naturalistic depiction. Alberti's short text on the "elements" of painting, as well as Piero della Francesca's writings, shows the extent to which the ability to construct numerous geometric figures was a key tool for artists.[53] And if we consider the correspondence of mathematics and ornament, we can view surface geometry as part and parcel of pattern and decoration.[54]

To take one contemporary of Alberti as an example, the Sienese painter Sassetta positioned the figures in his *Last Supper* (fig. 2.8) in an orderly, symmetrical, highly geometric fashion. This order, often called *ornamentum*, was a way of indicating honor. Icons had long displayed symmetry and measured proportion among their parts. Is Sassetta's style a throwback, the revival of a medieval mode,[55] or the imposition of a new idealism? In his ca. 1400 treatise—otherwise very terse on questions of design—Cennino Cennini recommended intersecting diagonal snapped lines to find the center of a composition.[56] The technique would also have enabled the orderly placement of figures. And the "decorative" geometrical approach to arranging figures is not exclusively "medieval": geometric pattern appears, for instance, in Pontormo's *Vertumnus and Pomona* at Poggio a Caiano, in which he shapes naturalistic figures into an utterly antinaturalistic pattern based on multiple triangles (fig. 2.9). Chronologically between these two are examples of geometric arrangement from Fra Angelico's San Marco altarpiece to Raphael's *School of Athens*. While little direct evidence of mathematical planning of such paintings exists, they simply could not have been made to look the way they do without the taking of measurements: to establish the center point and various distances from it, to define the heights of figures, to create order and symmetry.[57] In the *Divina proportione*, Luca Pacioli writes that painting delights the soul "through the appropriate distance and variety of colors" and concerns itself with "arithmetic and geometric" proportions.[58]

Many fifteenth-century paintings present their subject matter in an ordered geometric fashion while also employing perspective to map their space. John White, in *The Birth and Rebirth of Pictorial Space*, suggests an affinity, yet also a tension, between decoration and space in the early work of artists like Fra Angelico and Fra Filippo Lippi. He writes, apparently with some surprise, "The decorative artist is excited by the pictorial possibilities and not the scientific meaning of perspective."[59] But these "pictorial possibilities" might serve purposes relating not to natural vision but to the eyes of the mind.[60]

Artists did explore the internal mathematical logic of figures—something that can be observed most easily in the analogies made between bodies and buildings. Villard de Honnecourt had already built figures out of

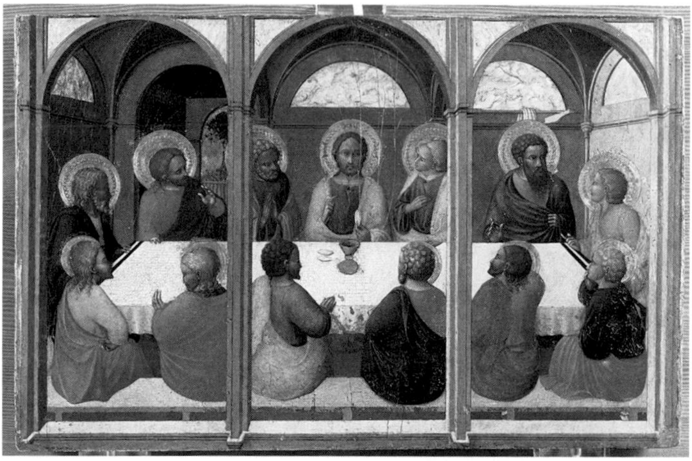

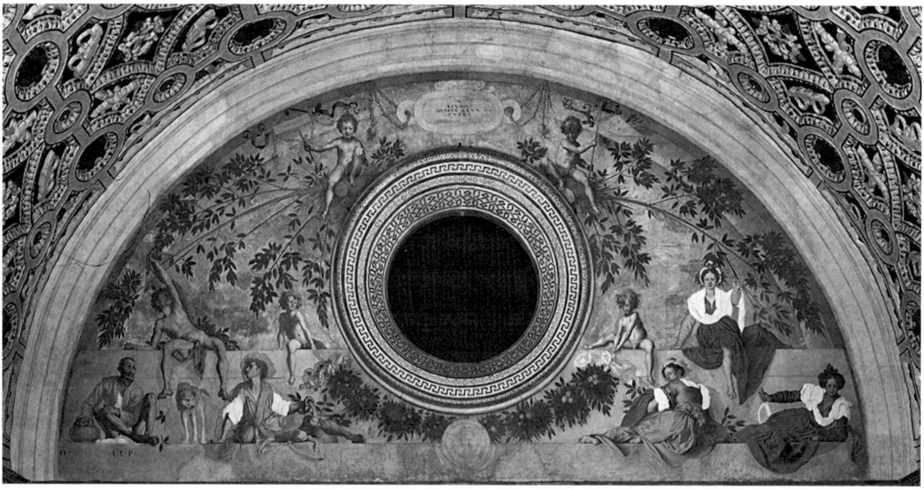

2.8 Sassetta (Stefano di Giovanni). *Last Supper*, 1423–1426. Tempera and gold leaf on wood panel, 24 × 38 cm. Pinacoteca Nazionale, Siena. Scala / Art Resource, NY.

2.9 Jacopo Pontormo. *Vertumnus and Pomona*, 1519–1521. Fresco, 461 × 990 cm. Villa Medici, Poggio a Caiano, Italy. Photo: Antonio Quattrone.

triangles in the thirteenth century. Various figures in the *Codex Huygens*, formerly attributed to Leonardo, appear with the proportions of their bodies defined by triangles in the manner of a ground surface mapped by a surveyor. Francesco di Giorgio drew many similar figures, and Dürer constructed figures from geometric shapes, studying their anamorphic transformations. This notion of design is fundamentally architectonic, like the division of a facade into appropriately measured parts.

If bodies and buildings "possessed" geometry, could the picture plane be understood as something like a body or building that itself should be ordered according to mathematical proportions? In the fifteenth century, no one would have argued that a painted panel, canvas, or fresco possessed "form" by virtue of its own design, circumscription, or composition; form was a higher property of created things or the idea in the mind of God. Alberti sidesteps the question of the artist's creative contribution—how he might order the scene he depicts, how an arrangement might be fictive, arbitrary, or decorative. What's more important, though, is that he seeks to keep painting in the realm of the material. His strong emphasis on the idea of *historia*—narrative subject matter—and the deep secularism of his writing suggest an understated resistance to the icon. The images he provides guidance on making are not miracle-working images.

There is, however, one element of fiction, or at least avowed convention, in Alberti's text. Alberti puts perspective on the firm footing of practical measurement, yet doubts persist about the reality or fictiveness of the drawing that is produced. The key to composing the panel, for Alberti, is the size and scale of the human body, something not mathematically required. The units of the ground plane, the *braccio*, not only are named for a body part—the arm—but are given by Alberti as one-third the height of the human body. A digression on Protagoras's famous line—"man is the measure of all things"—gives rhetorical cover to the move to a practical method in which the body is the principle of construction: it's the scale of the human body that allows the picture plane to be divided into parts.[61] On the quadrangle of the painted surface, the artist divides the ground line into *braccia*. He then places the intersection of the centric ray (what we call the vanishing point) at the same height at which the heads of the figures will appear. He draws lines from this point to each dividing point along the ground line. Alberti thus institutes human scale as the proper measure of perspectival construction: the entire construction is based on human height and may be understood as a series of lines drawn from a man's head to the ground. Along with Protagoras, he draws on ideas from the Roman architectural theorist Vitruvius, who viewed the proportions of the human body as the basis for architectural measurement. That human height was neither the only possible placement for the vanishing point nor a universally followed rule is made clear by the existence of numerous other schemas. A point that corresponds to human height is just as arbitrary—and hence potentially significant—as other placements.[62] Martin Kemp bluntly criticizes Alberti on this issue, writing that he "paraded half-explanations from Euclidian geometry and medieval optics, before outlining in some detail a practical means of constructing spatial illusion. He elusively failed to commit himself on the precise relationship of the visual pyramid to his constructional procedure, with no proof as to why the former results in the latter."[63]

Alberti describes his convention as an agreement, a "pactum." He writes: "it is proper for the position of this centric point to be no farther from the ground line than the height of a man to be painted, for by means of this convention [agreement, *hoc pacto*], the spectators and the depicted things will appear to meet [*adesse*] on equal ground [*aequali in solo*]."[64] We might relate this concern with equality and agreement to the famous "civic" character of early Florentine humanism, the notion of any ruler as merely *primus inter pares*. Thomas Golsenne and Bertrand Prévost have argued, in a similar vein, that Alberti's theories of painting reflect concerns about friendship and community.[65] There may also be an echo here of Alberti's thinking on symmetry in architecture, which he describes not with the term *symmetria* but rather with a notion of equal correspondence: "Care had to be taken to give openings a graceful appearance, and those on the right and the left would have equal dimension so that they balanced each other out [*coaequatis dextera sinistris respondeant*]."[66]

If this was not an absolute equality by any means, it certainly suggests the idea of coming together, literally, on an equal footing. Alberti's in-between social status—a brilliant scholar and illegitimate son who wrote sadly of the loneliness of scholars and the grand social entourages of rich men—may have made this notion appealing to him.[67] In contrast to the more famous metaphor of the window, these rhetorical choices suggest contact and proximity rather than distance and separation—the mirroring of viewer and work, their equality, their agreement. The viewer participates in the work; they exist on the same plane. Perhaps we might even think of the angled lines of the picture plane as a mirror image of the viewer's pyramid of vision—the picture's *own* pyramid of vision as it looks back.

Alberti's ambitions for art are modest. They are based in an understanding of visual veracity and perhaps suspicion of the claims made for religious images—anticipating the tensions over images in the time of the Reformation. But they also, significantly, present a social model of equality, similarity, and commensurability that will have unexpected resonance with political theory, to be addressed in chapter 4. First, however, in chapter 3, I introduce the uses that spiritual geometry made of the triangle, in which geometric arrangements could also serve as explicit devotional prompts.

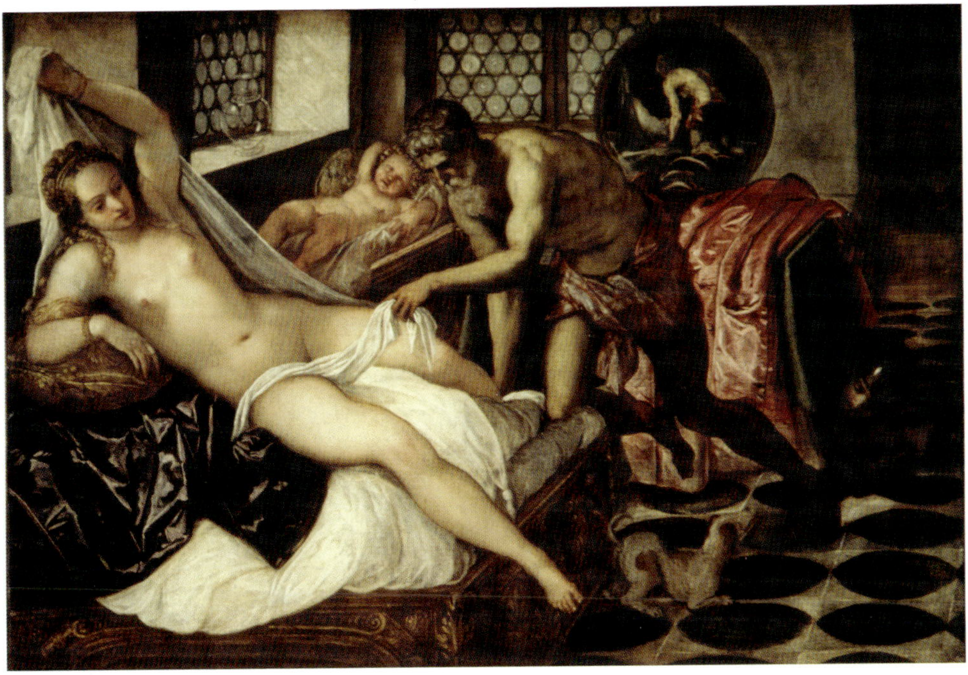

1. Jacopo Robusti Tintoretto, *Vulcan Surprising Mars and Venus*, ca. 1555. Oil on canvas, 135 × 198 cm. Alte Pinakothek, Bayerische Staatsgemäldesammlungen, Munich. Kavaler/Art Resource, NY.

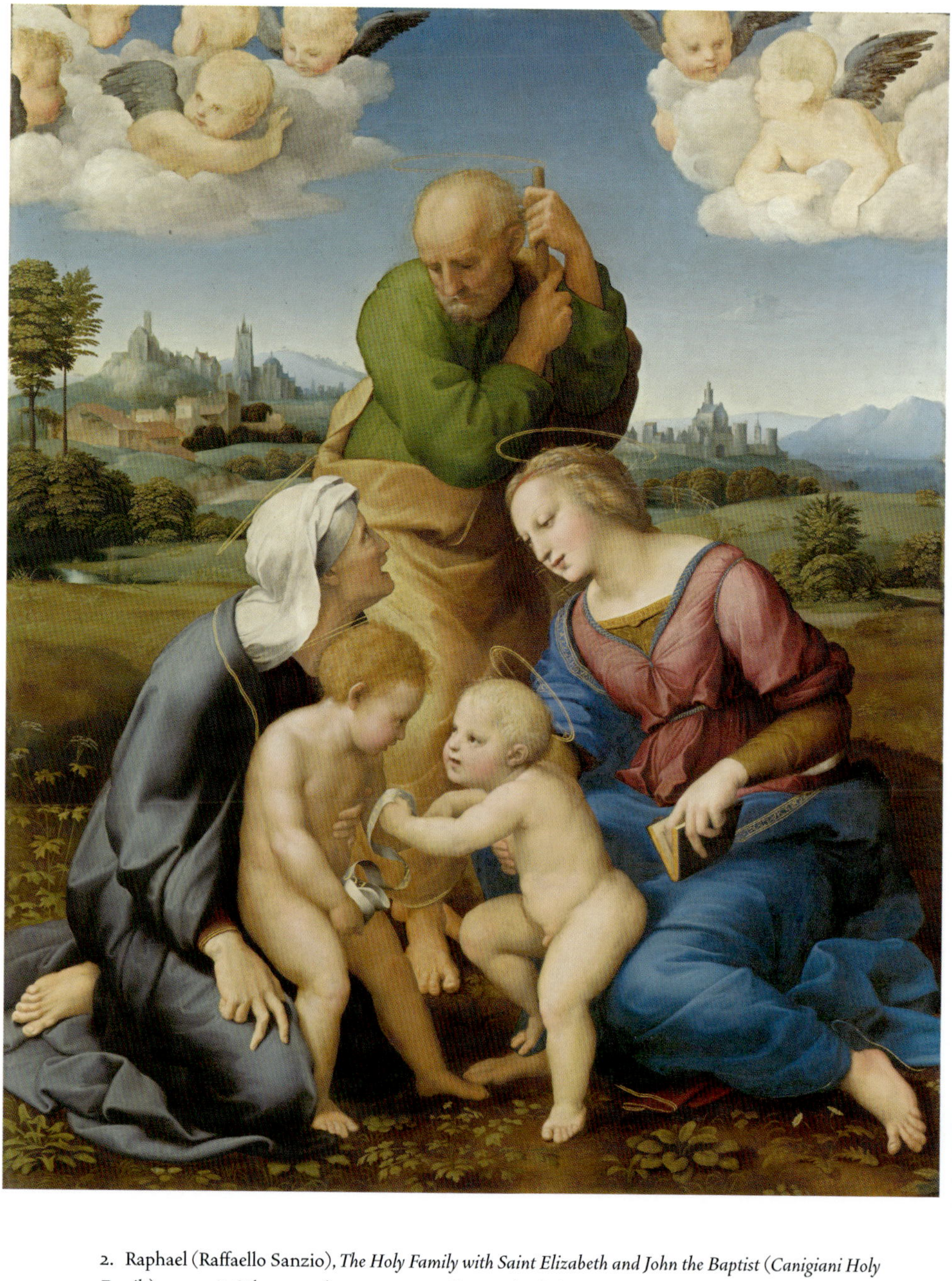

2. Raphael (Raffaello Sanzio), *The Holy Family with Saint Elizabeth and John the Baptist* (*Canigiani Holy Family*), 1505–6. Oil on wood, 131 × 107 cm. Alte Pinakothek, Bayerische Staatsgemäldesammlungen, Munich. Bildarchiv Preussischer Kulturbesitz / Art Resource, NY.

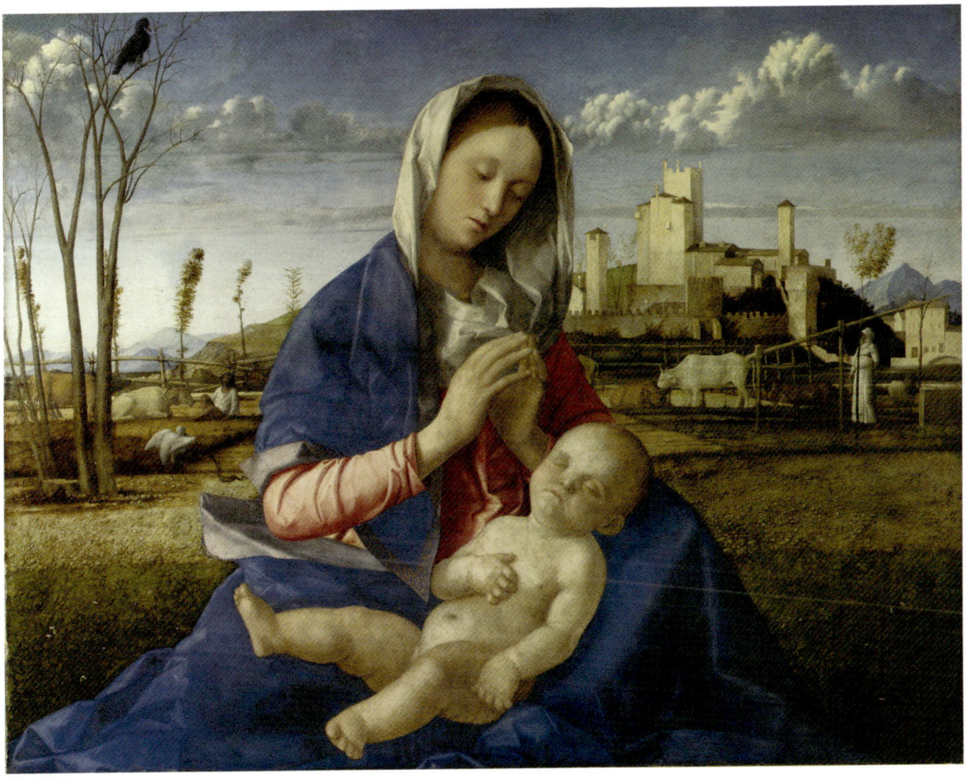

3. Giovanni Bellini, *Madonna of the Meadow*, ca. 1500. Oil and egg on synthetic panel transferred from wood, 67.3 × 86.4 cm. National Gallery, London. © National Gallery London / Art Resource, NY.

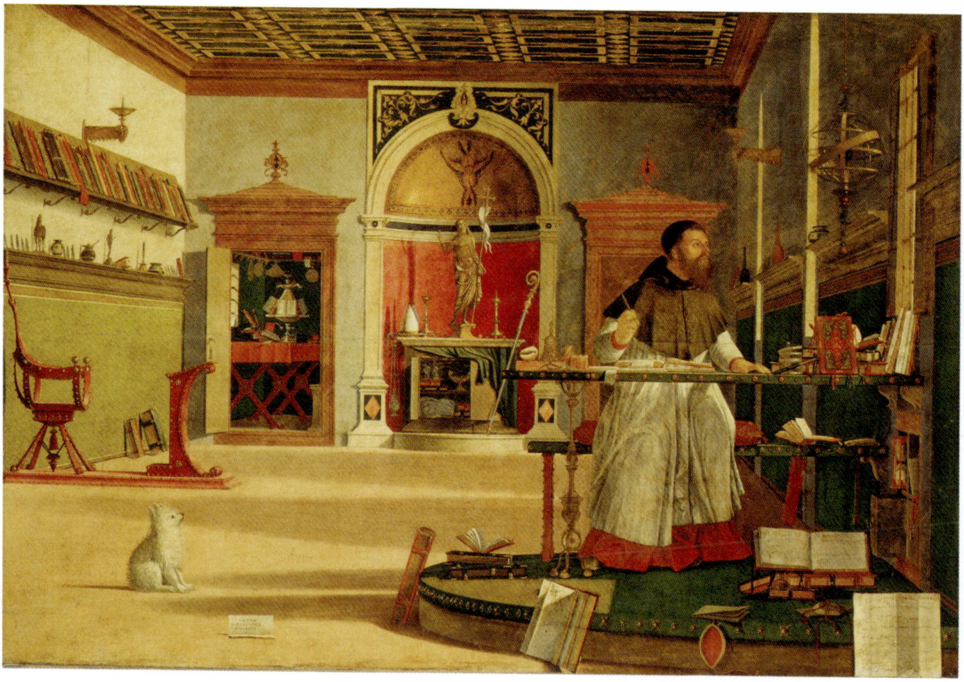

4. Vittore Carpaccio, *Vision of Saint Augustine*, 1502. Tempera on canvas, 141 × 210 cm. Scuola di S. Giorgio degli Schiavoni, Venice. Cameraphoto Arte Venice / Art Resource, NY.

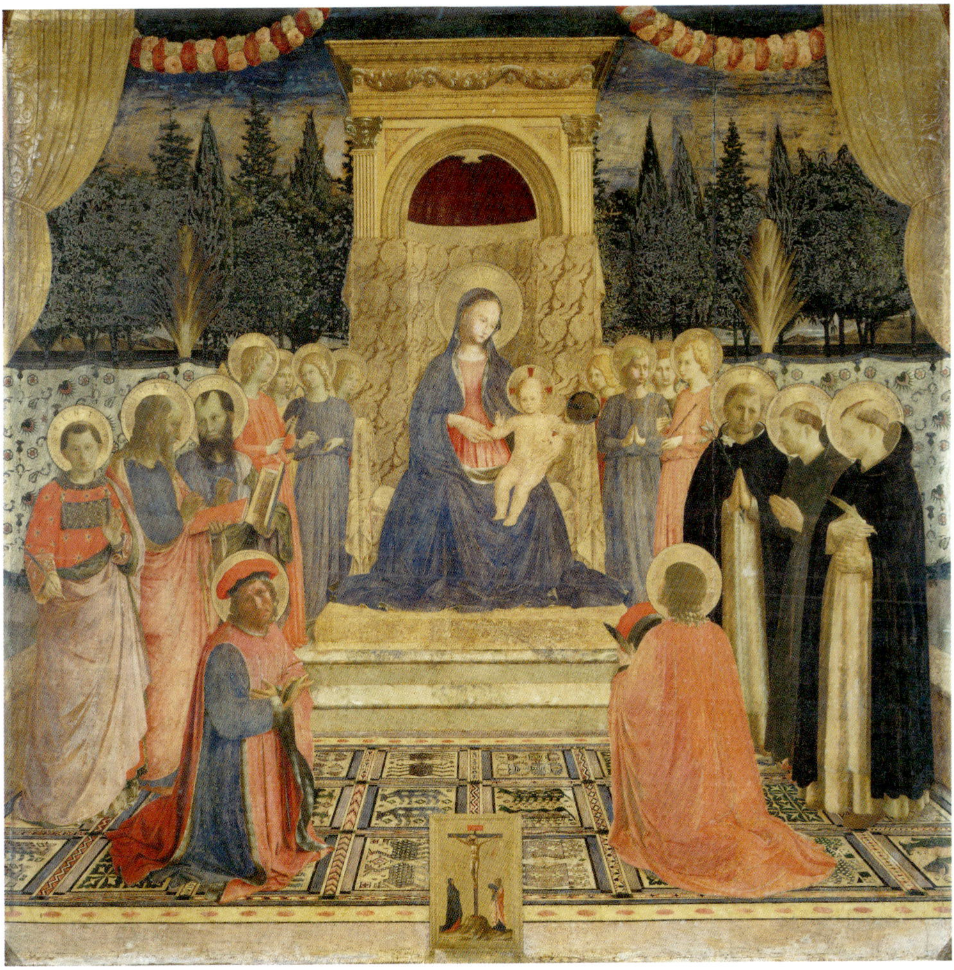

5. Fra Angelico (Giovanni da Fiesole), *San Marco Altarpiece*, ca. 1439–42. Tempera on wood, 220 × 227 cm. Museo di S. Marco, Florence. Erich Lessing/Art Resource, NY.

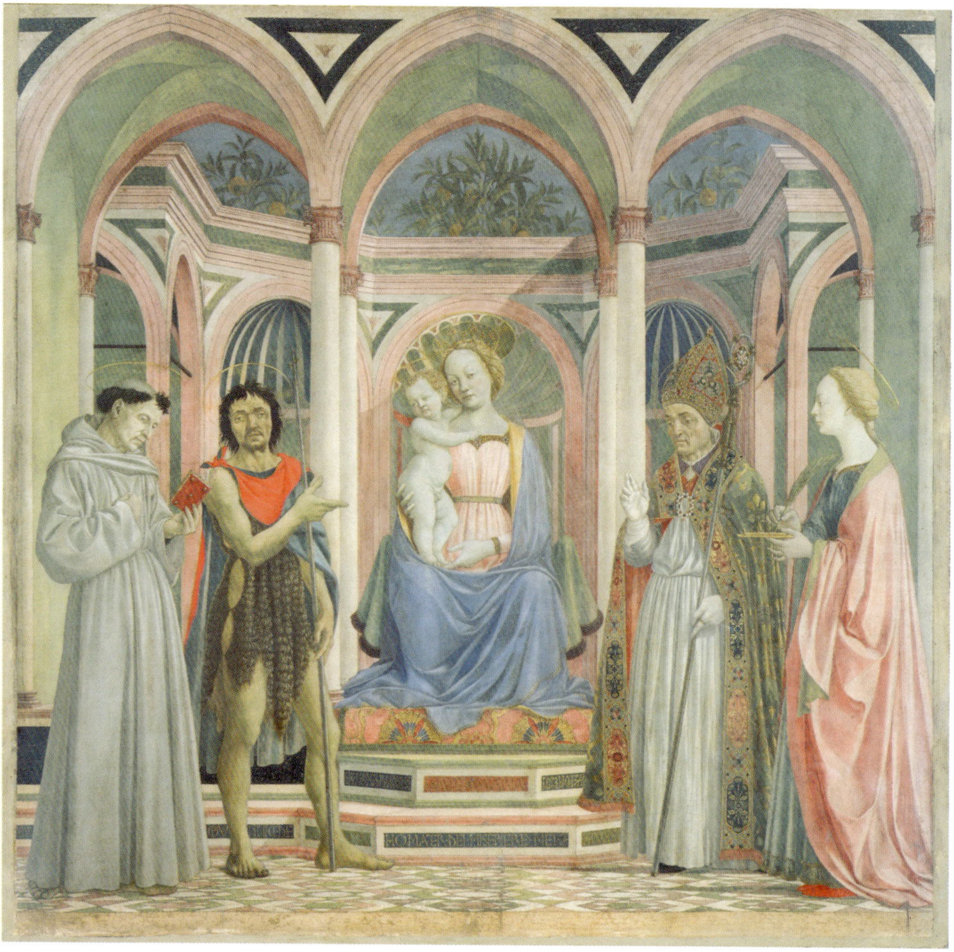

6. Domenico Veneziano, *St. Lucy Altarpiece*, ca. 1445. Tempera on wood, 210 × 215 cm. Uffizi, Florence. Scala / Art Resource, NY.

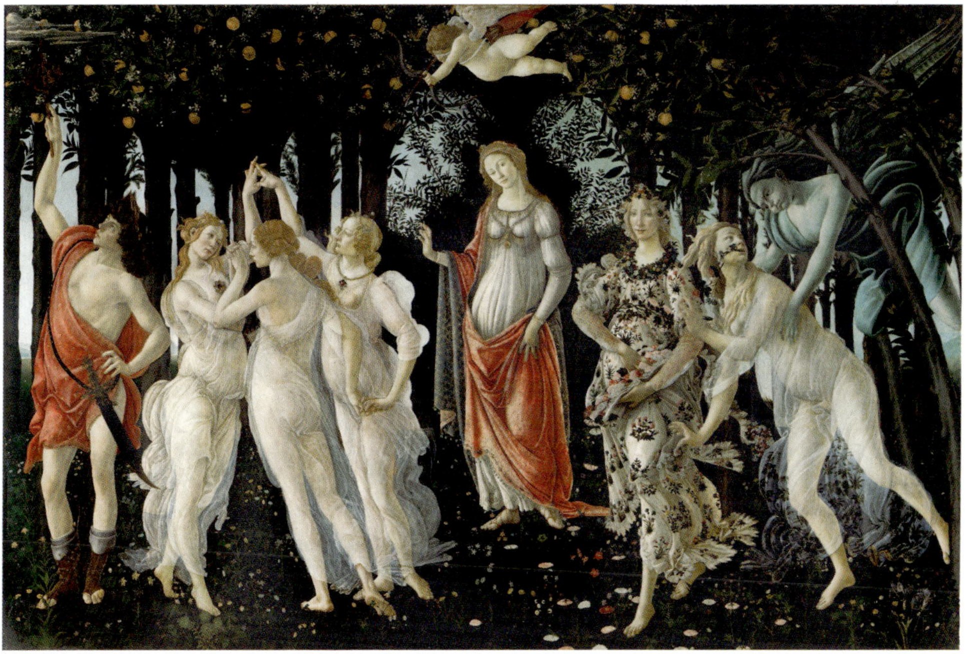

7. Sandro Botticelli, *La Primavera* (*Spring*), ca. 1477. Tempera on wood, 203 × 314 cm. Uffizi, Florence. Erich Lessing/Art Resource, NY.

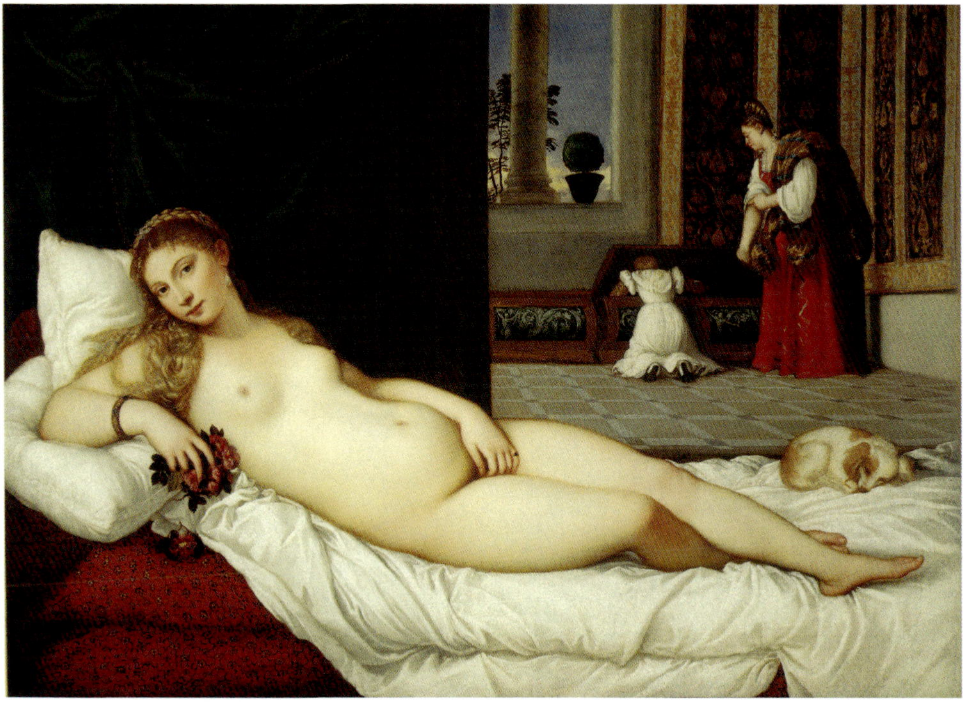

8. Titian (Tiziano Vecellio), *Venus of Urbino*, 1538. Oil on canvas, 119 × 165 cm. Uffizi, Florence. Scala / Ministero per i Beni e le Attività Culturali / Art Resource, NY.

3

The Triangle and the Trinity

The three figures—Mary, the angel Gabriel, and God the Father—in Francesco del Cossa's Dresden *Annunciation* (fig. 3.1) form a distinct triangle, constructed in perspective across three dimensions. The painting also contains a surprise, a snail that appears at its lower right corner. An iconographic argument holds that it is an allegorical figure for the Virgin, suggested by a belief that snails were fertilized by dew.[1] But the snail, as Daniel Arasse points out, is also out of scale for its position in the picture. Its shape and size, mirroring the tiny figure of God the Father, force the viewer to acknowledge this incongruity, even calling into question the painting medium itself as an arbiter of truth.

The depiction of God the Father contains another unusual element: he wears a triangular halo. There are other examples of the triangular halo in the fifteenth and early sixteenth centuries, but it is not typical until the Counter-Reformation. Del Cossa was a Ferrarese painter who created this work for the Church of the Osservanza in Bologna in 1470, and as it happens, representations of God the Father with the three-pointed aureole cluster thickly

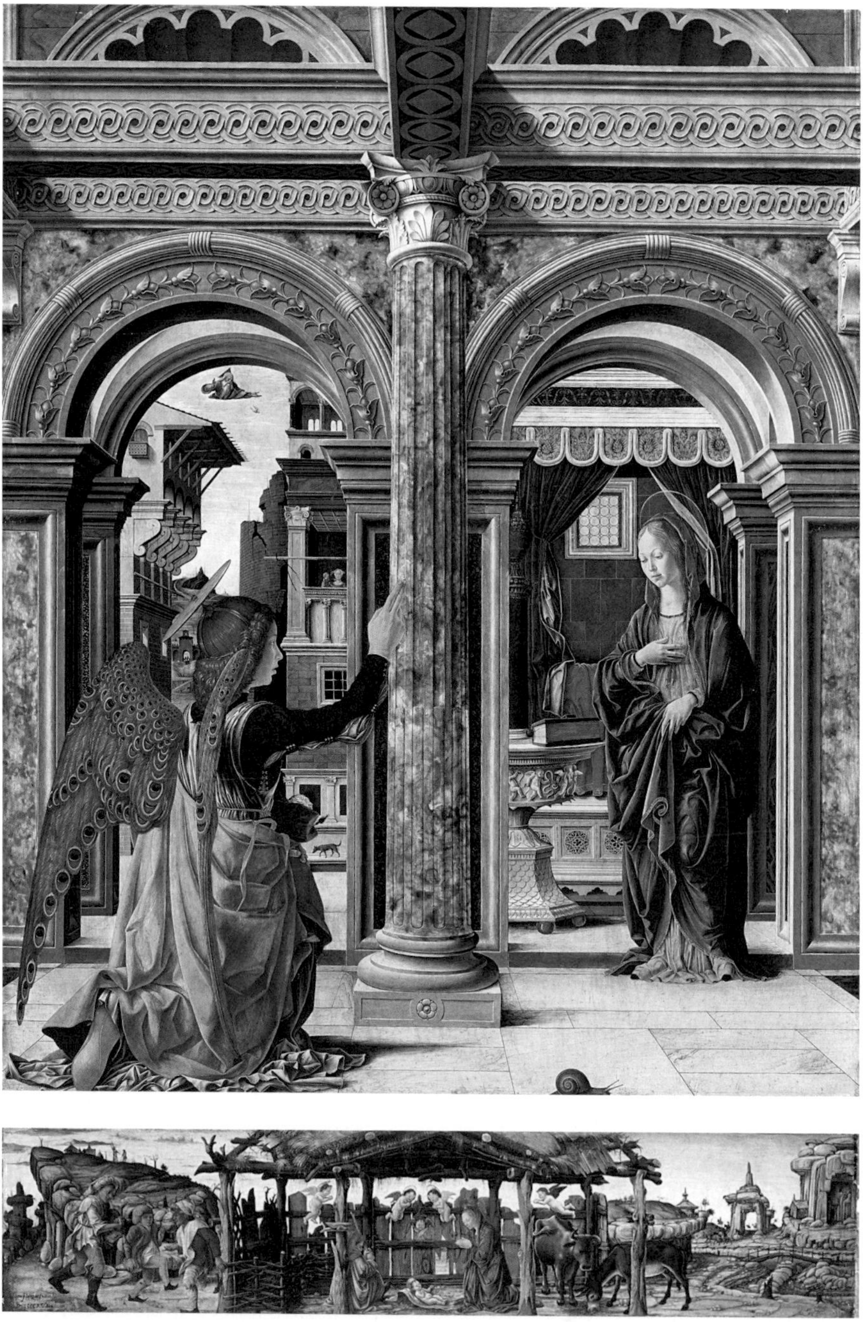

3.1 Francesco del Cossa. *Annunciation* and predella panel with *Nativity*, 1470/72. Oil and tempera on wood, 137 × 113 cm and 26.5 × 114.5 cm. Gemäldegalerie Alte Meister, Staatliche Kunstsammlungen, Dresden. Bildarchiv Preussischer Kulturbesitz / Art Resource, NY. Photo: Hans-Peter Klut.

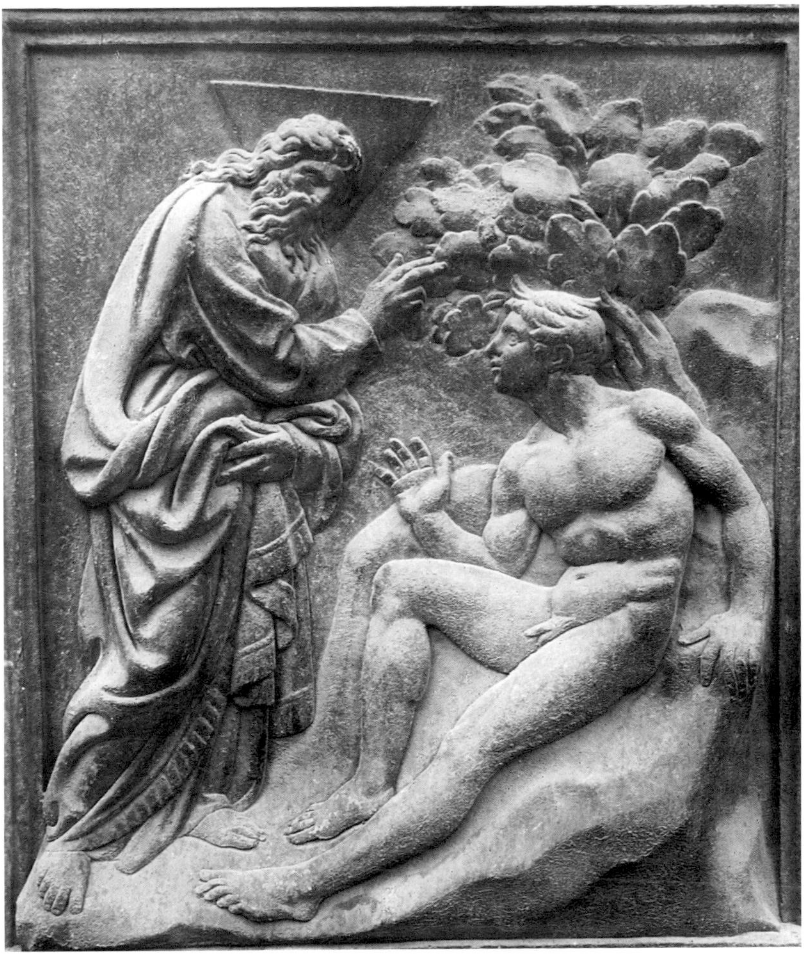

3.2 Jacopo della Quercia. *The Creation of Adam*, 1425–38. Marble facade relief, 85 × 69.2 cm. San Petronio, Bologna. Alinari / Art Resource, NY.

in the mid-fifteenth century around Bologna and Ferrara.[2] The triangle helps explain to the viewer that where God the Father is depicted as a single figure, "he" actually represents the Trinity. This idea seems to inform the images of God the Father in the creation reliefs that Jacopo della Quercia sculpted for the portal of the Church of San Petronio in Bologna around 1425–38, in which God wears a triangular halo (fig. 3.2).[3] These reliefs likely drew on commentaries on Genesis that insisted on an already-trinitarian God present at creation. According to Augustine and many others, evidence of the Trinity was found in the first-person plural of God's words in Genesis 1:26: "Let us make man in our image and likeness."[4] God as Creator is already plural, hence trinitarian. Some writers extended the trinitarian creation further. Rupert of Deutz argued for the direct resonance of the Trinity in the mode of creation itself: the Creator of earth, sky, and all their adornment (*coeli et terra, et omnis ornatus eorum*) must indeed be the Trinity by virtue of this triple act.[5]

Presumably inspired by such ideas, della Quercia's reliefs—public and celebrated images on an ongoing and highly visible building project—likely served as inspiration for many painters and miniaturists working in northern Italy and particularly Ferrara in the mid-fifteenth century. Many of them also used the three-pointed halo. Among these, a prime example is the magnificent Bible of Borso d'Este and other works by Taddeo Crivelli, along with del Cossa's complex painting.[6]

Did Jacopo della Quercia and his collaborators draw the idea from a visual source? They might have seen triangular diagrams representing the Trinity in manuscripts, but the most common iconography of the Trinity is not triangular at all: often called the "Throne of Grace" or *Gnadenstuhl*, it depicts God the Father holding the crucified Christ, with the dove of the Holy Spirit passing between them—Masaccio's *Trinity* is one famous example.[7] The triangle does appear sporadically as a symbol of the Trinity throughout the Middle Ages in devotional and scholarly texts.[8] Later—in the late sixteenth century and beyond—triangles often stand as emblems of the divine, because they provide resolution for Reformation-era conflicts over the use of images, particularly figurative ones. At that point, the geometric form is clearly a response to the risk of idolatry contained in embodied visual representations of God. Similar issues may have been at work in earlier periods that produced few explicit discussions of iconography; in this chapter I explore the relationship between the triangle and the Trinity, concentrating on the High Middle Ages up to the fifteenth and early sixteenth centuries, asking how the triangle supported the creative visualization of theology.

Representation and Abstraction

In the last chapter I suggested that Alberti's silence on religious imagery and the question of fiction was a significant silence, resonating with the conflictual history of Christian visual representation. In that history, the Trinity, as ineffable concept, poses particular kinds of problems. One might imagine that the simplest and most obvious use of the triangle—apart from the occasional representation of the musical instrument or the three-star constellation known as Triangulum—would be as a sign of the Christian Trinity. But it is not automatically the case that the Trinity could be represented this way, or at all. In one of his letters, St. Augustine wrote that it was a mistake to understand the Trinity as three monumental figures spatially proximate to one another—whether in a row or configured as a triangle. We must, he writes,

> not hesitate to overthrow a certain idolatry that the weakness of human thought tries to build up in our heart because of our familiarity with visible things and so that we do not dare to believe that the invisible, incorporeal, immutable Trinity, which we worship, is like three living masses, though very large and beautiful, each bounded by the limits of its own space and clinging to one another by close proximity in their places. It makes no difference whether one of them is located in the middle so that it separates the two joined to it on each side or whether, arranged like a triangle, each touches the others so that none is separated from another.[9]

The Trinity should not, Augustine believed, be represented as a physically bounded set of beings. It was a divine, incomprehensible mystery, and thus any visual representation of it was fraught. Over the course of the Middle Ages, depictions of God the Father, or of the Trinity, were less common than those of Mary or Christ, though not entirely absent. Augustine's hesitation about representing the Trinity, and broader concerns about images based on the Second Commandment, may have played a role in suppressing visual depictions of the concept.[10]

But visual images of God had their defenders. The defense of images by John of Damascus, a text that was widely read in Western Europe, explains that the incarnation of Christ is the key to the legitimacy of images: "Of old, God the incorporeal and formless was never depicted, but now that God has been seen in the flesh and has associated with human kind, I depict what I have seen of God. I do not venerate matter, I venerate the fashioner of matter, who became matter for my sake and accepted to dwell in matter and through matter worked my salvation, and I will not cease from reverencing matter, through which my salvation was worked."[11] The phrase "what I have seen of God" seems to imply a justification for images of Christ but perhaps not of God the Father. And John does not explicitly state that images of God the Father may be fashioned by human hands. Yet his Neoplatonic sanctification of matter seems to suggest an extension of the power of the image (as participation in, not only reference to, its divine prototype). Further, situating the artificial image in the context of definitions of Christ as image of the Father, John seems to open the door to representations of the Trinity.[12] That is to say, if one can make an image of the Son, who said, "He who has seen me has seen the Father" (John 14:8), could one not make an image of the Father?

Representations of God the Father and of the Trinity grew more common over the late Middle Ages, becoming nearly ubiquitous, though not universally approved, by the fifteenth century.[13] As Bernard McGinn writes, "The last two centuries of the Middle Ages saw a proliferation of experiments in Trinitarian iconography."[14] And, McGinn argues, theologians like Hildegard of Bingen, Joachim of Fiore, and Henry Suso took up this task enthusiastically. Suso suggested in the fourteenth century that although the Trinity is beyond thought or any visual image, one may selectively use images to "drive out one image by another" and get closer to divine truths.[15]

Augustine used his treatise *De Trinitate* to develop a more abstract and proper way to understand the difficult but crucial theological concept. Throughout the Middle Ages and well into the Renaissance, the treatise was both the most influential text on the Trinity and one of his own most widely read. It did more than simply describe the Trinity. It offered methods of interpretation: ways of reading the world and the human soul as triadic manifestations of the Trinity. It provided a template for understanding different conceptual and actual relationships. For St. Augustine, threeness had a particular valence as a trinitarian image in the world. This also appears in the writings of later interpreters, in particular Peter Lombard and the various authors who composed commentaries on his *Sentences*, such as Thomas Aquinas, Nicholas of Cusa, and Egidio da Viterbo. Augustine's theorization of the Trinity had a major, continuing impact.

The Trinity was a central article of faith for all standard forms of Christianity in the Middle Ages and Renaissance, but the doctrine of the Trinity was not a fixed and simple thing. From the sixth century onward, some Western liturgies began including the words *filioque* ("and from the Son") in their version of the Nicene Creed, the Christian profession of faith formalized in 325 CE at the Council of Nicaea. This became a major point of contention between the Western and Eastern Churches throughout the Middle Ages and has to this day never been fully resolved. The change made in the West emphasized the distinct importance of the Son within the Trinity. In the original creed, the Holy Spirit—the third member of the Trinity—is held to "proceed" only from the Father, to derive from him just as the Son does. According to the change, made in fits and starts in the West, the Holy

Spirit proceeded from *both* Father and Son. In the early Middle Ages, the *filioque* clause served a practical purpose as a counter to Arianism, which disputed the divinity of Christ. There is a certain mathematical logic to the Western view as well: if the Son and Holy Spirit proceed in exactly the same way from the Father, there is no distinction between them; the three persons of the Trinity are 1, 2, and 2, rather than 1, 2, and 3. But from the point of view of the Byzantine Church this was first and foremost a violation of Canon 7 of the Third Ecumenical Council, which forbade the promulgation of any other Creed. As a secondary effect, insistence on the original wording also underlined a hierarchical relationship between the Father and the other persons of the Trinity. Later, the attribution of an enhanced agency to the Son may have served the political agenda of monarchs: as earthly rulers, they played the symbolic role of "son" to the pope as father (pope, or *papa*, means "father"). A Trinity that emphasized equality, at least between Father and Son, might provide a model for a form of power-sharing.

If for Augustine the image of the Trinity cannot be found in any visual depiction, the concept of "image," as with John of Damascus, is nonetheless crucial: the image of the Trinity is in the human soul. Because human beings were created in the "image and likeness" of God, and God is triune, human essence must reflect the Trinity. The Trinity, Augustine argues, leaves both *images* and *vestiges* in the world. Only the angels and the human soul could boast of being "images" of the Trinity. The Trinity was "impressed" upon the soul, such that the soul operated in triads (e.g., of mind, knowledge, and love, or of intelligence, memory, and will).[16] Elsewhere in the world there exist *vestiges*, indexical signs, literally "footprints" of the divine. The Trinity acts on the world through numerous triadic relationships of which it is a template.

For many theologians who followed Augustine, the notion provides ways of reading the world in terms of triads. These might map the human soul—as in the triad of intelligence, memory, and will—or they might provide ways of imagining relationality, causality, order, and beauty. A primary explanation Augustine provides for the Trinity is that of love, which consists of the lover, the beloved, and love itself.[17] In the human mind, the highest part of man, the Trinity's image can be found in the mind, knowledge of self, and love of self, or (more commonly) memory, intelligence, and will.[18] While the three faculties always reflect the Trinity as vestige, they attain to the status of *image* only in their orientation toward God.[19] The image, thus, is understood as tending toward its exemplar, assimilating itself to it.

For Augustine, cognition and perception themselves were trinitarian. Especially significant for theories of psychology is the identity of the soul with its several faculties—that is, the notion that even if particular functions can be distinguished within the soul, they still form with it a single entity in the manner of God in three persons. The soul's faculties (intellect, will, and memory) and functions (mind, knowledge, and love) are, like the Trinity, unified yet multiple. That this diversity-in-unity of the faculties should rely on the *imago* of the Trinity stamped in the human soul is quite striking.[20] Augustine also sees a vestige of the Trinity in the operations of perception, where the perceiving power joins with the thing perceived, even though their natures are different: thus sight, the thing seen, and their joining together are a triad. He thereby produces a theory of vision based on the Trinity, privileging sight as the highest of the perceptual powers.[21]

Such triads seem to hint at the Hegelian dialectic of *thesis*, *antithesis*, and *synthesis:* Hegel was almost certainly influenced by trinitarian as well as Neoplatonic ideas in developing this notion.[22] More immediately, many medieval and Renaissance authors built on Augustine's triads. Thierry of Chartres and Alain de Lille both described substance itself as a trinity of matter, form, and their "spirit" or "joining."[23] Egidio da Viterbo in the sixteenth century described the first "person" as origin, the second as distinction, and the third as *ordo*, or order. Thus, the ordered existence of beings is itself trinitarian.[24] Where Egidio sees in the dyad the capacity for difference, Nicholas of Cusa saw equality, interpreting the Trinity as oneness, equality, and union;[25] he also illustrated it through the slightly less abstract concepts of fecundity, offspring, and love, as well as matter, form, and their union and, as with Augustine, the operations of the mind. In these examples we may incidentally detect a tendency of the triad to collapse into a dyad, figuring relationality itself: the third term is the relationship between the other two.

It may in fact be Nicholas of Cusa who contributed most to the theology of the triangle.[26] In what is perhaps his most famous work, the *De docta ignorantia*, he drew on the negative theology of such writers as Pseudo-Dionysius the Areopagite, by which only negative predicates can be attached to the divine: one can say what God is *not*, but not what God *is*.[27] He argued for an informed ignorance by which human beings come to knowledge in a negative fashion, by recognizing its opposite, ignorance, in themselves. He also argued forcefully for the value of mathematics for understanding divine truth. Mathematics has greater certainty than other forms of knowledge, he claims, because it exists almost entirely without materiality. Human beings must conduct their inquiry with material images; and materiality in the sublunar world implies changeability, an obstacle to any fixed certainty of the kind that must be the property of God. Matter being fickle, it is always ready to take a new form.[28] And even mathematics must be understood in the human imagination, which is composed of matter. But mathematical ideas are "more abstract than perceptible things, . . . very fixed and very certain."[29] In his text, Cusanus argues that human beings proceed to knowledge through created things that are "images" of divine realities. Mathematical objects are a special kind of entity: finite mathematical signs that humans can comprehend are the gateway to the *in*finite mathematical forms that both reflect and constitute divine truths. He also pursues a more visually oriented demonstration through geometry, including extensive discussions of the triangle. He suggests that an infinite triangle—in which all three angles and sides extend to infinity—is assimilable to an infinite line and an infinite sphere. All these figures approach God as they approach infinity. Taken together, these three figures—which are also one—constitute a "trine Maximum." The triangle (as opposed to the line or the sphere) has a special place as a symbol of the Trinity. It is the shape into which all shapes are divided; the "maximal" triangle, which corresponds to God, also "encompasses all polygonal figures."[30]

Cusanus elaborates in several texts on the practice of geometry as a spur to the mind's progress toward divine truths. In *De theologicis complementis*, written for Pope Nicholas V to explicate the symbolic applications of mathematics, he writes that "a geometer is not interested in lines or figures that are of copper or of gold or of wood; rather, he is interested in them as they are in themselves, although they do not exist apart from matter. Therefore, he views with his sensory eyes perceptible figures in order to be able to view with his mind's eye mental figures."[31] In other words, the mind of the geometer, beginning with figures that appear embedded in a material form, seeks to contemplate the figures "as they are in themselves." God as an infinite triangle might be understood, for example, by contrast with actual triangles.

> Moreover, the mind does not less truly behold mental figures than the eyes behold perceptible figures; instead, the mind views figures the more truly the more it views them in themselves as free of material otherness. Now, the senses do not at all attain to figures apart from otherness. For a figure receives otherness from its union with a material, which must be one material or another. On account of this union a triangular pattern in this floor differs from a triangular pattern in the wall, and the figure in the one is a truer [triangle] than [is the figure] in the other.[32]

If Cusanus believed mathematical understanding of the Trinity was superior, he nonetheless allowed for the common pedagogical use of material metaphors for God as three "persons." (To reject it would have been a very sharp, perhaps impossible, break with tradition.) He writes, "As for our most holy teachers having called Oneness, Father, Equality, Son, and Union, Holy Spirit, they have done so because of a certain likeness to these transient things."[33] The Trinity's effects in the world were also evidence for the triune God, of assistance in persuading nonbelievers. Defining the Trinity for a putative Muslim audience in his *Cribratio Alkorani*, he even used, among others, the more mundane analogy of spring, stream, and pond.[34]

Indeed, part of the appeal of Augustine's system was the possibility of finding meaning in the physical world. This extended to the shape of triangular things. In a twelfth-century mystic text, the *Vitis mystica*, the lily is described as having a triangular head that signifies the Trinity—one head, three angles.[35] The coat of arms of France—three lilies or fleurs-de-lys arranged in a triangle—was understood as trinitarian, and in medals minted during the reign of Francis I, the outline of the coat of arms was an equilateral triangle.[36] Similarly, the Medici arms with the six balls or *palle* were sometimes arranged to compose the triangular number six, with three rows of three, two, and one respectively.[37] And for Charles de Bovelles in the early sixteenth century, the heart is "sculpted" with the shape of a triangular pyramid. The pyramid, "simplest and first of all solid, bodily figures, serves as a pellucid and clear vestige of the divine Trinity"; thus it is that the heart's shape reflects the Trinity and embodies its image, the human soul.[38]

Bovelles's claims reflected earlier geometric understandings of the Trinity, which came to be applied to the problem of depicting the undepictable. The geometric triangle hovered between idea and reality; this provided possibilities for abstract ways of thinking and styles of representation. Thus many medieval authors associated the triangle with the Trinity, whether as mystical symbol or as a simple teaching tool.[39] Gerhoh von Reichersberg, in the eleventh century, wrote of the beautiful "mystery" of the number three and how the image of the triangle gives a form to (*informat*) faith in the Holy Trinity. The thirteenth-century Cistercian Garnerius of Rochefort developed an extended analogy between the Trinity and the equilateral triangle.[40] Alain de Lille wrote in the late twelfth century of the Trinity as an equilateral triangle of which all the angles are right angles. For him, the three equal lines constitute the persons of the Trinity; at the angles, they "kiss" one another.[41] Following Augustine's notion of the Trinity's "impression" on the world, Alain describes it as a specifically triangular effect. He writes, "God imprinted his triangle on the world . . . in [the world] the vestige of the Trinity reverberates."[42] He pursues this thought, inciting the believer to imprint the triangle similarly on his own mind, as a reference point for moral behavior: "Let us imprint this triangle on the mind. Let us have the power to resist vices, the wisdom to order our behavior, the beneficence to weather troubles."[43]

Roger Bacon, too, found in the triangle an apt analogy for the Trinity. In his *Opus majus*, he suggests using different kinds of geometric lines to express differing degrees of moral behavior. "It is impossible," he writes, "for the blessed Trinity and unity of essence to be represented by us more aptly . . . than by geometrical things. For in the triangle alone among all created things is found the unity of essence with the distinction of three things occupying that same essence."[44] Bacon's primary aim is to discuss mathematics as one among other practical sciences necessary to grasp the literal sense of difficult passages of the Bible. With the triangle, however, he also develops an account of the symbolic uses of geometry.

Texts like these were not necessarily designed to provide tips for artists. But the symbolic triangle did appear in a variety of medieval illuminated manuscripts. Abstracted, without figures, it appears as a divine symbol or mystical vision in manuscripts made for aristocratic patrons or monastic use—such as the Uta Codex, the Breviary of Charles V, and the Rothschild Canticles.[45] In some cases the triangle is constructed from the outlines of other objects: a suggestive absence, itself undepictable. Along with these devotional examples, the Trinity also appears as a triangle in diagrams in scholarly texts. In his twelfth-century *Dialogues*, Petrus Alfonsi, a converted Jew, finds the Trinity in the Hebrew Bible and produces a diagram of the Tetragrammaton (the divine name YHWH) as a triangle. In his diagram, each point of the triangle has a syllable formed from two of the letters of IEVE (the Latinized version): IE, EV, and VE.[46] This image was copied in many manuscripts.[47] Joachim of Fiore adapts this version of the Tetragrammaton to his own diagrammatic image of a truncated triangle, a ten-stringed psaltery that represents the Trinity.[48]

An image that became widespread, the *scutum fidei* or "shield of faith," arranged Father, Son, and Holy Spirit as a schematic triangle, set in relation to one another and to the general concept of divinity (fig. 3.3).[49] The scutum combines image with text, explaining the relationship among the three members through their distinction from one another (*non est*) and their general identity with divinity (each one "est Deus"). The iconography of the shield was articulated in the early 1230s by Hugh of St. Cher, who explicitly referred to it as a *triangulum*.[50] It became popular in France, England, and elsewhere, and had a long life, later appearing in printed French and English books of hours well into the sixteenth century (fig. 3.4).[51] Perhaps

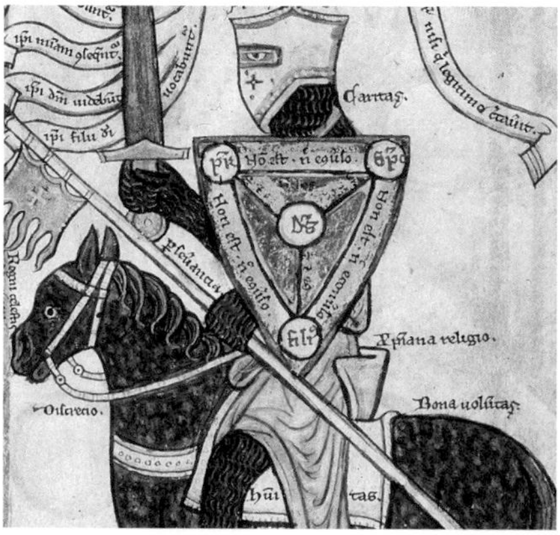

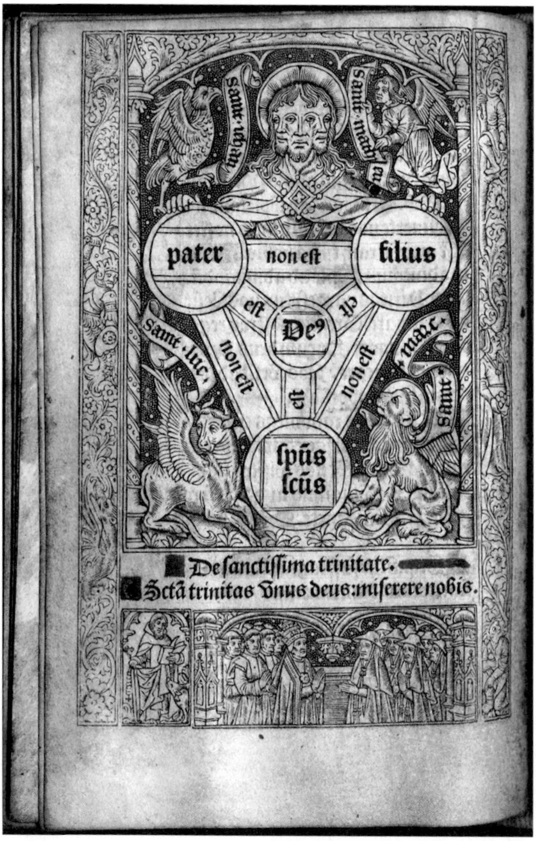

3.3 Anonymous. *A Knight's Battle with Vices*, detail of *scutum fidei*, from William Peraldus, *Summa vitiorum*, between 1255 and 1265. Illuminated manuscript. British Library Harley Ms. 3244, f. 28r. © All Rights Reserved. The British Library Board. Licence Number: DEPART02.

3.4 Anonymous. Shield of faith in *Horae Beate Marie Virginis ad usum Leodiensem* (1500), f. o2v. Woodcut. The Newberry Library, Chicago. Vault Wing ZP 539.K464. Photo courtesy of John M. Wing Foundation.

because of this strong visual association, the "shield-style" triangle was the dominant orientation for the triangle until the mid-fifteenth century. The shield is also a functional object, a defense against unbelievers. It shares its overall outline with certain depictions of the *arma Christi* as a shield, a coat of arms of the Passion of Christ that is often also subjected to geometric arrangement.

In some cases, the trinitarian triangle is depicted in combination with anthropomorphic figures. In a manuscript of Dirk van Delft's devotional *Tafel van den kersten ghelove* (fig. 3.5), the representation of a diagrammatic Trinity also contains four small emblems that suggest the positions of the *scutum fidei*: Father, Son, Holy Spirit, and, in the center, *Deus*. The emblems consist of a cross with a crown of thorns, a blessing hand with a cruciform halo, and a dove. The center image is the Holy Face—the face of Christ as imprinted on the cloth of Saint Veronica. Each of the three persons is indicated by absence: alongside the conventional (but clearly metaphorical) dove, the cross appears without Christ's body on it, and the blessing hand represents the agency but not the body of the Father. The image thus represents a juxtaposition of presence (the face) and absence (the other figures). Finally, the stark whiteness of the lines of the diagram emphasizes the immateriality of the relationships they indicate.

Geometric diagram and anthropomorphism are integrated in a fifteenth-century engraving attributed to the Master of the Berlin Passion (fig. 3.6), in which the head of Christ, the body of the dove, and the head of God the Father together plot an equilateral triangle. In a historiated initial in an early sixteenth-century lectionary made in France, the persons of the Trinity are "doubled" by a diagrammatic, shield-style triangle held by Father and Son (fig. 3.7). Combinations of figurative and geometric modes of representation present two ways of approaching the divine in a single image, as if to suggest meditative movement from one to the other, or an address to different faculties or audiences, or a reminder not to trust too deeply in visual appearances.

The direct meeting of visual representation of the Trinity as triangle in the context of theological writings finds its Renaissance culmination in the work of Charles de Bovelles. In chapter 5 I will deal more directly with Bovelles's theories of the human mind, closely linked with his views on the Trinity. Here, I address how he brought together the triangle and the Trinity in theoretical ways that extend and elaborate on the medieval discourse. Bovelles was notable for the numerous quirky, and often triangular, diagrams that appear in his texts.[52] Many of his diagrams used the triangle to

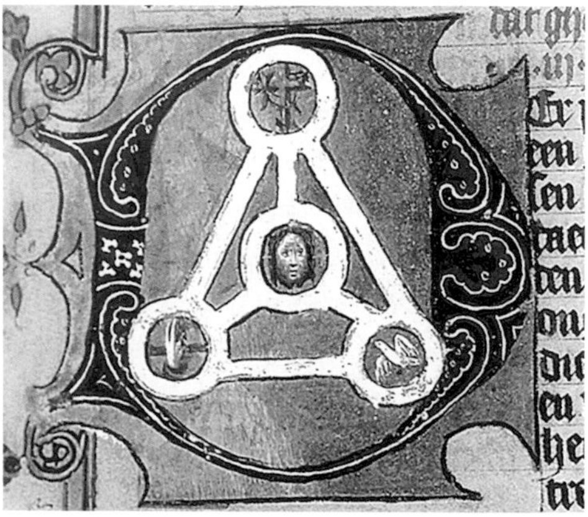

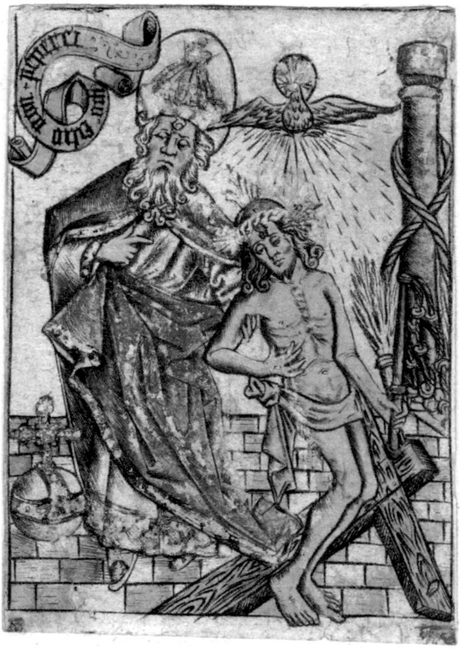

3.5 Anonymous. Trinitarian diagram in Dirk van Delft, *Tafel van den kersten ghelove*, early 15th century. Illuminated manuscript. Walters Ms. W.171, f. 2r. © The Walters Art Museum.

3.6 Master of the Berlin Passion (German or Netherlandish). *The Trinity*, 1450–1500. Hand-colored engraving with traces of gold, 8 × 6.1 cm. British Museum, London. © Trustees of the British Museum.

illustrate aspects of the universe and the workings of the human mind—such concepts as physical change, causality, and reason. He wrote a series of basic texts on mathematics and anthropology; in his works on the senses, the intellect, and the wise man, Bovelles keeps ever present the view that "mathematical symbols are the most appropriate reflection of the divine realities; they are indispensable for eyes that seek the divine sun from its reflections."[53] He also followed Cusanus in viewing mathematics as a middle place between the physical and the metaphysical,[54] opening onto both sen-

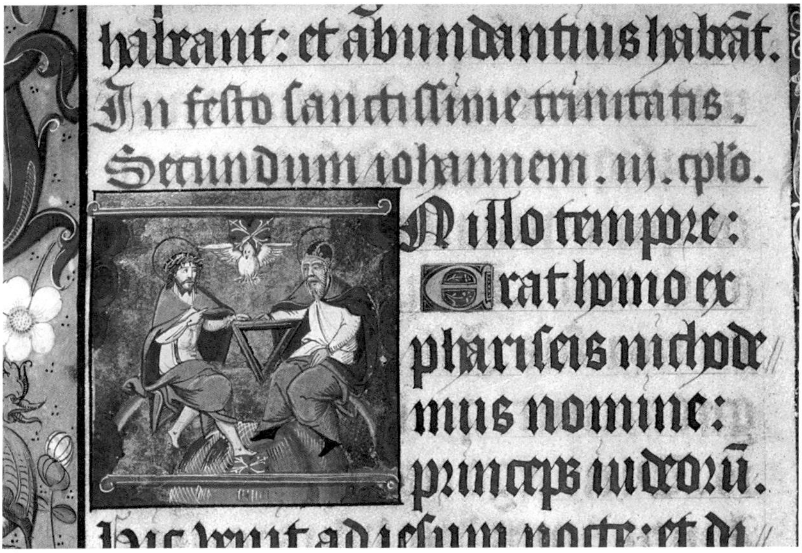

3.7 Anonymous. Trinity in initial I, "In illo tempore . . ." (John 3:1), *Gospel Lectionary*, 1500–1520. Illuminated manuscript. Princeton University Library Garrett 33, f. 153v. Courtesy Manuscripts Division, Department of Rare Books and Special Collections, Princeton University Library.

sible and intelligible realms. The trine power of the triangle, Bovelles writes, "proposes a mathematical analogy. We leave this analogy for fathoming by those who can grasp hidden things [*arcana*] from sense-based and mathematical signs and, thanks to a lofty bent of mind, the sublime truths of intelligible things."[55]

Cusanus's own manuscripts contain a sprinkling of diagrams, but nothing like the quantity in Bovelles's printed texts. For Bovelles, these writings must have suggested possible justifications for the diagrammatic rendering of theological ideas and conceptual relationships, as well. For instance, Cusanus describes the hierarchy of existent beings as a series of more acute angles toward God as God draws beings close to himself: some merely exist (an obtuse angle), some live (a right angle) and some have understanding (an acute angle). Bovelles's version of this idea is expressed in an elaborate table (which adds the function of sensing): rocks merely exist, plants exist and live, animals exist, live, and sense, and humans exist, live, sense, and think. In a corresponding section, men sunken into various vices are paired with each "lower" form of being.[56]

Diagrams provided a pared-down mode of representation that got at the "essential." Indeed, they constitute an antinaturalistic thread running through medieval and early modern visual representation. Bovelles drew on precedents in manuscript illumination and prefigures the popularity of diagrams in the print culture of the Reformation and Counter-Reformation, in which some artists sought to present doctrinal matters in a more abstract, less sensual way. Part of what gives Bovelles's diagrams the aspiration to depict abstract truth is their connection to mathematics. He rarely fails to emphasize the mathematical character of his images and concepts. Bovelles's teacher Jacques Lefèvre d'Etaples envisioned what Jocelyn Hillgarth calls "another logic, higher than that of Aristotle, one by which men could reach divine truth above conjectures."[57] Lefèvre differentiated between rational and intellectual philosophy: the rational was based on Aristotelian logic, and the intellectual in the mathematical theology of Cusanus.[58] He was, in fact, editing Cusanus's complete works (published in 1514) as Bovelles was beginning to write.

Like many others, Bovelles was inspired by Augustine's *De Trinitate* to find triads in nature and the mind. Bovelles explains that Reason operates as a triangle because unlike the eyes (which see in the daytime) and unlike the imagination (which sees, Bovelles suggests, only at night or in the darkness), Reason sees all the time. Physical causality, too, operates via a *triangle* of matter, actual form and potential form. In later chapters we will see how Bovelles imagined the soul itself, perception, and contemplation as triangular operations. Starting with the Augustinian detective procedure of examining the natural world for triadic vestiges of the Trinity, he adds to it Cusanus's mathematical theology and generates copious illustrations for these concepts. Bovelles thus provides visual models for geometric theology, making him a key figure for the relationship of idea and image in the Renaissance.

For Augustine, human beings could approach God by enlivening the triune character of the human soul—continually forming themselves into God's image. A medieval trinitarian hymn calls upon the Trinity itself to remold the human soul into its image: "Trinity light of lights / illuminate the hearts of men / remolding your likeness into your image."[59] Such notions might help explain the triangular shape of Albrecht Dürer's famous 1500 Munich *Self-Portrait*, which Franz Winzinger argued in a 1954 essay was based on a triangle inscribed within a circle.[60] When artists shaped their images into geometric forms, were they aspiring to a more abstract depiction of God—one beyond the material? Or were they instead demonstrating the effects of the Trinity on the material world? In the case of Francesco del Cossa (fig. 3.1), perhaps it is significant that the triangular halo finds echoes in the triangular formations of the composition—not only the triangle the angel and Mary make with God the Father but also the one they make with the snail. In the predella panel beneath, earthly trinities play out in the Holy Family (including Joseph) and the three dancing shepherds. More than simple "ensign" (to use Damisch's formulation), the trinitarian triangle shapes the world it inhabits. Structured by triads and triangles of differing scales, connecting beings from the lowly snail through the human to the divine, del Cossa's painting gives us a world vibrating reassuringly with the divine.

4

Equality, Hierarchy, and the Birth of Triangular Composition

Scholars and survey texts often connect idea of triangular composition with the High Renaissance; certainly the artists of the first decades of the sixteenth century put the triangle to uses that were both emphatic and innovative. But they worked from a convention that had been firmly consolidated in the mid-fifteenth century. By the time of Domenico Ghirlandaio's 1483 *Madonna and Child with Saints* (fig. 4.1), the triangle was a convention, not an innovation. In this chapter I focus on a classic moment in the history of Florentine Renaissance art: the development, in the 1430s and 40s, essentially by Fra Angelico and Fra Filippo Lippi, of the single-field altarpiece—the *tavola quadrata*—with its figures arranged in a triangle.[1]

As we saw already, John White, in *The Birth and Rebirth of Pictorial Space*, identified both affinities and tensions in Quattrocento artists' approach to decoration and the representation of space. Twentieth-century scholars often viewed perspective as *conquest* of space, and White is no exception. To him, Fra Angelico's San Marco altarpiece (plate 5) finally resolves the tension between decoration and space in favor of the latter: "space is dominant, whilst the sur-

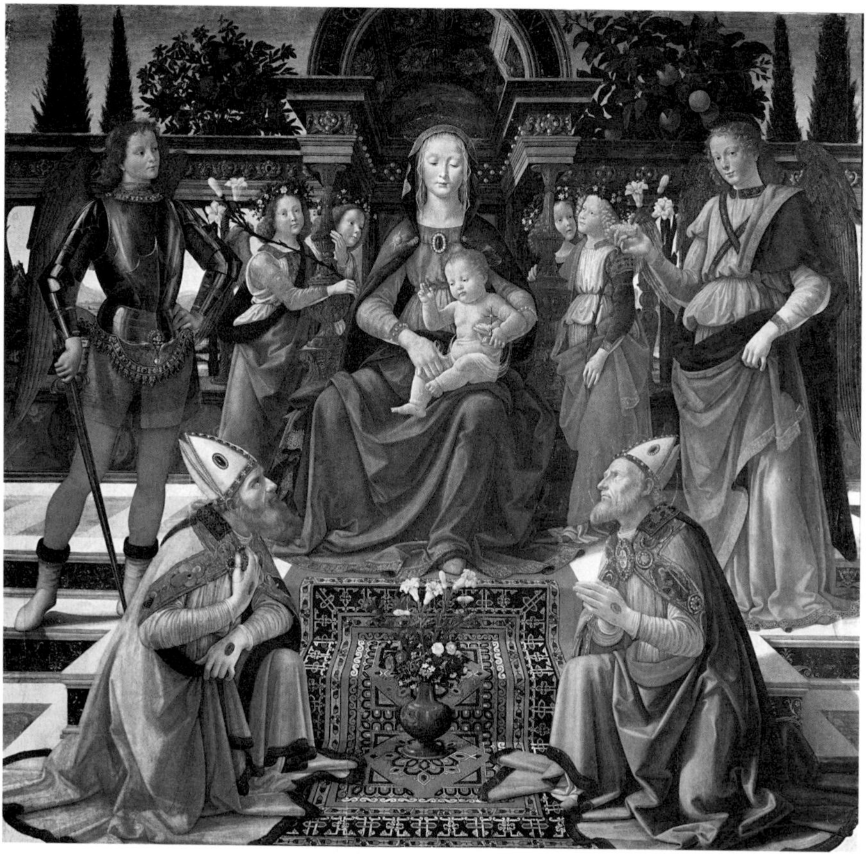

4.1 Domenico Ghirlandaio. *Madonna and Child enthroned with Saints Michael, Justus, and Zenobius, and the archangel Raphael* (San Giusto altarpiece), ca. 1484. Oil on wood, 190 × 200 cm. Uffizi, Florence. Erich Lessing / Art Resource, NY.

face filling is achieved by purely decorative elements such as curtains and garlands."[2] Ornamental geometry is indeed evident in the backdrop to the holy scene—but not only. The arrangement of the figures creates a well-defined, nearly equilateral triangle on the surface of the panel, rhyming with the perspectival lines of the carpet.

Megan Holmes and others have explained the development in terms of Brunelleschi's preference for chapel altarpieces to be *tavole quadrate* and Alberti's assumption of a quadrangular panel as a paradigm of the painting surface.[3] As Holmes notes, perspective may also have played a role. Both the compositional convention of "triangular composition" and perspectival geometry produce triangular forms.

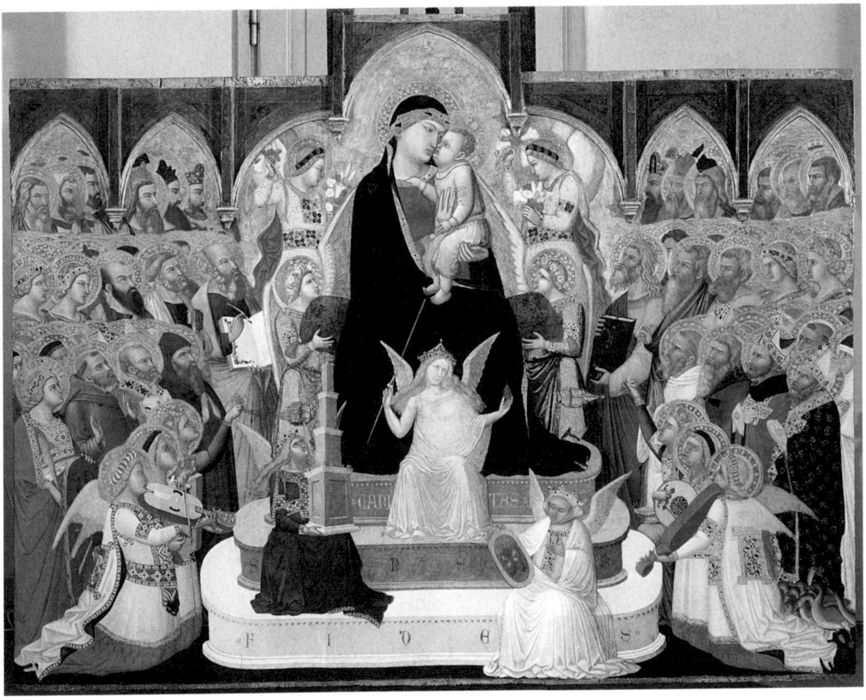

4.2 Ambrogio Lorenzetti. *Maestà*, ca. 1335. Fresco, 155 × 206 cm. Palazzo Comunale, Massa Marittima, Italy. Scala / Art Resource, NY.

Triangular arrangement became a convention in fifteenth-century *tavole quadrate*. What I will suggest is that the compositional triangle emerges from a very particular political (as well as artistic) moment and became a well-understood convention by the second half of the fifteenth century. These paintings required the artist to manage the arrangement and attitudes of the figures in a distinctive way, unlike that of a triptych or polyptych. The flanking figures in Fra Angelico's and Ghirlandaio's paintings are not *simply* lower, or smaller, than the Virgin; this would not in itself be remarkable. What's notable is that they incline toward her in such a way as to form the lines of the triangle with their bodies, or the perspectival construction of the image is such that a triangle is emphatically described—or both.

This type of arrangement is not without precedent in fourteenth-century painting.[4] Ambrogio Lorenzetti's *Maestà* in Massa Marittima, which may have been influenced by Augustinian ideas, is perhaps the strongest example (fig. 4.2).[5] Its pronounced triangular composition finds an echo in

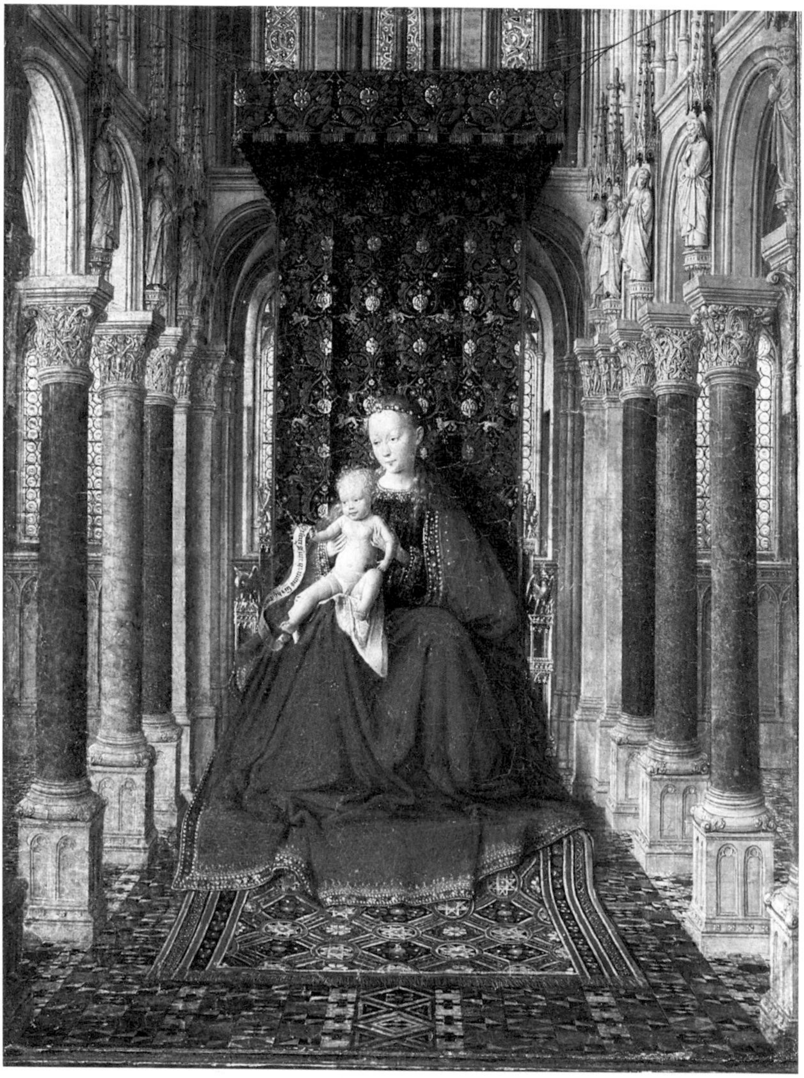

4.3 Jan van Eyck. *Madonna and Child, Saint George, and Saint Catherine (Dresden Madonna)*, 1437. Oil on wood. 33.1 × 27.5 cm (center panel). Gemäldegalerie Alte Meister, Staatliche Kunstsammlungen, Dresden. Bildarchiv Preussischer Kulturbesitz / Art Resource, NY. Photo: Hans-Peter Klut.

the three theological virtues (Faith, Hope, and Charity) arrayed on the steps of Mary's throne. A fifteenth-century northern parallel is found in the work of Jan van Eyck. Van Eyck's tiny portable Dresden triptych (fig. 4.3), with its strikingly triangular Virgin and Child, is known to have been in Italy by the end of the sixteenth century, and may have been there earlier; Florentine painters certainly had examples of northern works.[6] Features specific to fifteenth-century usage are the combination of two main properties: a unified perspectival space with naturalistic figures that all appear at the same scale, and a markedly triangular form. What's significant for this period is how decisively the shape caught on, especially in Florence. I will argue here that it did so for reasons that reflect political theology as much as, or more than, form. By the end of the first half of the fifteenth century, such triangles had become a standard way to express hierarchy through geometry.

Diagrammatic Thinking: Heymericus's Triangle and the Council of Basel

The orientation of triangular composition contrasts sharply with another triangular image that emerged from the same political moment as these Florentine altarpieces. In the early 1430s, at the Council of Basel, a rather obscure academic named Heymericus de Campo devised a triangular image he called the *sigillum eternitatis*, the "seal of eternity." He created a version with colored lines and solid black triangular forms (fig. 4.4) and several in monochromatic ink (fig. 4.5). In this image Augustine's relational notion of the Trinity, the diagrammatic tradition, and the *scutum fidei* come together. Heymericus's ambitions for what this triangular image can do provide new insights into the import of geometric arrangement in Quattrocento painting.

Heymericus was a theologian from the Netherlands who studied in Paris and taught at the universities of Cologne and Louvain.[7] At Cologne he was elected rector, and it was in this capacity that he was sent to the Council of Basel from 1432 to 1435. The Council of Basel was the first church council following the Council of Konstanz, which had ended the late fourteenth- through early fifteenth-century schism in the Western Church, established Martin V as pope, and proclaimed regular council meetings. Basel built on the example of Konstanz to argue for its own supremacy over the pope—by that time no longer Martin but Eugenius IV. The council was charged with seeking peace, in particular to attempt to win over the Hussites and to reunify the Eastern and Western Churches. Its members enacted the radically democratic practice of "one man, one vote," giving that vote to members of the council at all levels of the church hierarchy. The council must have appealed to academics like Heymericus in that it gave the same suffrage to a lowly academic—a mere canon with a negligible benefice—as to a cardinal. Antony Black has argued that such political theories derived ultimately

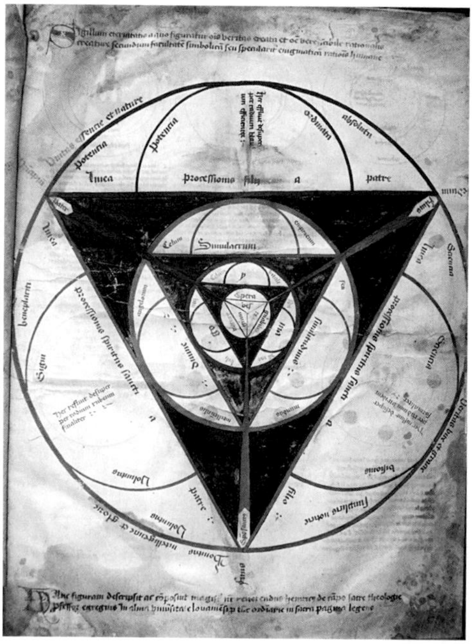

4.4 Heymericus de Campo, *Sigillum eternitatis*, ca. 1432. Illuminated manuscript. Tongerlo Abbey, Belgium, V 325, f. 1r. Courtesy Tongerlo Abbey Archives.

from the tradition of democratic self-governance in the universities, whose faculty and graduates formed most of the council's members.[8]

The lengthy meeting took its toll: Heymericus sent dispatches back to Cologne, telling his correspondents that he had spent ten florins and needed more money, that he was embarrassed in front of his colleagues from the University of Vienna who were paying his expenses, and that he would keep better accounts of his expenditures next time.[9] Heymericus debated with the Hussites and helped defend the orthodoxy of the *Revelations* of St. Birgitta (Bridget of Sweden). A diligent scholar, he also took the opportunity to copy numerous manuscripts for his university.

At the council, Heymericus created his *sigillum*, an equilateral triangle inscribed in a circle, with three lines extending from the vertices of the triangle to meet at its center. Its colors were blue (celestial substance), green (earthly generation), and red (fire, and the Holy Spirit).[10] It is not known exactly how he distributed or discussed the image at the council, but he must have tried to garner an audience for it among the council members. For Heymericus, this diagram was a means of both indexing and performing the possibility of human knowledge of the divine—in much the same way as Ramon Llull's diagrams had done in the late thirteenth and fourteenth centuries. The *sigillum* refers to God in his existence *in se* (the circle), in his three per-

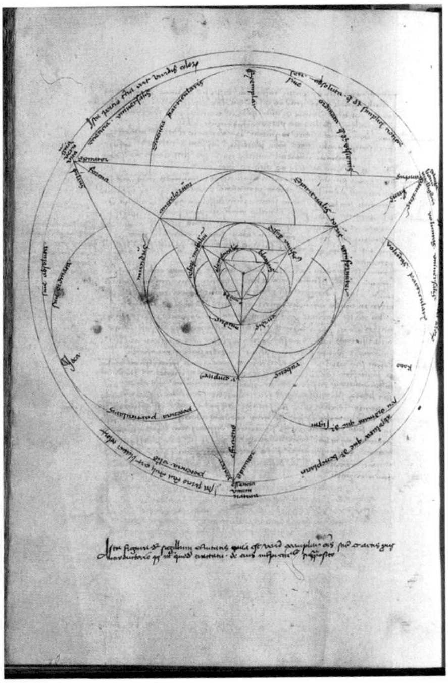

4.5 Heymericus de Campo, *Sigillum eternitatis*, ca. 1432. Ink drawing. St. Nikolaus-Hospital Bernkastel-Cues, Codex Cusanus 106, f. 186v. Courtesy St. Nikolaus-Hospital Bernkastel-Cues.

sons (the triangle), and, through the three lines, in his threefold causality (as exemplary, efficient, and final cause of things in the world). Heymericus suggests the *sigillum* as an image of how universal knowledge might be made proportional to the human intellect, which requires species or "phantasms"—often risky in their materiality—to understand divine truth. This is the problem, indeed, for mathematics. Nicholas of Cusa too suggested the triangle as a symbol for God, viewing mathematics as proportional to the mind. Heymericus also had ambitions for the political relevance of his figure, hoping it would help reconcile the church during the council's meetings.

Heymericus's seal makes concrete what in Augustine's treatise on the Trinity remains fairly abstract: the stamping of the Trinity on the world. Indeed, Heymericus uses Augustine's term *vestigium* to describe his seal; but he also uses the term *imago*, suggesting a deeper resemblance of the type Augustine reserves for the human soul. It serves as both description and demonstration, allowing him to answer all kinds of theological questions. This image, he explains, is "proportional" to the "lower part of the intellect," which is to say, the triad of reason, imagination, and sense.[11] Existing in perceptible form, it allows the human faculties preliminary access to divine truth: it is, to be precise, the means by which "eternal truth translates its image into human reason."[12] Heuristically, it was a flexible figure, capable of being interpreted in multiple ways as particular situations demanded.

Heymericus's triangle is laid out "shield-style," like the diagrams of the Trinity and the *scutum fidei* we saw in chapter 3. The *scutum fidei* had made an appearance at the Council of Konstanz too, as a shield to repel Hussite heresy, and Heymericus may have been inspired by it.[13] The shield-style orientation also appears in certain fifteenth-century images of the coronation of the Virgin by the Trinity, as we will see in chapter 5; with the Virgin forming a triangle with Christ and God the Father, these images resonate with one of Heymericus's versions of the *sigillum*, in which the church (Ecclesia, often associated with Mary) takes up the lower point of the triangle.[14]

Heymericus presumably designed the seal with the council's deliberations in mind, and he quickly extended its applications to political discussions. The relevance to Basel came in particular from the triangle. The equality of the sides and angles of the equilateral triangle, which typically remains a matter of abstract symbolism, here inspires concrete political applications. Heymericus mapped key figures in the conflict onto the persons of the Trinity: the pope was the Father, the council the Son, and consensus the Holy Spirit.[15] (The standards for the attribution of consensus to the Spirit were too loose for Ambrogio Traversari, who reported that "whatsoever this raging mob [*furiosa multitudo*] decrees is ascribed to the Holy Spirit."[16]) Heymericus recognizes the pope's authority but limits it by asserting the need to come to consensus with the body of the church as represented by the council. To Heymericus this was a middle position—indeed, later he would give qualified support to the pope while still hoping for a conciliar solution.[17] But from the pope's point of view it must have appeared quite a radical limitation of his (potential) power.

Arguing for the papal side was, notably, the pope's in-house theologian, the Dominican Juan de Torquemada, uncle of the more famous Inquisitor General Tomás de Torquemada.[18] To defend church hierarchy, Torquemada mobilized writings on celestial hierarchy by the sixth-century author who wrote under the pseudonym Dionysius the Areopagite (now known as Pseudo-Dionysius).[19] Torquemada used the Neoplatonic doctrine of emanation to claim that all power flowed from above. The pope is thus all-powerful; any power held by those below him derives only from him.[20]

A diagram from a sixteenth-century French edition of Boethius (fig. 4.6) shows how an emanationist hierarchy might be imagined. A hundred years after Heymericus's *sigillum*, this diagram represents what by that time was

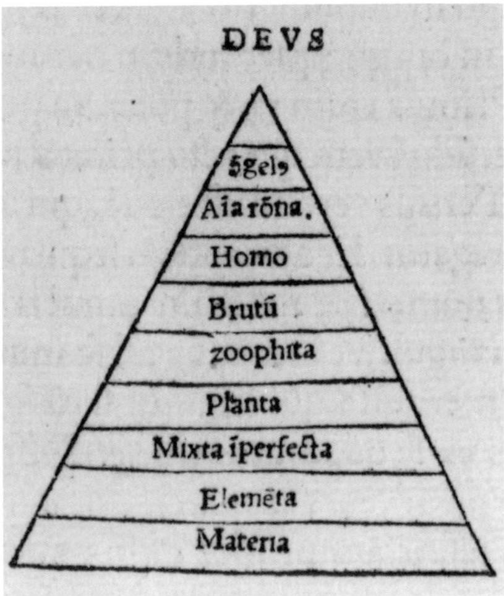

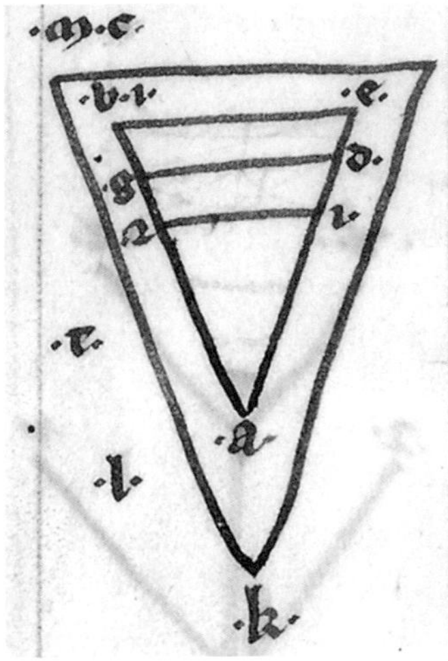

4.6 Anonymous. Triangular diagram of existents in Boethius, *Arithmetica* (1521). Houghton Library, Harvard University, Typ 515 21.224 F. By permission.

4.7 Anonymous. Triangular diagram in Aristotle, *De memoria*, 13th century. Ink drawing. Biblioteca Apostolica Vaticana Ms. Vat. Lat. 2083, f. 196v. © Biblioteca Apostolica Vaticana.

the typical orientation for triangles. The contrast between these two triangles is not trivial. The shield-style triangle was the more typical way of representing the Trinity as a triangle prior to the establishment of the hierarchical convention. It was also simply the dominant way of representing a triangle: for instance, most late medieval manuscripts of Aristotle's *On Memory and Reminiscence* orient the text's triangular diagram shield-style (fig. 4.7), sometimes with rounded lines that actually make it resemble a shield. Over time, the hierarchical type ("mountain-style") gained prominence.[21]

By the mid-sixteenth century when we see a triangle representing the Trinity, it is almost invariably hierarchical. As we have already seen, the shield-style version can be viewed as an image of equality, with the equal relation between the two top figures represented by their joining in the third. While this sense of equality is implicit in the *scutum fidei* (and reserved for God), it is fully articulated by Heymericus and made applicable to human affairs. Given the discussions on the procession of the Holy Spirit (the *filioque* clause) destined to form the bulk of discussions at the Council of Florence, it is worth noting that Heymericus accepts the position of the Western Church. The persons of the Trinity are not identical (God the Father does not proceed from either of the other persons). But he extends the relevance of the co-procession of the Spirit to the realm of politics.

Heymericus did not dispute the existence of church hierarchy; he accepted, in part, the view derived from Pseudo-Dionysius. But he believed it must be balanced, limited, by another understanding of the church. Ingeniously, he took *literally* a distinction made in the Nicene Creed's description of the church: the phrase "apostolic and catholic church."[22] For Heymericus, this meant two overlapping institutions were established within the one church. The apostolic church observed hierarchy. But the *catholic* or universal church was constituted by the council in its existence qua council and was founded on egalitarian principles—as demonstrated by the equal sides of his own equilateral triangle. In its catholicity, the church could legitimize, or delegitimize, the power of the pope.

The notion of ready translation between a diagrammatic trinitarian triangle and political theory is not unique to Heymericus. Egidio da Viterbo, the Augustinian prior influential at the papal court in the early decades of the sixteenth century, imagined twelve earthly rulers constituted as tetrahedrons of divine love by the passionate flames of the triangular Trinity.[23] A political triangle also appears in manuals written for François d'Angoulême, the future King Francis I of France, by his preceptor, Jean Thenaud. In his *Cabale metrifiee*, also inspired by Pseudo-Dionysius, Thenaud brings together the two orientations to different purposes, establishing a shield-style trinitarian triangle of "glory" (representing God) and a second one, mountain-style, which he explicitly describes as "hierarchical," below it (fig. 4.8). This is the triangle of "triumph," which represents the hierarchy of the angels. Further triangles, alternating in their orientation and imbricated in one another, emanate downward in sequence.[24] Another text by Thenaud, the *Troys resolutions et sentences*, an attempt at refuting gloomy astrological predictions, models the relationship among members of the royal family directly on his two trinitarian triangles. In a vision, his pilgrim sees "two triangles . . . in which there were six shields with *fleurs-de-lys*. From which resulted the most perfect more than pure and silver quadrangle of love and inseparable royal union."[25] Francis and his family members are mapped onto a shield-style trinitarian triangle, with Francis as the Father, Louise de Savoie (his mother) as the Son, and Marguerite de Navarre (his sister) as the Holy Ghost.[26]

Heymericus's diagram may not have been politically efficacious, but it seems to have had aftereffects in the work of one of his colleagues from Cologne, with whom he shared many interests—in Ramon Llull, in geometric theology, in Rhineland mysticism. This colleague was Nicholas of Cusa, who would later devise his own mystical *symbolum* consisting of a triangle,

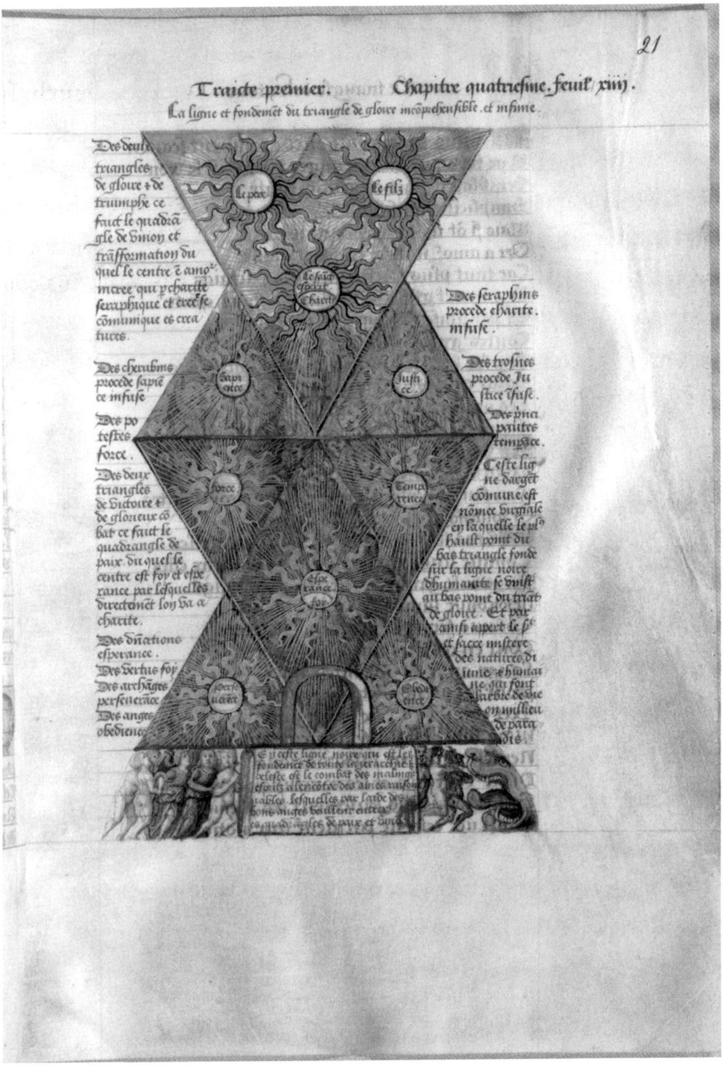

4.8 Anonymous. Triangle of glory in Jean Thenaud, *La saincte et très chrestïenne cabale metrifiée*, 1519. Illuminated manuscript. Bibliothèque nationale de France Ms. fr. 882, f. 21r.

circle, and line and would also expound at great length upon the mystical significance of what he called the "maximum triangle" (one in which all the angles are infinite, making it equivalent to an infinite line or infinite circle).[27] Cusanus attended the council as an emissary of a claimant to the archbishopric of Trier; famously, he there switched sides from the conciliarists to the papalists, becoming a strong advocate of the pope.[28] Cusanus hoped for union between the Eastern and Western Churches, and it seems he became convinced that the best vehicle for union was a strong central authority. (That authority was, it must be said, more aspirational than real; Eugenius had been run out of Rome by his opponents there.) In a sense, we might view this as a political triangulation among the positions of the council, the pope, and the Eastern Church.

The effective work at Basel came to an end after Eugenius IV called for a change of venue to Ferrara to accommodate Greek representatives coming from Byzantium. The councilgoers declined to move; Eugenius proclaimed the council dissolved; some refused to go and elected an antipope, but more of their number left for Italy, where the council convened first in Ferrara and then in Florence. While Heymericus had already left Basel by the time the council moved, Cusanus played a major role at Ferrara and Florence, accompanying the Greek emissaries to Italy and working on behalf of the pope at the reconstituted councils.

In Florence, the council finally—temporarily—negotiated union with the Eastern Church. In the discussions aimed at repairing the split between East and West, to create a united front against the Ottomans, the *filioque* clause was a main area of dispute. Western church representatives argued that as a development of an idea already existing in the very definition of the Father—the Son's identity with the Father—this was not actually a "new" statement. The Easterners did not forcefully dispute this definition of the Trinity but held strongly that the Creed could not be altered. A compromise, in which different phrases ("from the Son" and "through the Son") were declared compatible, was short-lived; after the council it did not win the approval of the Byzantine synod of bishops. But one consequence was that the pope advanced the cause of central authority. The event helped give both practical and philosophical support to a hierarchical trend that would intensify over the fifteenth century both within the church and in political bodies throughout Europe.

Mathematics and Its Meaning

Given that triangle, line, and circle are the three components of Heymericus's *sigillum*, it is suggestive that Cusanus defined divinity as an infinite triangle that is also an infinite line that is also an infinite sphere. He seems to have alluded to Heymericus's ideas in his *De docta ignorantia*.[29] But Cusanus's notion of the triangle is more abstract, not so concretely applicable to political events or papal power. Separated from the human realm, it is a way of accepting human limitations in a quest for spiritual ascent. Less charitably, we might see it as a way of evading certain difficult political questions.

Like Heymericus, Cusanus uses geometry as a way of explaining divine truths. The divine triangle is absolute and beyond all comparison, unlike material triangles, which have relative differences among them. By recognizing this incommensurability, one can begin to contemplate God. Thus, the varieties of matter allow for diversity in materializations of the triangle. Some are "truer" than others, but none is a true triangle in the mathematical sense, that is, a triangle present only in the mind. The mind is not composed of matter: it is "free of perceptible material." Not only is the mind thus empowered to view figures without material otherness, and hence in their true form, but the mind itself *is* their form. "Therefore, whatever [figures] the mind views, it views in themselves."[30] Cusanus suggests a specific place for the human mind in the cosmos, one defined by mathematics.

This idea forms part of a broader theory of mathematics as spiritual tool that Cusanus developed, drawing on Plato and in dialogue with another friend and collaborator, the cardinal Basilios Bessarion, who helped negotiate union at Florence on behalf of the Greeks.[31] Bessarion was one of the fifteenth century's great champions of Plato and a personal vehicle for the arrival of Greek manuscripts, including many mathematical texts, in Italy. The view of mathematics as a middle place was picked up by Bessarion in his defense of Plato (*Adversus calumniatorem Platonis*), written in 1459. In it he argued that mathematics offers a midpoint between the material and the immaterial that allows the mind to become abstracted from material things.[32] Both Cusanus and Bessarion described mathematics as a contemplative practice that promised to train the mind to "sequester" itself from matter and aspire to the spiritual. It was a purgative mental discipline.

For Cusanus and Bessarion, taking the word *mens* (mind) as connected to measurement (*metior*, *mensura*), mathematics had a specific identity as the language of the mind, a midpoint between the earthly and the spiritual. Bessarion cites Plato's argument that "one can find nothing better suited to the illuminating of the human genius than the mathematical disciplines, which, abstracting from the material, seem to bring themselves up to the divine condition and offer a sort of step by which the soul can ascend to divine things."[33] Continuing his reasoning, he asserts that the more the human mind is absorbed in "the senses and visible and imagined things," the more it cannot reach the immaterial. Bessarion wrote,

> With mathematics, which are, with a middle nature, constituted partly as material, partly as immaterial, [philosophers] trust it can be accomplished, that the mind can more easily attain its goal. Naturally it is necessary for the triangular figure or line or point to be contained in matter. But it is known to be possible for the mind to be abstracted by the use of mathematical reason without any material support. . . . Thus we become accustomed to departing from matter and we exercise our minds to consider form in itself without any material condition and thus we are more easily brought back, penetrating to divine things through this very apt medium, or, we might say, vehicle.[34]

Mathematics is different from other forms of human knowledge; it can serve as a kind of contemplative exercise that allows the mind to reach a more spiritual state. With the idea of mathematics as a *median* realm comes the notion that it is a *medium*: it is what allows movement between opposites. In this way mathematics is also like an image. Bessarion pursues his argument with a visual metaphor:

> To be sure, mathematical things seem to be images of intelligible things in about the same way as we see the images of natural things represented in shadows or water or mirrors. Those who delve into their study do not remain forever considering these sorts of images, but are stimulated finally to those things of which they are images. . . . Little by little they sequester the sense and act by the sole judgment of the soul concerning these things, which we call mathematical intelligibles, ... by which alone man can truly be called divine and a god.[35]

He does not liken these "images of intelligible things" to painted or drawn images but rather to those created naturally: shadows and reflections. But he does use, of mathematical "images," language that resonates with debates over the proper use of artifactual visual images in worship. Worshipers, similarly, were expected not to "remain forever considering these sorts of images" but to be "stimulated finally to those things of which they are images." Following this practice, one could begin by contemplating the image of the divine through mathematics and (Bessarion emphasized) eventually refer one's contemplation to God himself, not remaining "stuck" in the image. This was standard Christian image theory, in which devotion was to be referred to the holy prototype (Christ, the Virgin, saints) rather than resting in an image.

Though Cusanus, like Bessarion, is influenced by Plato, his view of the mathematical forms that human beings use in thinking is distinct from the ancient philosopher's. He argues that these forms are *specific* to the human reason and intellect. In *De theologicis complementis*, he suggests that when the mathematician draws a polygon—the example is a triangle—he imagines its "infinite exemplar" before bringing it into finitude in his own drawing.[36] He writes in *De beryllo* that "Plato is seen wrongly to have concluded—when he says that mathematical entities, which are abstracted from perceptible objects, are truer in the mind—that therefore they have another, still truer, supra-intellectual being."[37] Instead, they are truest, most "at home," in the mind itself. This makes mathematics an especially good way for humans to think—it is the *proper* tool, even the material constitution, of the human mind. In his *De theologicis complementis*, Cusanus describes the way the mind perceives geometric figures in their existence "without otherness." That is, they are present in the mind "as in their own form"; the mind "stands in relation to mathematical figures as being their form."[38] This corresponds with his statements in *De coniecturis* that number is the means by which reason operates; its use of mathematics is its own self-fashioning.[39] In assimilating the perceptual images into intelligibility, the intellect actualizes itself:

> Now, since the perfection of the intellect is its actual understanding (for when the ability to understand passes into actuality, it is perfected), the intellect is its own fecundity: it makes intelligible (by means of its own resources) that which comes into it. The intellect's descending unto perceptual images is the perceptual images' ascending from the conditions of their contractedness unto less contracted simplicities. Therefore, the more deeply the intellect enters into these images, the more they are absorbed by its light, so that, at length, the intellectual otherness, having been resolved into the intellect's oneness, finds rest as in its own goal. Therefore, the oneness of the intellect is made more perfect the more it proceeds from potentiality into actuality.[40]

One might assume with Plato that mathematical order exists infinitely in the mind of God and preexists as "ideas" in the human mind—or that, in Aristotle's view, it exists in the world and enters the mind through a process of abstraction. Cusanus takes an original, middle view by which mathematics *is* the idiom of the human intellect.

How, then, can we have knowledge of divine ideas, or forms? We have, he suggests, oblique knowledge of them by analogy with mathematical knowledge. We can understand the immaterial figure of a triangle through our knowledge of its material manifestations (the figure of a triangle conjoined with the material of a wall, for instance). In his *De possest*, he writes, "If we know something about [divine works], we surmise it by likening a figure to a form. . . . From figure, which gives being in mathematics, [we make an inference about] form, which gives being: just as the figure of a triangle gives being to the triangle, so the human form, or species, gives being to a man."[41] Thus, contemplating physical shapes of things, we have an analogy for divine ideas. He concludes this same set of reflections by asserting that human thought requires images to operate at all: "Everyone who understands must behold images."[42]

Aristotle had used the metaphor of the plan of a building in the mind of an architect to convey the notion of form; Cusanus renders this analogy more accessible with the figure of the triangle, but he also extends it to make mathematics the basis of all human knowledge, from the empirical and scientific to the theological and speculative. One implication of these ideas is the notion that philosophy is properly conducted through the drawing (and rigorous construction) of figures; visual art, as visual demonstration and argument, could thus have a special place in philosophy.

In contrast to the view that all of nature is written in the language of mathematics—associated with Galileo but present in philosophy since the Neoplatonists—Cusanus places mathematics specifically *in the human intellect* as the mediating place between nature and God. He thus attributes to human beings a unique contemplative role in the cosmos. We might speculate, following these ideas, that when painters attempted to "mathematize" their pictures, it was in order to make visual arts that were more proportional, more "appropriate," to the powers of the mind, specifically the intellect.

The Language of the Mind and the Language of the Body

What might an image that speaks this "language of the mind" look like? For Heymericus, it was surely his own *sigillum*. But for some of his contemporaries, it might have been the geometry of painted altarpieces. Cusanus could contemplate the relationship between a geometric triangle and a triangular pediment and imagine what further purgation might be necessary to move beyond geometry altogether into divine contemplation. As we saw in chapter 1, Ernst Cassirer argued that the union of mathematics and art produced the concept of the "necessity" of nature—the philosophical basis of Renaissance naturalism.[43] Rather than assuming that the mathematization of images equals naturalism, we might consider this process as a method for enhancing the aptness of images as contemplative prompts. If paintings come to be composed geometrically, as triangles, perhaps it is in order that we might contemplate more elevated triangles in the mind—to move beyond, to even more sublime things.

Indeed, can we understand paintings as an effort at producing a representation proportional to the human intellect—and eventually a theological truth—not a capturing of reality at all? If most artists did not seek to be theoretical mathematicians or profound theologians (at least not most of them), still, I argue that one of the ambitions of artists in the fifteenth and sixteenth centuries was to tell devotional and other stories in *the language of the intellect*—pictures that speak the language of the mind, that is, mathematics.[44]

What might this mean? The idea that the picture should be written in the language of the mind suggests a kind of mirroring between viewer and object. This appears, too, in the analogy Alberti makes between painting and the love of Narcissus. He wrote, "The inventor of painting, according to the poets, was Narcissus. . . . What is painting but embracing, by means of art, the surface of the pool?"[45] The structure of equality and participation suggested by Alberti's remarks on perspective is radicalized here as a kind of merger between viewer and viewed, or painter and painted. But Alberti—even as he celebrates painting—seems to hint at a certain danger in this assimilation. For him, painting should represent what can be seen. Like Masaccio's *Trinity*, Fra Angelico's San Marco altarpiece is a major painting of Alberti's time that depicts things *not* seen—exactly what Alberti states should not be the concern of painters. The Virgin is enthroned in a celestial company of saints. Details of the setting allude loosely to the heavenly Jerusalem—the throne and trees (of life?), and the river (perhaps that of paradise) that flows in the background. Mary delicately veils her infant son's genitals; he holds a globe that bears an actual map of Europe, Asia, and Africa, with the Holy Land indicated as a bright spot at its precise geometric center. Cloth of Eastern origin—the Anatolian carpet and Bursa silk hangings—echoes this geography. While it may be constructed of elements based in visual observation, it is clearly not a scene presented to the eyes of the body.

To address this painting further, let's return to Florence and the years 1439 to 1443, when Pope Eugenius was in residence in Florence, in quarters in Santa Maria Novella, as the council was meeting. The pope had for several years already been exerting influence over the city's political and religious affairs. He helped Cosimo de' Medici return from exile in 1434 and encouraged him to undertake an energetic building campaign of monasteries and churches. In 1436 he intervened directly at San Marco, ordering the then-resident Silvestrines to exchange residences with the Observant Dominicans, the order to which Fra Angelico and Juan de Torquemada—who also argued the case for the switch at Basel—belonged.[46] San Marco had been granted to the Dominicans by Martin V, Eugenius's predecessor, but they had not yet been formally installed there. Two years later Cosimo began financing the renovation and expansion of San Marco, where he also acquired patronage rights to the high altar on which Fra Angelico's painting would stand.[47] Connections between the Medici and the pope tightened even further as Cosimo bankrolled the council of Florence.[48] Cosimo also shared Torquemada's interest in Pseudo-Dionysius; the same Ambrogio Traversari who had denounced the council as a mob was also, in the 1430s, translating Pseudo-Dionysius's complete works into Latin under Medici patronage.[49]

In these same years we can identify the beginnings of a convention: the quadrangular altarpiece that displays triangular composition. As we have seen, triangularity in itself is not especially new. In many central triptych panels, Mary sits enthroned and, with her flowing robes, forms a steeply sloping isosceles triangle. If we compare this panel with standard polyptychs with saints in a row to either side of Mary, we might imagine that as the central panel extends outward to subsume the wings into itself, the base of the triangle broadens to include additional figures. It becomes, essentially, an equilateral triangle. And certainly one element in creating this convention, as Holmes suggests, must have been the development, over these same years, of geometric perspective. Perspective can provide a schema that "supports" or emphasizes the compositional triangle; it might also support theology—as Daniel Arasse argued—and ecclesiology, rather than just describing space. With the erosion of the polyptych format, triangularly composed square altarpieces subsume all figures into one space. The saints come to join a hierarchically posed Virgin in a space that she defines.[50]

Classically, we would say that this is a function of the new, perspectival mapping of reality. It is also, I suggest, a visual metaphor for the papal doctrine of church hierarchy—ideology that may also be at work in another monument of early Renaissance art in Florence, the Brancacci chapel at Santa Maria del Carmine.[51] If all power flows from above, the "unified field" might work to deny the existence of separate, distinct sources of power. The acute viewer might see, behind the triangular arrangement of the figures, the geometric triangle that can prompt contemplation. But any viewer could get the message of relational hierarchy. This is not just any triangle; it is one that metaphorically turns Heymericus's *sigillum* upside down.

Does it do so consciously? It is likely that Torquemada had seen Heymericus's *sigillum*; it is almost certain that Nicholas of Cusa had. Moreover, the Dominican archbishop of Florence, Antoninus, as we saw in chapter 3, took an interest in "spiritual geometry." Fra Angelico was not just any painter but an Observant Dominican himself and a resident of the convent who would soon go on to work directly for the pope.[52] He may have been especially attuned to mathematical symbolism: Arasse has argued, for instance, that his Cortona and San Marco *Annunciations* represent the manifestation of the Trinity in the Virgin's body through Mary's triple-arcaded dwelling.[53]

We cannot know how directly ideas about mathematical contemplation might have influenced developments in the arts. There is most likely no "smoking gun." The point I want to make is not that Fra Angelico was thinking about overturning Heymericus's triangle. I am interested in the politics around the painting, but even more in the politics that surrounded the establishment of the triangular convention. When we consider such paintings in the light of Heymericus's triangle, their statement of hierarchy seems less inevitable—more a *statement* and less a simple fact. The altarpiece, in other words, comes to aestheticize a political tendency toward hierarchy.[54] The development of the convention of hierarchical triangular composition in the years of the council's presence in Florence might be an indication of just how decisively the road of Basel was not taken.

Heymericus is not only important for the orientation of his trinitarian triangle. He is significant because he suggests a sense of the power of visual representation, an alternative way of thinking about the "geometric" image. His *sigillum* visualizes theology—but doesn't merely diagram theological truths; it seeks to bring the political world into accordance with them. If we accept the force of his egalitarian emphasis, it can defamiliarize triangular composition. To those of us fully accustomed to this convention, it makes the hierarchical orientation less obvious. The triangle is not, by nature, a hierarchical entity; it is so when the choice is made to present it in a certain way. What I want to stress is this: the repetition of the exercise of power is not the same as its naturalness or inevitability. Each instantiation makes its own intervention, and the power expressed here *was* contested by at least some of its contemporaries. Too often, in assuming the inevitability of the outcomes that actually occurred, we lose sight of the conflicts, real or imaginary, that resulted in them. Reaction is reaction to something; history presents multiple possibilities.

To sum up these three triangles, then: Nicholas of Cusa's triangle is completely abstract—it's through *distance* from the visible that one can imagine the spiritual. In this way, it downplays the significance of specific doctrinal programs or political allegiances. Heymericus uses a more concrete conceptual mode, showing that it was possible to depict and enact theology visually—in his case, to contest papal hierarchy. His triangle represents a symbolic use of geometry, a clear effort at producing an image proportional to human faculties that will tell divine truth and be efficacious in contemporary political affairs. Meanwhile, Fra Angelico's rigorously triangular, "mountain-style" composition represents a distinct shift toward geometric hierarchy. At a crucial moment in the formation of the triangular compositional convention, the altarpiece presents a highly geometric arrangement keyed to contemplative use and strongly supportive of hierarchy.

The Body of the Virgin

It remains to consider, however, the fact that the top point in the hierarchy is expressed through the body of Mary: that is, the Virgin's body is the vehicle for the instantiation of hierarchy. Is the abstraction of geometry troubled by the presence of this physical maternal body? Fra Angelico's use of the figure was in one sense typical: a seated Virgin and Child was certainly conventional subject matter for early fifteenth-century Tuscan altarpieces. If he or his interlocutors imagined a mathematical contemplative process for the learned viewer, what was the place of Mary in this process? To be honored, but also, perhaps, to be transcended in favor of a purer, less material (but still geometric) form of contemplation?

In his treatment of the figure of Mary, Fra Filippo Lippi's altarpieces of this period can be seen as presenting an alternative to Fra Angelico's. In his Barbadori altarpiece in the Louvre (fig. 4.9), which also composes the three main figures in a triangle, the figure of Mary comes radically forward from the position she holds in the San Marco altarpiece—or we might say, since Fra Angelico likely knew Lippi's painting when he painted his, that the Dominican painter *removed* her from the viewer's space. For Lippi, Mary and Jesus are more immediate, intimately embodied presences. The composition is also dynamic, with moments of slight asymmetry, the triangular composition echoed in reverse by the two saints' croziers, and the slightly swaying stance of the Virgin rhyming with that of the "stepping" Christ child. Judging from the smaller scalloped half-dome topping the niche behind them, what we might have here is a statuette of the Virgin striding out of its niche

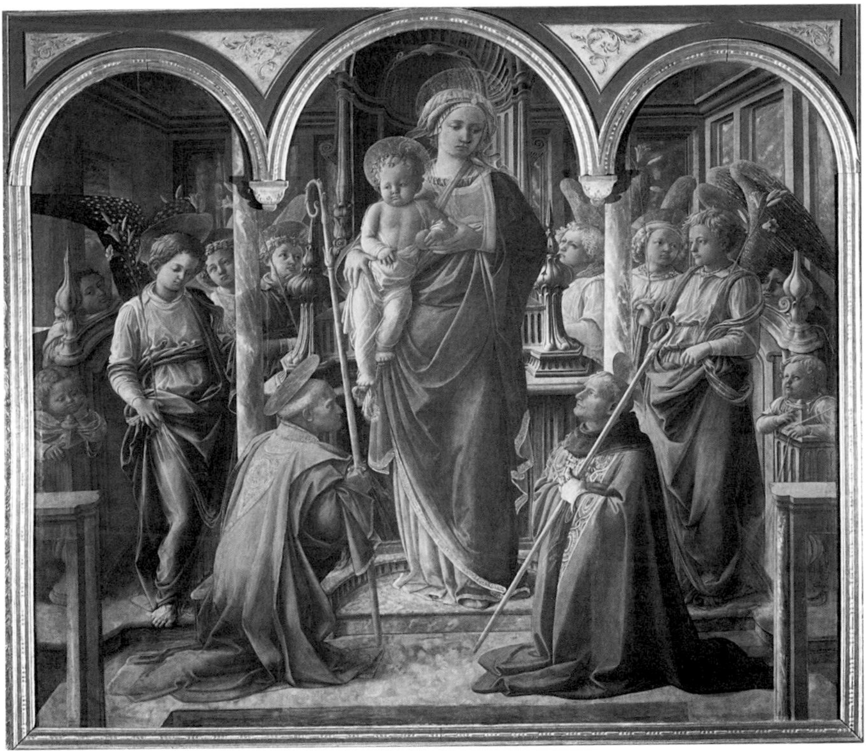

4.9 Fra Filippo Lippi. *Madonna and Child with Angels Saints Frediano and Augustine (Pala Barbadori)*, 1437. Tempera on wood, 208 × 244 cm. Louvre, Paris. Erich Lessing/Art Resource, NY.

and into the figures' space. In these two paintings and similar ones, the figure of Mary seems to have been the object of geometric reflection in terms of perspective as well as composition.

And it is striking how frequently—especially for paintings created soon after Alberti wrote his *De pictura*—artists place the vanishing point of a painting at a significant spot on Mary's body. As we saw in chapter 2, Alberti suggests it be placed at the height of a man. But artists relentlessly disregarded this precept and placed it within the Virgin's body. In the San Marco altarpiece, it is at her heart. In Botticelli's Sant'Ambrogio Madonna, the vanishing point is positioned at her belly.[55] In addition, the Virgin's blue robes, parted over her heart, create a triangle of red cloth bounded at the bottom by her girdle—perhaps emphasizing, indeed reflecting, the painting's triangular order.

Compositional triangles also echo the one formed by the perspectival system; as in the San Marco altarpiece, human bodies and abstract geometries respond to one another. If we think of perspective as made up of such triangles, we may also think of it as a compositional tool. If perspective and compositional geometry echo one another, without being exactly the same, it might suggest that one represents the eyes of the body, the other the eyes of the spirit. The saint who looks out at us—Cosmas-Cosimo—gestures us toward the devotional experience his companion, Damian, is in the process of receiving. Damian faces away from us, facilitating the viewer's potential identification with his position.

Taking into account the relational, bidirectional conception of vision in this period, we might say that the hierarchical triangle, or the Trinity, in fact *emanates* from the body of the Virgin. Mary is symbolically associated with Ecclesia, the female personification of the church. She serves as a support for the expression of many different theological concepts. She is also described in various devotional texts (as we shall see) as the temple or receptacle of the Trinity.[56] On the threshold of Fra Angelico's San Marco *Annunciation*, she is described (in a quote taken from the theologian Adam de St.-Victor) as the Trinity's "triclinium," which in medieval Latin designated a room divided into three parts.[57] The Virgin might express the generative power of God, her body representing the humanity of Christ. Her position, too, expresses concepts of majesty and hierarchy. Through the iconography of the triangle, emanation and generation can be seen as comparable, even allied processes: birth as metaphor for divine creation. This would provide some of the visual and symbolic grounding for the triangular compositions that were to become so prevalent in representations of the Holy Family—most notably those of Raphael and Leonardo.

The bodily focus of the vanishing point in many fifteenth-century paintings has been an issue of curious confusion. Scholars have puzzled especially over the placement of the vanishing point in the main panel of Domenico Veneziano's St. Lucy Altarpiece, which also violates Alberti's precepts (plate 6). In the main panel are the Madonna and Child flanked by saints, an example of the iconography now known as *sacra conversazione*. A near-exact square, the shape implies a conscious use of geometry, echoed by the coordination of spaces within the composition and the multiple triangles that appear in the architectural forms.

In *The Renaissance Rediscovery of Linear Perspective*, Samuel Edgerton gives the vanishing point of this painting special prominence by placing it on the cover of the book, with lines marking its perspectival construction. The problem he addresses is the fact that the painting's centric point is lower than it ought to be according to Alberti's prescriptions; it is well below the heads of all the figures. Edgerton argued that Domenico used the idiosyncratic placement of the centric point, between the Virgin's legs, to avoid the effects that a higher placement would have had on the relationship of the figures to what we might call the decorative aspects of the architecture. By placing the centric point low, he suggests, Domenico created a sort of architectural halo for the Virgin and the four saints who surround her. While Thomas Puttfarken takes issue with this view, he suggests that the unusual placement of the vanishing point is the result of the painting's original position above the viewer's head—but this is not Albertian either. James Elkins does point out that it might have something to do with the Virgin as bodily origin of the Savior, saying, "The vanishing point . . . is certainly lower than the position Alberti prescribed, but not arbitrarily so: it coincides with the Virgin's womb, the spiritual and dogmatic center of the painting."[58]

But this is not quite right. Presumably the Virgin's womb is quite a bit higher than the centric point delineated on this image. The triangle framed by the primary perspective lines is a shield-style triangle, like the trinitarian diagrams we have seen. Tracing these perspectival lines from above, one finds them framing a pronounced triangle of cloth between the Virgin's wide-open legs.[59] Its centric point might more properly be associated with the very base of her pelvis—that is, her genitalia, or perhaps we might say, her intact hymen—perhaps suggested, too, by the triangle of cloth between her knees.

This triangle of cloth, echoed in the stark black triangles of the spandrels above, reminds us that there are things we are not meant to see. The three steps of the Virgin's throne move from one with a legible Latin inscription, to pseudo-Hebrew mystery, to complete veiling.[60] The body encloses an unseeable mystery; the cloth might refer to the hymen but also blocks our view of this mystery of mysteries. The viewer is offered access to the world within Domenico's painting, but only to a point. In my reading of Alberti I suggested that the *intercisio*, as blockage, is as significant a metaphor for perspective as the view through the window. Here we seem to have both, an invitation to enter but a conviction that some things will remain hidden. Perspective is neither *only* penetration nor *only* its refusal. And if we consider the lines that radiate outward from a central point, we might consider it also to be about generation. For as much as our gaze "penetrates" the image, the image radiates outward toward us, just as salvation radiates outward from the Virgin's body.

The question of vision and visibility is also brought into play by the presence and actions of St. Lucy, the eponymous saint of the church for which the painting was made. Before being martyred, the early Christian saint was held to have torn her own eyes out to avoid being the object of attraction to suitors. Here Lucy certainly possesses at least one intact eye with which to gaze upon the Virgin—but it may be that it is a spiritual eye, since her two glutinous eyeballs rest like fried eggs on the platter she holds in one hand. The raw materiality of the eyeballs on the platter asks us to set aside our mortal eyes and see something more abstractly spiritual.

To bring this back to Fra Angelico, there is one more detail to consider in the San Marco altarpiece: the small, fictive, archaizing panel of the crucifixion that appears front and center. It functions as a *repoussoir*, insisting on its own presence—a made object—within the viewer's reality and the corresponding immateriality, as divine vision, of the scene beyond. The scene beyond, unlike the small panel—this would be the picture's argument—overcomes its own imageness.

The status of the whole panel as a divine vision can explain its strictly ordered geometry. It certainly calls into question the assumption that perspective was employed to depict real space. And yet the picture is still in the world; hence the internal struggle, within this emerging convention, over the status of the body of the Virgin. Geometric figure for Angelico, real embodied presence for Lippi; both at once, in tension, perhaps, for Domenico. The physicality of Lucy's eyes is what she seems to be manipulating with her penlike, or perhaps paintbrushlike, palm frond, on a tablet that could be a painter's palette or sketchpad. The eyes themselves might be understood as the medium of the painter: St. Lucy's name, too, refers to light. She holds a palm leaf—"like a pen," Hellmut Wohl pointed out, but one might even say, "like a paintbrush."[61] Here is something we could not see with mortal eyes without painting to bring it to us.

5

A Flood of Tears

The triangular arrangement of Fra Angelico's altarpiece presents a philosophical argument for hierarchy grounded in mathematics, while Filippo Lippi and Domenico Veneziano present strong alternatives that foreground the body of the Virgin. But even these differences are exploded by the stark compositional alternative that appears in Paolo Uccello's *Flood*, in the cloister of Santa Maria Novella, which Harald Keller called the "most daring" of Quattrocento instances of triangular composition (fig. 5.1).[1] Keller writes, "The sides of Noah's ark—which is depicted twice in a symmetrical arrangement—converge to make up the diagonals of a triangle enclosing a wild rout of panic-stricken and dying human beings."[2] This frescoed lunette is one of a series of Old Testament narratives in the cloister (called the Chiostro Verde) of Santa Maria Novella, in Florence. Noah's Ark appears on the left, at the onset of the deluge—and then again on the right after the waters have withdrawn. In the triangular field between the two arks, people (and animals) drown, fight, and desperately scrabble to get inside. The scene is all

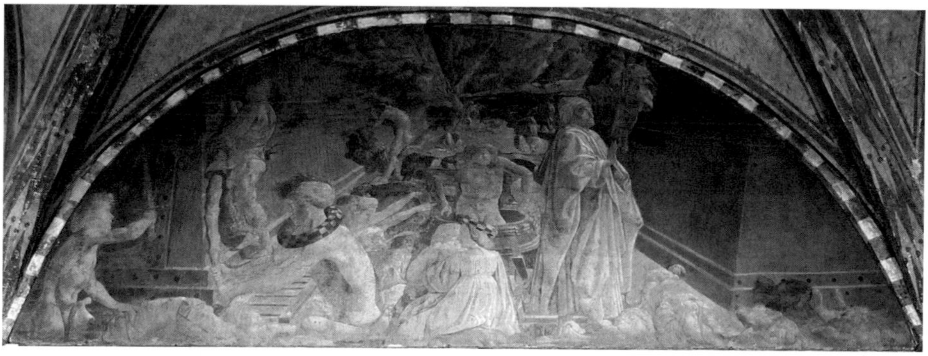

5.1 Paolo Uccello, *The Flood*, 1447–8. Fresco, 215 × 520 cm. Santa Maria Novella (Chiostro Verde), Florence.

the more poignant for the fact that we know what the participants don't—how the story will end for them.

The mathematically punctilious Uccello was famously criticized by Vasari for his obsessive interest in perspective. The critique—as it often does with Vasari—points to excessive devotion to an otherwise acceptable (even necessary) component of the painter's art. What is striking is not only that Uccello devotes singular attention to perspective in his paintings but also that he makes perspective part of the subject matter of the work. Composition, narrative, and perspectival system infiltrate one another. In the *Battle of San Romano* panel in London, a grid of broken, fallen lances rhymes with the perspectival scheme. In Uccello's *Flood* too, perspective infuses subject matter. The arks' smooth surfaces and straight lines—their geometry—suggest perspectival triangles. They serve as a sharp contrast to the writhing mass between them. Likely inspired by the ark on Ghiberti's *Gates of Paradise*, they are represented as pyramids.[3] Yet this geometry is difficult to follow, and not only because it can't be understood as the snapshot of a single moment. The right ark, set at a very slight angle with the picture plane, makes the perspective difficult to understand at a glance. We have the hint of a few orthogonals on the left that can help us find a vanishing point; it corresponds approximately to a point where lightning strikes in the background. Following Alberti's precept, Uccello places this point at the approximate height of those figures that stand on the ground. But most of the figures, swept up in the rising waters, cannot actually stand.

Finally, another problem emerges. It appears the two arks represent different moments—"before and after"—as the waters rise and then as they

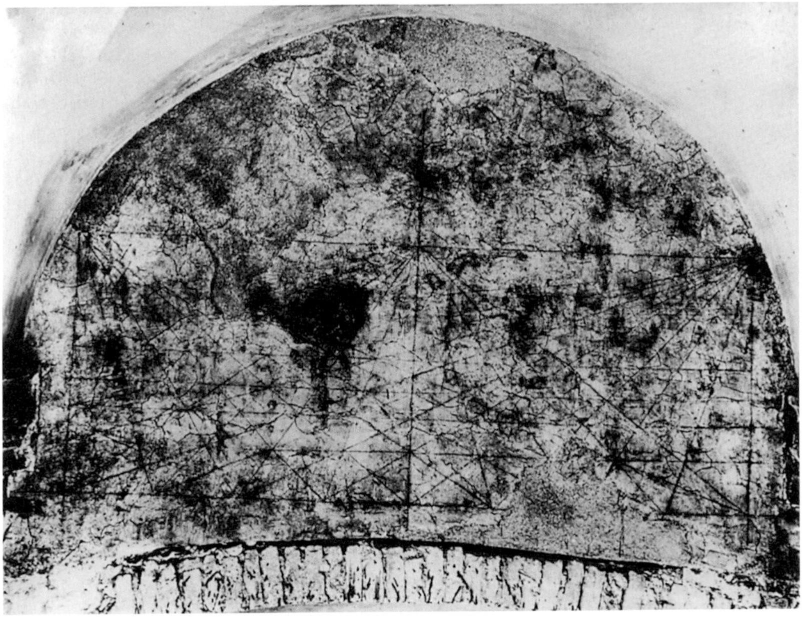

5.2 Paolo Uccello, *Nativity* (sinopia for the *Nativity of San Martino alla Scala*), ca. 1446. Uffizi, Florence. Courtesy Soprintendenza Speciale per il Polo Museale di Firenze.

recede from a devastated world. This sharp temporal divide—and even moral state—of the two sides casts any notion of "unified" perspectival space into confusion. This painting seems specifically to make a problem out of its temporal disjunction, creating a mathematically impossible view.

In a sinopia drawing for a *Nativity* attributed to Uccello (fig. 5.2) the method that Robert Klein called the "bifocal" or "horned" perspective appears.[4] Here there is a central vanishing point, but contrasting orthogonals also run to two points on either side of the painting. Theoretically, it seems to acknowledge that most viewers have two eyes, not the single point of vision assumed by Alberti's theory. In practical terms, it works in a mathematically equivalent way to the distance point method: it is as if a distance point were placed at the edge of the painted field, positioning the viewer at a distance half the painting's width. Scholars have questioned Klein's assertions about the prevalence of this procedure in Quattrocento workshop practice, but it is clearly at work in the sinopia.[5] In *The Flood*, we might see the two arks as solid manifestations of the "horns" of the perspective system.

Thus, in the same physical space but distant in time, a serene standing figure on the right and another—Noah himself?—climbing out of the ark, represent those who have already touched ground at the conclusion of the long drama. The turbulent scenes inside the triangle constitute a shocking depiction of humanity at its lowest ebb, in a cultural moment often held up as a shining exemplar of faith in humankind. Here and elsewhere, geometry plays a role in delineating relationships between order and disorder. The flood, like excess and emotion, blurs boundaries. What was the moral order of geometry up against?

As we saw in the last chapter, the hierarchical triangle established itself as a convention in the fifteenth century, amidst a few cross-currents. By 1500, Neoplatonic writers were convinced of its implantation in the soul. Uccello suggests already that the middle position of the human being, aspirational for Cusanus, could in fact be an uncomfortable place between the celestial and the catastrophic. For writers and artists around 1500, the triangle would constitute a privileged figure for exploring this position.

Figuring the Human Soul

As we saw, Nicholas of Cusa and Basilios Bessarion played key roles in introducing the taste for Plato to Italy in the mid-fifteenth century. By the late fifteenth century, interest in the ancient Greek philosopher had developed into a broader cultural movement: intellectuals like Marsilio Ficino and Giovanni Pico della Mirandola studied Plato along with other thinkers and combined these pursuits with their own peculiar version of Christian theology. Ficino elaborates extensively on mathematics as a source of knowledge and had a major impact on later thinkers. A doctor, scholar, and Medici courtier, ordained a priest at the age of forty, he energetically translated Greek texts into Latin, using Plato as the starting point for a syncretic neo-antique philosophy that draws on many Greek authors.

This set of views and interests is sometimes known as Neoplatonic, although this causes confusion with the late antique Neoplatonists from whom Ficino and others also drew many ideas. Further confusion arises from the fact that in modern art-historical scholarship we have come to understand Neoplatonic interpretation of pictures as a process of idealization, often requiring sexual sublimation, in which the body is transcended in spiritual experience. In the work of twentieth-century art historians, with a few exceptions, this often meant that the body's sensuality was utterly erased and the viewer was required to discount the evidence of her or his eyes.

Renaissance interest in the philosophy of Plato was not, however, an essential ingredient for a process of spiritual sublimation. Late medieval devotional thinkers proposed a similar process of beginning with the material world and the body and ascending by steps to a spiritual plane. What is different, perhaps, is the degree of outright eroticization that appears in some Renaissance paintings: the use of all the resources of the new naturalism to—apparently—unabashedly sexual ends. The recourse to Platonism on the part of modern scholars may be explained by the fact that Plato theorized a chaste, spiritual love that grew out of, but left behind, immediate sensual experience.[6] But Ficino's writings are diverse. They also explore sensual realities, the physical world in its materiality and in its connectedness to the spiritual realm.[7] At its most optimistic, Neoplatonic emanationism held that divine sparks, present all the way down the celestial and material hierarchies, had the potential to light up and turn toward God, no matter how refracted and attenuated they might be.

Along with the interest in Plato came continuing interests in the symbolic and metaphorical uses of geometry. Like Bessarion and Cusanus, Ficino thought of mathematics as a middle realm between the material and the spiritual and developed numerous philosophical implications of mathematics—both real and symbolic.[8] He also writes, following Cusanus, that the Pythagoreans viewed what existed above the soul as unnumbered and "united" while all things below the soul were numbered and "discordant." The soul joins them, being in between: "like mathematical objects, souls occupy the middle position between the natural forms and the divine forms."[9]

Karl Popper suggested that the difficulties Greek mathematicians had with irrational numbers in arithmetic led to Plato's higher valuing of geometry as something more powerful and closer to truth.[10] This would account for the specificity of the fabled inscription over the doorway to Plato's Academy: what is required of the initiate is not mathematics in general, but specifically geometry. Many of Plato's own writings supported the idea of a key role for both geometry and arithmetic; among them were the *Republic*, *Timaeus*, *Theaetetus*, *Meno*, *Laws*, and *Epinomis* (the latter now considered of doubtful authenticity). The *Timaeus*, as we have already seen, presented a theory of matter as composed of triangles; the character Timaeus also refers to the right triangle with sides 1, 2, and the square root of 3 as the "fairest" of triangles.[11] The *Republic* was of particular interest to Renaissance thinkers, with its notion of the "nuptial number," a special number governing procreation and allowing for the birth of the best possible offspring for the proper governance of the polis. Ficino believed that the key to this number was found in the basic Pythagorean 3-4-5 triangle.[12] George Hersey, in his book *Pythagorean Palaces*, suggests that Renaissance architects and architectural theorists, following Ficino, thought of architectural forms as the product of copulation among numbers.[13] In Plato's notion, "the ideal society's leaders ought to copulate at moments best suited to a specific rhythm of fertility and gestation, and in accordance with a corresponding cosmic rhythm established by the muses."[14] As Hersey describes Ficino's extension of the idea, "not only do social leaders copulate in accordance with these cycles, so do an endless host of numbers and shapes . . . that pervade every cranny of existence. Everything is structured for good or ill around this dense, endless mass of mathematical coition."[15]

Many ideas about Plato held by fifteenth- and sixteenth-century scholars actually derived from late antique commentaries such as Proclus's commentary on the *Republic*. It seems to have been from Proclus that Ficino took the idea of the human soul as a triangle.[16] Proclus's own mathematical speculations are quite difficult to understand; in promiscuously deploying the insights of other commentators, he often makes his own position unclear. He ascribes the notion of the soul as triangle to Xenocrates, basing this on an analogy Xenocrates made between types of triangles and types of beings. While most devotees of the triangle chose a primary type or conflated all types together, Xenocrates articulated a specific cosmic typology. The equilateral triangle is like God; the scalene triangle represents human beings; the isosceles triangle, midway between the two, represents the hybrid being of daimons.[17] Proclus extends this idea to suggest that the soul itself is a triangle, and Ficino follows Proclus in this, also citing the Pythagoreans.

Thus in his *Platonic Theology*, Ficino writes that the Pythagoreans "think that the soul, depending on the divine unity, pours itself into matter as into duality, just as a triangle similarly descends from the angle at its apex to end in the two angles at its base; and thus the soul is always estranged and perturbed in the composite body. With justice then they figure the soul as a triangle."[18] He comments elsewhere, "The triangle accords with the soul; for just as the triangle from one angle extends to two more, so the soul, which flows out from an indivisible and divine substance, sinks [*labitur*—slides down, lapses] into the entirely divisible nature of the body." He emphasizes the powers of the soul as multiple, derived from a single point, and in some sense fallen:

> In this we can compare the soul . . . to the triangle, because the triangle is the first figure of those figures which consist of many lines and are led forth into extension or in a right angle. Similarly, the soul is the first of all to be divided up into many powers—powers that are subjected in it to the understanding—

> and it seems to be led forth into extension when it sinks from divinity into nature. In this descent it flows out from the highest understanding down into three lower powers, that is, into discursive reasoning, into sense, and into the power of quickening [*vegetandi virtutum*], just as the triangle too, having been led forth from the point [*signum*] is drawn out into three angles.[19]

As Michael J. B. Allen suggests, if the world is made up of triangles and subject to triangular powers, it makes sense that human beings would be subject to the triangle.[20] As we saw in chapter 1, Giovanni Pico della Mirandola and Gérard Roussel also saw the soul as triangular. But Ficino extends its significance to a more pessimistic view of the human state. In his elaboration of Proclus's triangle, he changes its orientation and emphasis. Proclus describes the triangle of the soul as a right triangle, laterally oriented.[21] Ficino's triangle of the soul flows downward ("sinks" or "descends") into materiality from a single point at its apex.

The hierarchical Quattrocento triangular composition had become the dominant form of the triangle by the later fifteenth century, and its prominence may have been what led Ficino to his own understanding of the hierarchical triangle. Perhaps Ficino took for granted that the triangle of the soul would have a hierarchical orientation because he saw this orientation displayed in devotional paintings. Giovanni Pierio Valeriano Bolzani, in his *Hieroglyphica*, later gave a similar reading of the ideas Proclus had derived from Xenocrates. For Valeriano Bolzani, different isosceles triangles carried different moral messages depending upon how broad the base was in comparison to the height: a narrow base displays reason, while a wide base displays dissolution.[22] Physical extension means a falling away from an ideal state: if dimensionality is a descent into materiality, greater extension means a further descent, a moral slippage. With respect to his immediate philosophical predecessors, Ficino rematerializes the triangle: it is no longer purely abstract like Cusanus's, but it is the opposite of the moderate egalitarianism of Heymericus. It marks quite a different, more pessimistic way of thinking the triangle. The base of the triangle, rather than simply "grounding" a visual composition or providing the "basis" for ascent, actually represents the soul being dragged down to earth.

Measuring and Fecundity

Compared to Ficino's literary speculations, Charles de Bovelles's diagrams, at first glance, may appear rather pedestrian. But it's perhaps in these kinds of images that the significance of mathematical symbolism to visual media becomes most apparent. Ficino's ideas had an important and immediate influence on Bovelles, who, like Heymericus, particularly pursued the visual representation and instantiation of philosophical and theological ideas. Bovelles's diagrams have been given surprisingly little attention, despite the fact that Cassirer and others give him a central place in the development of Renaissance anthropocentrism.[23] For Cassirer, Bovelles's speculations on cognition and contemplation and his notion of a perspectival viewer with the "God's eye" position made him both a typically Renaissance thinker and a harbinger of modernity.

Originally from Picardy, Bovelles (or Bouelles, in Latin Bovillus) published a plethora of Latin treatises on theological, philosophical, psychological, and mathematical topics in the first half of the sixteenth century. He studied at the University of Paris with Jacques Lefèvre d'Étaples but left without receiving his degree to embark on an ambitious publishing campaign. Like Lefèvre, he was reform-minded but did not become a Protestant. His writings include practical, mystical, and pious works as well as letters that document a famous quarrel in which he denounced Johann Trithemius, author of the cryptographic *Steganographia*, over practices he believed involved black magic.

In 1511, Bovelles published a compendium on mathematical, theological, and anthropological topics (I refer to it as *Que hoc volumine*, the opening words of its makeshift title—"what [is contained] in this volume"). In doing so, he entered into a moment of particular mathematical curiosity attached to the year 1500. The text was profusely illustrated with woodcut diagrams. The blocks used to print them, as in the case of most illustrated books of the sixteenth century, were cut by artisans whose names are now lost to us; they were based on designs that must have been provided or supervised by the author. (What we do know about Bovelles's diagrams is that the mathematician Oronce Fine designed and supervised the illustrations of Bovelles's 1542 *Liure singulier*.)[24] Diagrams were common in medieval devotional, theological, and scientific manuscripts, and printing gave them a much broader audience. In Bovelles's diagrams, the triangle appears as a technique for measuring the world and thought—for thinking about thinking itself. Even in his more strictly technical treatises, he does not limit himself to an abstract presentation of numbers and theorems but also makes value judgments, mentioning the nobility of certain numbers, angles, and shapes.[25] Mathematics, he contends, helps one to discover the "secret properties of all natural things."[26] Bovelles was pivotal in the reception of Cusanus, Llull, and Ficino and was perhaps most distinctive in giving schematic visual form to many of the significant philosophical concepts of his time.

In one of the included texts, *De sapiente* (On the wise man), Bovelles carefully diagrams wisdom with visual demonstrations. Man's nature is perfected in wisdom, Bovelles argues, and the *sapiens* or wise man is accomplished (*consummari*) in the "number" of its principle, the Trinity.[27] Knowledge is itself a trinity of perceiver (*spectativum*), object, and medium of perception.[28] The "natural man" (monad) and the "man of art" (dyad) combine to form a triad in the triple *sapiens*.[29] The wise man's practice is one of abstraction through elevation: the wise man abstracts himself from the world. Bovelles uses a metaphor of posture: while an animal or infant crawls close to the ground, the fully realized human being stands. This places the wise man above sensation (associated with the animal) and puts him closer to the angels and, eventually, to God.

For Bovelles, mathematics in general and triangles more specifically regulate the body, the soul, and relations between the two, demonstrating and enacting the lifting of the human mind up toward the divine. In his diagrams, triangles help define the "human." At some points they seem to be merely arbitrary taxonomies; elsewhere they are put to philosophical ends. For instance, Bovelles set up a hierarchy by which the various levels of being, from the most spiritual to the most material, corresponded to geometry: point (God), line (angel), triangle (man), cube (world).

The senses are also ruled by the triangle. In an intriguing set of images, whose production Bovelles must have closely supervised, the senses are mapped according to a triangular schema (fig. 5.3). "The two ears are further apart than the eyes," he writes, "the eyes more than the nostrils, and the nostrils more than the tongue, which is joined and indivisible. Thus let these four senses situated in the head . . . define a not unattractive, though sideways [*eversa*], pyramid."[30] The ears are "higher" because for Bovelles objects of hearing are simpler and less material than the objects of vision.[31] According to Bovelles, the power of each sense is proportional to the *size of the angle* it forms with the imagination. Hence the obtuse, right, and acute angles indicated on the diagram: the sense of hearing is the "highest" of the senses, followed by sight, smell, taste and touch. The ears, farther apart than the eyes, produce an obtuse angle with the imagination, which receives all the species delivered by the various senses, and the wider angle represents not only greater powers of perception but also increased capacity for

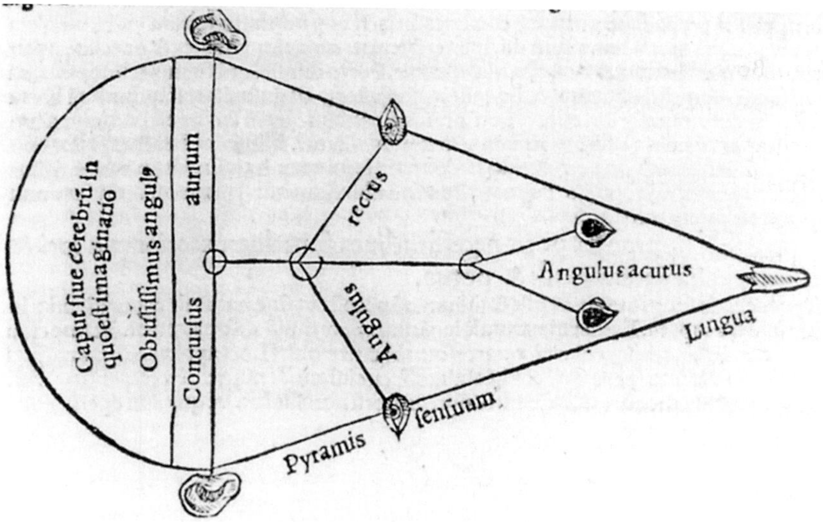

5.3 Anonymous. Triangular schema of the senses in Charles de Bovelles, *Liber de sensu* in *Que hoc volumine continentur* (1511), 33r. Woodcut. University of Chicago Library Special Collections Research Center. By permission.

abstract understanding. In the *Art et practique de Geometrie*, Bovelles writes, "Thus I say that these five high and honorable senses [touch is excluded but the imagination is included], situated and organized in the head, are not exempt from geometric figure. For the order according to which they are naturally placed and situated in the head, keeps and observes the beautiful triangular figure."[32] The image looks like a kind of reversed perspectival diagram, its angles seeming to imply an interior space, with the tongue as a grotesque parody of a vanishing point. And for Bovelles, the wise man's mind operates like a perspectival system, using measurement and proportion. He is not only the mirror of the world but also a careful measurer of himself and the objects of his senses.

Apprehending the world involves a movement of progressive simplification and dematerialization toward the mind of God, and this too can be understood as a triangle, or a cone, in which the world forms the base and the mind, or eye, forms the vertex.[33] In *De sapiente* Bovelles provides an illustration showing the intellect of the wise man as a kind of all-seeing eye, vertex of a triangle, set at a distance from the base that represents the world and all the things in it (fig. 5.4). It is a kind of perspectival image. Man, Bovelles argues, is the universal spectator, outside of things and viewing them from a distance, a function within the world that is able to set itself apart and see the world—self-consciousness, in Cassirer's Hegelian terms.[34]

As we have seen, for Bovelles the wise man is trinitarian. Knowledge is a trinity—the looker, the object looked at, and the action of the two; in the soul, similarly, a triad is formed by the intellect, the memory, and contemplation.[35] But in a sense, the highest ambition for man is to become a point, to disembody himself as he approaches the divine mind. This sets the human apart from the world, enabling aspiration toward disembodiment, the disembodied eye/I. The figure of the triangle conjoins the material and the spiritual, symbolizing real or conceptual transit between the extended base and the infinitesimal but commanding point of the vertex. We might also refer to Pacioli's diagram of the proportions of the human head (fig. 5.5). Though presented as a matter of proportion in creating figures, a sense of viewpoint is loosely suggested: if we imagine the diagram as an interior one, not only exterior, it seems to imply intake and abstraction, as the triangle diminishes toward the back of the head. Pacioli writes: "In the most noble of the exterior members, that is, the head, if you look carefully you will see that it is formed with the form of the first figure made of straight lines, that is, the equilateral (or isopleuros) triangle, placed as the foundation and principle of all the other books of Euclid in the beginning of his first book."[36]

Bovelles provides a thicker description of "the world" in its relation to geometry in his vernacular *Geometrie practique*, written for the layman. He wrote specifically for literate artisans with a particular need for geometrical knowledge, having produced the text, he writes in his preface, on the request of some "handymen" (*auturgis* and *manu operariis*).[37] Here, as he presumably thought proper in addressing laborers, he downplays the metaphysical implications of the science. This is where he describes geometry as subordinate to arithmetic: whereas the numbers of arithmetic are reclining (*gisans*) and situated in the soul, "geometry considers the measures, quantities, and corporeal dimensions that are placed and situated in the body, and in any solid and material thing."[38]

Although in his *Mathematics of Bodies* he deals with qualities normally associated with physical bodies, Bovelles's "mathematics of bodies" is not to be understood as addressed to the human body. While carefully defin-

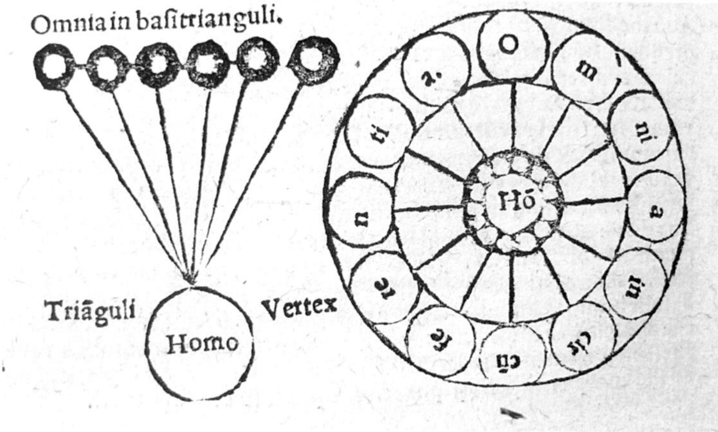

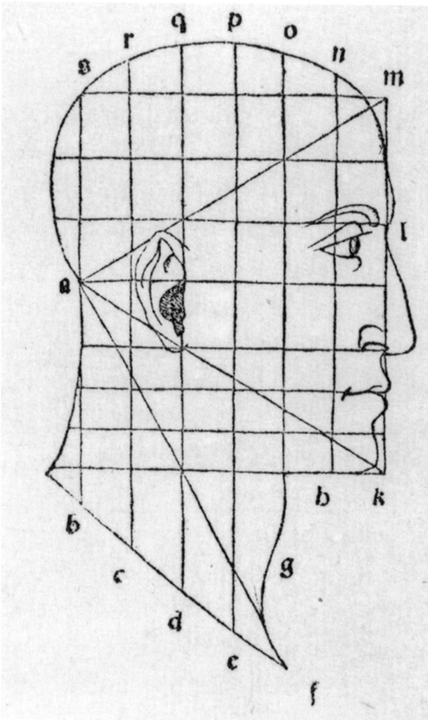

5.4 Anonymous. Diagram of man as universal spectator in Charles de Bovelles, *Liber de sapiente* in *Que hoc volumine continentur* (1511), 132v. Woodcut. University of Chicago Library Special Collections Research Center. By permission.

5.5 Anonymous. Proportions of the human head in Luca Pacioli, *De divina proportione* (1509), f. 25v. Houghton Library, Harvard University, Typ 525.09.669a F. By permission.

ing the geometric body as distinct from the human body, he characterizes the body's condition as one of servitude to the soul, an emphasis that might also relate to his audience's subordinate social position: "The body is taken in Geometry, not for the substance of the human body subject to and serving the soul, but for any corporeal measure with three dimensions, that is length, breadth, and depth."[39] This seems similar to Alberti's caution that the point in geometry is not the same as the point that represents it visually in art. Bovelles makes no cautionary remarks about geometric *surfaces* being understood as *fields*, in the sense of plots of land; he seems comfortable with their similarity.[40] This accords with the traditional usage of geometry for surveying, and with Bovelles's account of its very origins.

At the beginning of the text, Bovelles tells us that geometry was invented in order to survey land in Egypt. In his story, which draws on accounts by Aristotle and Herodotus, geometry arose historically in Egypt because the Nile flooded annually, providing fertilization to the fields but also confounding boundaries. Bovelles omits Herodotus's account of equal division of land among all subjects of the Egyptian pharaoh.[41] He focuses on how geometry allowed fertility to operate in a properly regulated way:

> Instead of rain, for the fertility of the fields, by the providence of God, the Nile overflows every year in summertime, and waters the fields, and for a certain amount of time remains on the fields. Then when it draws back, the edges and boundary markers of the fields are troubled and confused, from which great annoyances and questions arose in ancient times among the Egyptians. Because of this, in order to reduce popular controversy, the Kings of Egypt ordered that some art should be found by the priests, who were idle and paid no tribute to the Kings, to measure and limit the fields, so that through the annual flooding of the Nile, the fields would no longer be confused or troubled.[42]

With his mention of the fact that the priests paid no tribute, Bovelles is embellishing upon Aristotle's account in the *Metaphysics*, in which the priestly class developed geometry because it had the leisure time in which to do so. Aristotle makes no mention of tribute or the lack thereof. On the one hand Bovelles is reminding us of the reason a king would want careful measurement of the fields: he needs to know who is responsible for the taxes on them. But one might presume that the priests or scholars attached to a royal house might well be asked to devise such a scheme whether or not they already paid tribute. Perhaps this suggests a different kind of exemption: the priests' lack of tribute highlights their celibacy, their existence outside patrilineal agrarian social structure. They are idle with respect to the fields because they do not cultivate them. They do not produce agricultural wealth for tribute, as they correspondingly do not engage in procreation (the human parallel to the fertility of the fields). Geometry begins, if not with sexuality per se, then with fecundity and its troubling confusion of boundaries, and with the restoration of structure imposed by a sacred class.

Field boundaries and fecundity are also connected in an unusual story told later in the sixteenth century by Bernard Palissy. According to Palissy, it was Priapus who invented the division of the fields: the phallic god "invented the division of land, such that each would cultivate his own share."[43] The French Protestant author may have been inspired by the function of garden terms, ancient statues that consisted of a block of stone with a head and erect penis as their only anthropomorphic attributes.

Geometry, "measuring of the earth"—in Bovelles's tale, this is a matter of dividing property and regulating fertility. An analogy is found in Lorenzo Lotto's small *Allegory of Virtue and Vice* (or, I would suggest, *Moderation and Excess*) in the National Gallery of Art in Washington (fig. 5.6). While the satyr on the right indulges his lusts and expends his wealth under a stormy sky—where a ship founders in the sea and the dead side of the tree marks the sterility of his pursuits—the mathematical tools deployed by the child on the left make a tree flourish, clouds part, and the fields glow with wheat.[44] Flourishing fields and trees might be connected to the proper application of measurement, whereas dissolution and flowing excess lead to wrack and ruin. The function of the Platonic nuptial number is to regulate fertility, a theme that also appears in visual images—for instance, in an engraving (fig. 5.7) by Jacopo Raibolini Francia (ca. 1523–24), a chaste and proportionate Venus bears a geometric tool—the square—contrasting with the disordered, disproportionate pair of lovers to her right.

Trinity, the Family, Fecundity

We saw in chapter 3 that Bovelles mapped elements of the physical world as triangles and trinities. With his lists of triadic structures, it sometimes seems Bovelles is simply intent on viewing all concepts as triangles, however tenuous the connections might seem. In one diagram he presents, for example, a triangle of sun, moon, and earth—a view that did not correspond to the astronomy of his time (fig. 5.8). Other scholars held stricter definitions of which triads or triangles "worked" as vestiges or images of the Trinity. Bessarion suggested that the Trinity's stamp could be found in such disparate triads as power, wisdom, and goodness; substance, virtue, and operation; number, weight, and measure; and unity, appearance (*species*), and order.[45] But he argued against the notion that the three spatial dimensions, width, length, and height, could be understood as trinitarian. This view had

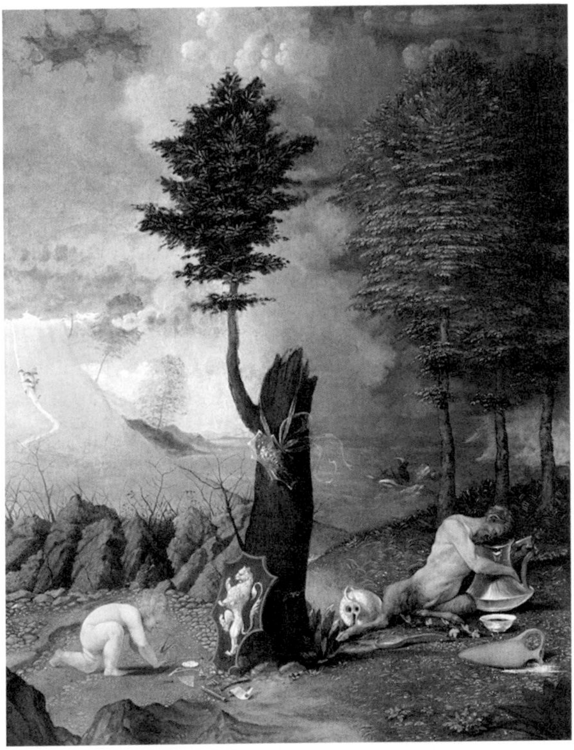

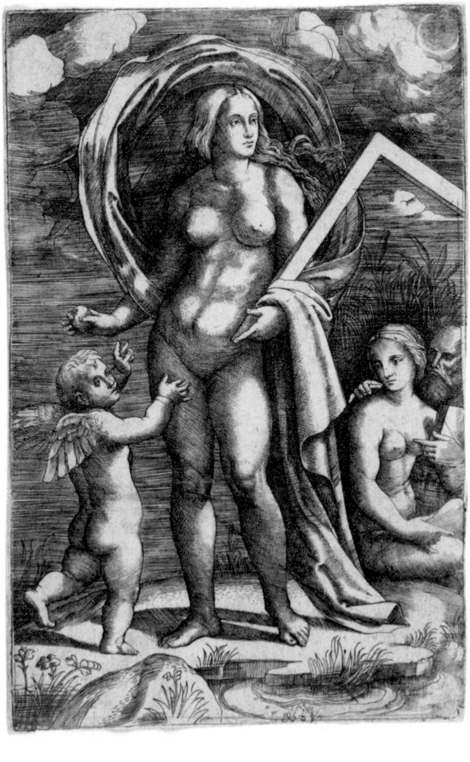

5.6 Lorenzo Lotto. *Allegory of Moderation and Excess (Allegory of Virtue and Vice)*, 1505. Oil on wood, 56.5 × 42.2 cm. Samuel H. Kress Collection, National Gallery of Art, Washington. Image courtesy of the Board of Trustees, National Gallery of Art, Washington.

5.7 Jacopo Raibolini Francia. *Moderate and Immoderate Love*, ca. 1523–4. Engraving, 22.9 × 14.8 cm. Minneapolis Institute of Arts, Minneapolis. Gift of Herschel V. Jones 1920. By permission.

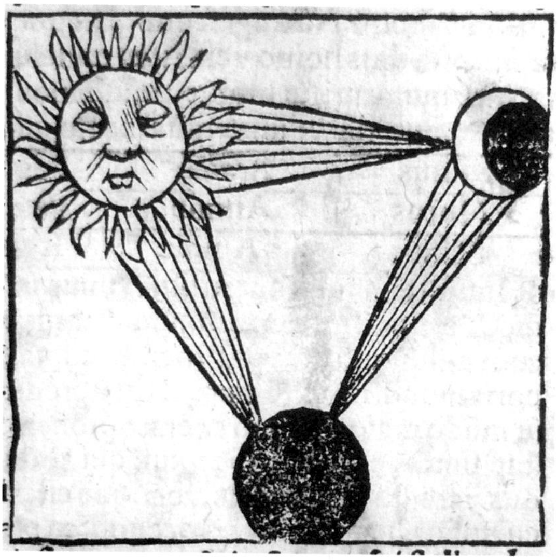

5.8 Anonymous. Sun, moon, and earth, in Charles de Bovelles, *Liber de sapiente* in *Que hoc volumine continentur* (1511), 133r. Woodcut. University of Chicago Library Special Collections Research Center. By permission.

been put forward by Christian interpreters of Aristotle, to suggest that the philosopher already knew of or at least anticipated the Holy Trinity. For Bessarion, whose interest was in defending Plato and critiquing Aristotle, this triad of physical attributes did not provide an apt analogy. Their concreteness and commensurability meant they could constitute, for him, only a "sordid" Trinity—a *sordes Trinitatem*. (Later, the idea of dimensions as trinitarian reemerges in Luca Pacioli's notion of the "divine proportion"—the golden mean. Pacioli argues it is coherent with the divine through the Trinity, because of the three terms involved in the proportion.)[46]

Another problematic triad presented itself as well: father, mother, and child. Could the family constitute a vestige of the Trinity? Theologians frequently proposed this idea in their accounts of the Trinity and its manifestations in the world—most often to refute it summarily. In Aquinas's *Summa theologica*, he writes, explaining the concept of *procession*, "So also did Abel proceed immediately from Adam, inasmuch as Adam was his father; and mediately, as Eve was his mother, who proceeded from Adam; although, indeed, this example of a material procession is inept to signify the immaterial procession of the divine persons."[47] Augustine wrote: "We must reject the idea that the Trinity is imaged in the triad of man, woman, and child."[48] Nicholas of Cusa, in *De docta ignorantia*, allows it but in a rather backhanded way, suggesting that the triad of mother, father, and child can be held to be a trinitarian vestige in arithmetical terms merely because all three are human. They are so no more or less than a collection of three donkeys.[49]

But the abstract, dialectical trinitarian notions of many commentators are often suggestive of procreative "production," in that they posit a first concept, an opposite one, and the product of the two. Renaissance authors also had the example of Plutarch, who wrote, drawing on Pythagorean ideas, that the joining of concept and material, form and matter, to produce the world could be described as father, mother, and offspring. To him, this meant that the family constituted a right triangle with sides of three different lengths (that is, a scalene right triangle).[50] One might suggest in fact that the need to *refute* the trinitarian family speaks to the appeal of the idea. Barbara Newman writes of the long history of this "tempting, but forcefully resisted, familial metaphor."[51] Folk wisdom might argue for a human trinity of the family by analogy, but also by contrast, with the divine Trinity. And while most authors were not willing to associate the human family with the Trinity, Charles de Bovelles did. He declared that the human species demonstrated the Trinity through the triad of Adam, Eve, and Abel, which he diagrammed as a triangle.[52] He also associated the primal family group, through a chain of analogies, with contemplation and knowledge. One of the triads that interest him is that of human wisdom. Bovelles defines the wise man, his ideal, as a "man of art," recipient of the gifts of nature.[53] Similarly, he describes human contemplation as an "Abel" born of Adam and Eve. Abel corresponds to the form of contemplation undertaken by the wise man: where Adam corresponds to intellect and Eve to memory, Abel is the union of the two. Contemplation, like knowledge itself, is thus the result of a kind of procreation.[54]

Bovelles places emphasis on the triangle itself as *fecund*—which might express obliquely this sense of the familial triad as a model or reflection of the Trinity. Nicholas of Cusa, though ambivalent about the family as a representation of the Trinity, does write of the fecundity of the Trinity, and specifically of its expression through the triangle. The late antique Neoplatonist Proclus wrote in his commentary on the first chapter of Euclid's *Geometry* that "the Pythagoreans assert that the triangle is the ultimate source of generation and of the production of species among all things generated."[55] Near-equilateral triangles were used to represent trees of consanguinity—that is, family trees—in medieval manuscripts of Isidore of Seville's *Etymologies*.[56] Boethius expounded upon the fecundity of the triangle in his Latin adaptation of Nicomachus's *Introduction to Arithmetic*, which was widely read as a basic textbook. The first triangular number—one—is "mother of all the numbers"; it "produces the sums of all the figures and is involved in the procreation of them all."[57]

Bovelles follows in the tradition of this Neoplatonic reflection on the triangle, fecundity, and the Trinity—and extends it. In a short treatise, the *Mathematical Bodies* (*De mathematicis corporibus*), contained in his 1511 compendium, Bovelles associates geometric shapes directly with procreation.[58] In the text, regular plane figures give birth to solids, specifically the regular Platonic solids, tetrahedron, cube, octahedron, dodecahedron, and icosahedron. The language of fertility and procreation (*generare, procreare, gignere, gestare, congressio, coitus*) is insistently present in this short treatise.[59]

Of the three plane figures, according to Bovelles, the triangle is the most fecund (*fecundissimus*) because it produces three solids, not only one. The square gives birth to the cube; the pentagon gives birth to the dodecahedron; but the triangle produces the tetrahedron, the octahedron, and the icosahedron. The excellence of the triangle thus also appears in its fecundity. It therefore includes its own identity within itself because it creates three solids from triangles alone.[60] "From these things," Bovelles writes, "it is clear that the power of the equilateral triangle is trine . . . it has three powers. Thus it is marked with the name of the Trinity; indeed it is triply fecund and gives birth to three things."[61] *De mathematicis corporibus* is described by P. M. Sanders as "the presentation of a mathematical guide by which the mystery of the Trinity can be contemplated."[62] But Bovelles also brings together the Trinity with the idea of fecundity through the geometric properties of the triangle. Moreover, the "potency" (*potestas*) of all the figures is triangular, because the sum of points, lines, surfaces, and bodies (the number of bodies being one in each case!) is divisible by three.[63] And the triangle is more powerful than the square or pentagon (it is *fecundissimus*) because it can create, or rather procreate, shapes with them as well.[64] It is one of these shapes, a "rhombicuboctahedron," that appears as a glass "ball" hanging on the viewer's left in Jacopo de' Barbari's highly geometric portrait of Luca Pacioli at the Museo di Capodimonte in Naples.

Another way that the triangle links with fertility is in images of the Virgin Mary, as we saw in the last chapter. In some images Mary appears with the Trinity as a kind of shadow member of it, perhaps to express its fecundity. As we have seen, few figurative images of the Trinity actually "inhabit" a triangular geometry. Yet there is one type of trinitarian iconography that uses triangles, and in a rather striking way. This is the Coronation of the Virgin by the Trinity, an iconography with no particular biblical grounding. In these images, we witness the coronation of the Virgin by a Trinity whose two anthropomorphic members form a triangle with her. This triangle thus expresses the Virgin Mary's special relationship to the members of the Trinity. These images differ sharply from representations that show Christ crowning a submissive Virgin whose pose (similar to that of the Virgin Annunciate) places her in a clearly subordinate position.

The triangular coronation images may be related to the *Buch der heiligen Dreifaltigkeit* (book of the holy Trinity), written in the German vernacular by a Franciscan named Ulmannus from 1410 to 1419 and circulated at the Council of Konstanz. The book proclaims the affinity, even identity, of Mary with the Trinity. In the text, Mary is described as an image (*Gestalt*) of the Trinity; she is also represented as one of its three persons, replacing the Holy Spirit.[65] In her book *God and the Goddesses*, Barbara Newman suggests that this text is the basis for the iconography of the coronation of the Virgin by the Trinity; at any rate the illustration of this motif in a manuscript of the text constitutes one of its earliest instances.[66]

In trinitarian coronations, it sometimes appears that Mary takes the place of the Holy Spirit. In Enguerrand Quarton's *Coronation of the Virgin* (fig. 5.9), not only do an identical God the Father and Christ form a triangle with the Virgin, but each figure, with its flowing robes, forms a kind of triangle, and they all intersect in the midsection of Mary's body. In an example from fifteenth-century manuscript illumination, an image by an Olivetan master (fig. 5.10), Mary seems to be a member of the Trinity, with the Holy Ghost in the midst of the three figures rather than taking its place at the vertex of the triangle, as if to represent the relational space of the *scutum fidei* that sometimes holds the cross. A number of fifteenth-century woodcuts depict a coronation by the Father and Son with Mary forming the spatially subordinate vertex, much as the Holy Spirit does in "shield-style" trinitarian triangles. If one recalls the *scutum*, this might also suggest generativity through the genealogical character of the shield, since the coat of arms itself was a sign of birth and parentage. The sacred heart, too, often appears in "shield" format but encloses the Christ child with the *arma Christi* in a uterine way that suggests the metaphor of the "womb of the heart."[67]

Images like these are not necessarily to be understood as doctrinally rigorous, but they give us some ways of thinking about how devotional concepts could be elaborated visually. Mary's special relationship to the Trinity—one that is not without theological problems—is reflected in late medieval German poetry, as well as Latin theological texts, as Peter Kern has shown in his 1971 book *Trinität, Maria, Inkarnation*. Sometimes viewed

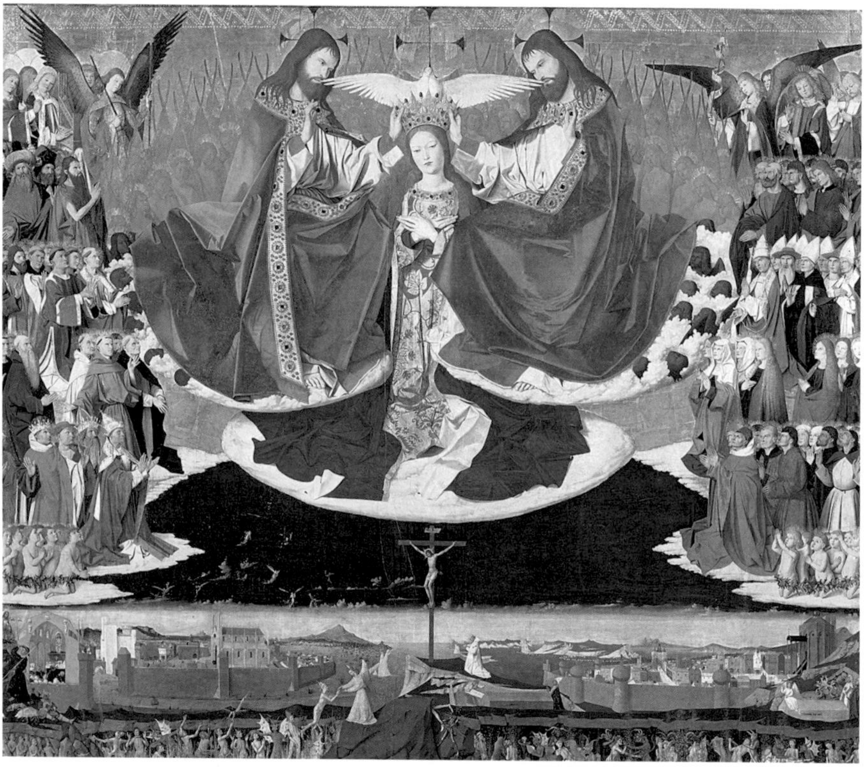

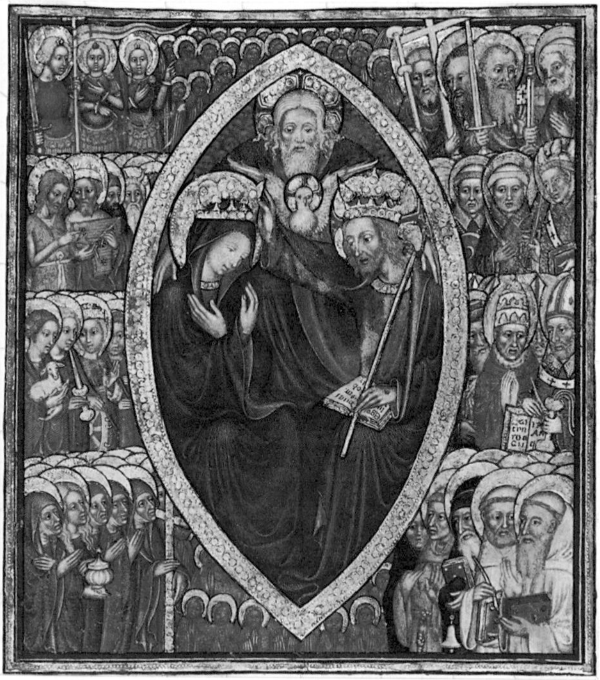

5.9 Enguerrand Quarton. *The Coronation of the Virgin*, 1454. Oil on wood, 183 × 220 cm. Musée Pierre de Luxembourg, Villeneuve-les-Avignon, France. Photo Giraudon/The Bridgeman Art Library

5.10 Anonymous Olivetan Master. *Coronation of the Virgin with the Trinity and Saints*, ca. 1440. Miniature on vellum, 15.7 × 14.3 cm. Rosenwald Collection, National Gallery of Art, Washington. Image courtesy National Gallery of Art, Washington.

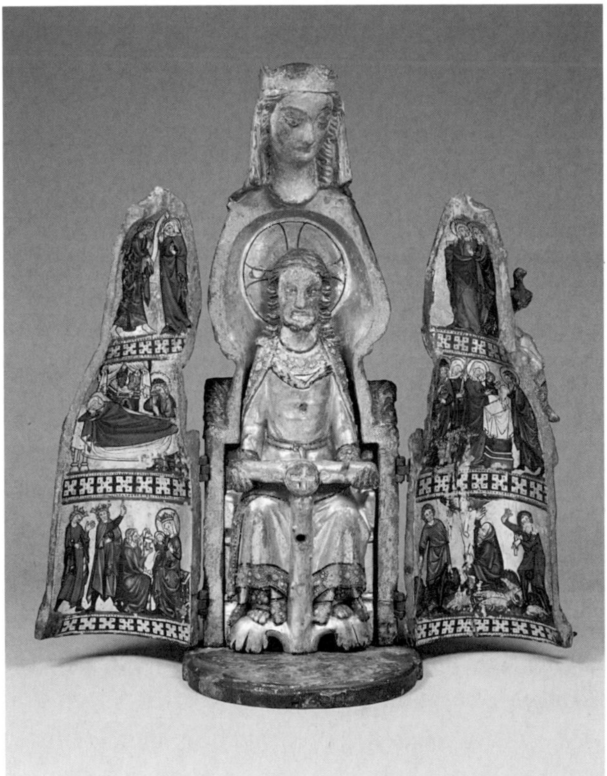

5.11 Anonymous. *Shrine of the Virgin*, ca. 1300. German Rhine Valley. Oak linen covering polychromy gilding gesso, 36.8 × 34.6 × 13 cm (open). Gift of J. Pierpont Morgan 1917. The Metropolitan Museum of Art, New York. Image © The Metropolitan Museum of Art/Art Resource, NY.

as "fertilized by the Trinity" (*fecundata Trinitate*)[68] or inhabited by it (*inhabitatio Trinitatis*, and *arca*, *templum*, *claustrum*, or *triclinium Trinitatis*),[69] Mary is at times even identifiable with the Holy Spirit itself, where the latter is understood as the love between Father and Son.[70] The triangle was itself a fertile form, and perhaps its visual appeal—the possibilities it provided for creating arrangements—could prompt new reflections on theology.[71]

The idea that the Virgin might enclose the Trinity was also present in the fifteenth century in the famous *Vierges ouvrantes*, small hinged devotional sculptures of the Virgin that opened to, in some cases, reveal a depiction of the Trinity. Jean Gerson, the influential canon of Notre Dame and chancellor of the University of Paris, was so incensed by this doctrinal irregularity that he had one such object destroyed.[72] The shape of many of these objects, when opened, was roughly an equilateral triangle (fig. 5.11). Opened, the Virgin's body is triangular and triadic—a triptych. As One opening onto Three, the structure of the object itself, not only its interior iconography, alludes to the Trinity. The Virgin's triangularity, then, might suggest, somewhat subversively, that she actually embodies the Trinity. And with Raphael's triangular compositions depicting the human family of Christ, or even Diana Mantuana's *Birth of Apollo and Diana*, a common theme suggests itself: motherhood, family, and generation represented by the symbolic triangle.

Measuring the Maternal Earth

I would now like to look more closely at Bellini's *Madonna of the Meadow* (plate 3), a more complicated painting than it appears at first glance. I first mentioned this painting because in it the Virgin's body, like Christ's face in the *Sancta Facies* paintings of Dürer and van Eyck, produces an equilateral triangle.[73] We might see in it a Christian version of the rhyme between earth and maternal body that appears so often in images of reclining nymphs in Venetian landscapes (in paintings by Giorgione, prints by Campagnola, and others). But there is more to this strange, sad, lyrical painting.

Given Mary's contemplative pose, we might liken her triangular form to Bovelles's triangles of contemplation. She sits at the edge of a rocky stretch of land, perhaps a road, with leafy plants growing to her left and right, praying over her sleeping infant. The nearly imperceptible shift from shimmering vegetation to rocky ground suggests greenish-brown veined marble. Mary's robes, though oddly stiff, do not seem to provide enough support for the Christ child—he is more like a stone statue than a real child, and his legs, especially, appear suspended in the air. Sleep in this instance seems, as often in nativity imagery, to be a premonition of death; the figure is a Pietà in which the body of Christ has been miniaturized. Like Poussin's *Arcadian Shepherds*, this painting connects the pastoral to loss.[74] In the background, a cowherd leans contemplatively against a post, and a white-robed figure stands, in the shadow of a fortified village or monastery, with a pair of goats. A tiny drama unfolds between a wading bird and a serpent, while a black vulture observes from a tall tree.[75]

Bellini's painting is a variation on a type called the "Madonna of Humility" and participates in an iconographic shift toward the equilateral triangle discernible from the end of the fourteenth century. The phrase *Madonna dell'humiltà* (*umiltà*, in its modern spelling) was associated with the iconography of the Madonna seated on the ground from the fourteenth century.[76] Worth remembering is the fact that the Latin word *humilitas* derives from *humus*—the earth. This version may also be related to northern or Italian images of the seated Madonna and child (figs. 5.12, 5.13) or, especially, to

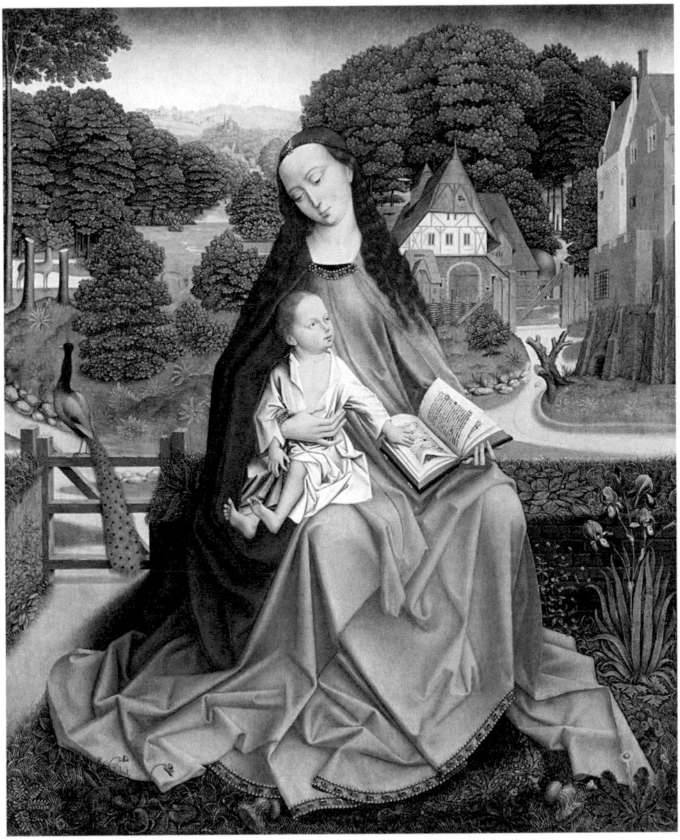

5.12 Anonymous (possibly Master of the Madonna Grog or Aert van den Bossche; formerly Master of the Embroidered Foliage). *Virgin and Child in a Landscape*, ca. 1492–1498. Oil on wood, 104.8 × 87 cm. Minneapolis Institute of Arts. Gift of the Groves Foundation by exchange, The Miscellaneous Works of Art Purchase Fund, The Paintings Acquisition Fund, The Painting Deaccession Fund, The Putnam Dana McMillan Fund, and unrestricted fund balances. By permission.

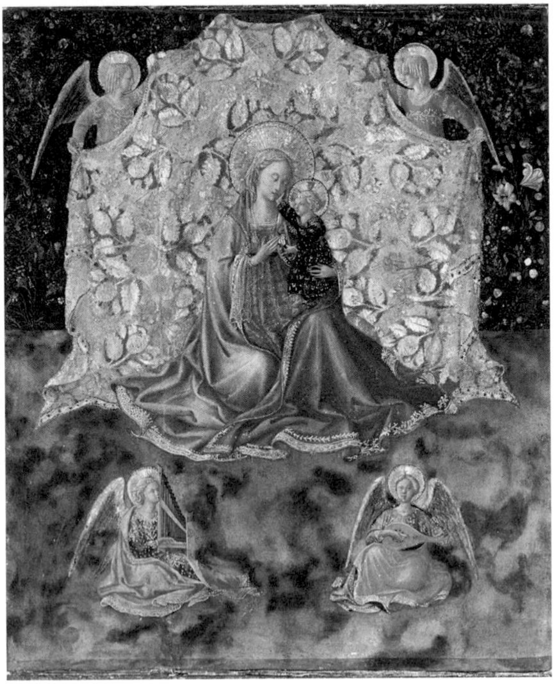

5.13 Fra Giovanni da Fiesole (Fra Angelico), workshop. *Madonna of Humility with Angels*, ca. 1440–1460. Tempera on wood, 34.7 × 29.4 cm. Accademia Carrara, Bergamo, Italy. By permission.

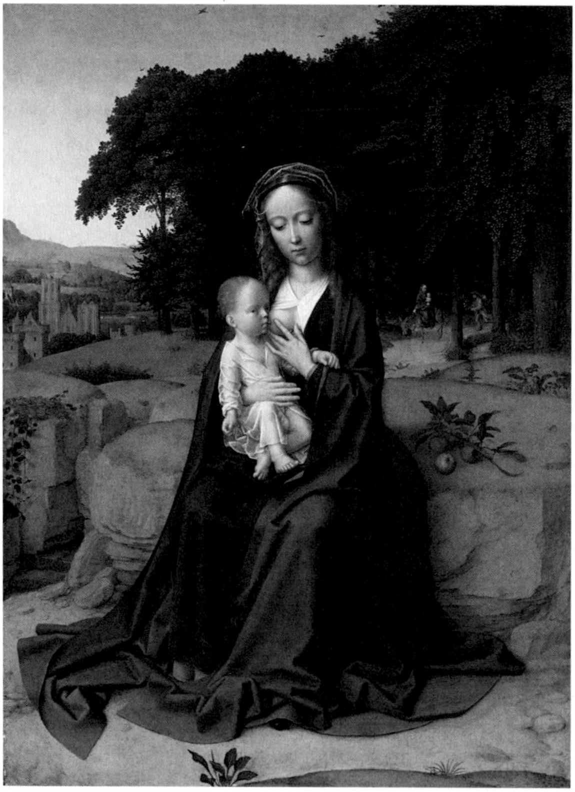

5.14 Gerard David. *Rest on the Flight into Egypt*, ca. 1512–15. Oil on wood, 50.8 × 43.2 cm. The Metropolitan Museum of Art, New York, The Jules Bache Collection 1949. Image © The Metropolitan Museum of Art / Art Resource, NY.

Gerard David's many panels depicting the *Rest on the Flight into Egypt* (fig. 5.14), the story of the Christ child's human family and their flight from Herod. If David marginalizes the figure of Joseph by making him a tiny spot in the background (part of a different moment in the narrative), we might say that Bellini effaces him entirely. A trace of that same story might be seen in the cultivated fields of the middle ground. In one episode of the apocryphal infancy narratives, a miraculous growth of wheat produced by the passage of the Holy Family helps them to evade Herod's soldiers. This story seems to have grown up and flourished in the European High Middle Ages.[77] In a brief written account of a passion play in 1431 in Paris, the grain miracle is given the same degree of prominence as major episodes: "the Annunciation, the Three Kings, the innocents, and the good man who was sowing his wheat."[78] The story in the play is situated as one of the episodes of the Life of the Virgin, which Pamela Berger suggests associates the story with earlier legends of a grain-protecting goddess.

The depicted animals may also be particularly meaningful. The vulture is often associated with Egypt. The wading bird fighting a serpent is most likely a crane; the crane and snake were considered traditional enemies. In one story, Moses helped Pharaoh defeat the Ethiopians by sending cranes ahead to eat their snakes.[79] G. P. Valeriano Bolzani's *Hieroglyphica* tells us that "the crane . . . carries on an assiduous battle with serpents: because indeed this type of animal is thoroughly terrestrial, such that it creeps on the earth, always sticks [*adhaerescat*] to the earth or hides itself in the most secret lairs of the earth. Thus the soul which recoils from the pleasures of mortals, thoroughly averts and keeps at a distance the unrestrained passions of the earth, through the hieroglyph of this bird, which tears the serpent to pieces, is aptly signified."[80] We cannot read this text directly into Bellini's painting, but its emphasis on the earth is suggestive. The bodily orientation of vertical crane fighting horizontal serpent suggests the same tension of elevation and extension that one sees in the triangular form itself.

Barely perceptible is the fact that the field is gridded, marking the measuring and dividing of property, or the "underlying fictive pavement" that Jean Pèlerin suggests for the representation of a country landscape.[81] This grid is telling: the perspectival lines created by it meet in the Virgin's body, specifically at her heart. In the Florentine devotional paintings addressed in the last chapter, in which the Virgin's body served as the site of the vanishing point, it was also recessed in space. There her position at the apex of the imaginary lines of the perspectival construction was clear. But here the lines recede behind her body and yet must be imagined to come to a point exactly at her heart. A contrivance closes the gap between foreground and background—insisting on the fusion of surface and space. It is as if the grid pierces her heart. And given the Virgin's mournful air, the lines suggest another analogy: an echo of the iconography of the seven swords that pierce the Virgin's heart, each representing one of the sorrows of the Virgin; in many such *Mater Dolorosa* images she also creates a triangular form on the ground (fig. 5.15).[82]

The Virgin's triangularity might figure heavenly ascent—but also generation and its corollary, death. The triangle reaches toward the heavens, but as in Ficino, it is pulled to the ground, with the Virgin as a melancholic Mother Earth. If contemplation operates through a triangular form, the eye hovering above the world, it is the body that, in the end, must join the two. Elsewhere, the geometric schema may imply order and hierarchy. Here, we might think instead of the figural type of melancholy, a body that appears as a triangular form, with a gaze both spatially and emotionally downcast—earthbound.

Like Bellini's Madonna, figures of Melancholia (fig. 5.16) are neither horizontal nor vertical but fall somewhere between the two, their heads drawn down, their bodies spreading at the bottom. Viewed purely in terms of shape, they also parallel the emerging Renaissance convention of the half-length portrait, another type of triangular arrangement. The implicit (or explicit) mortuary character of portraiture might, too, connect it to melancholia. In this regard, it is noteworthy that the Mona Lisa, an especially paradigmatic case of the triangular half-length portrait, is specifically connected to melancholy in Vasari's account of the production of the painting.

Melancholy was also associated with the earth—elementally both were cold and dry (understood in relative terms). The weightiness of the melancholy humor produced constancy and a firm capacity of the memory. It was associated with mathematics, the imagination, contemplation, and the habit of a downcast gaze. Ramon Llull associates melancholy with both earth and water, and writes:

> By the nature of ponderousness [melancholics] are constant and firm in their appetites and speech and naturally look down toward the earth and have a good memory, because water is restraining, greedy and impressionable; and they love mathematical figures and the images of the fantasy [*species phantasticas . . . et mathematicas*]. Earth too is a dense medium, in which the impressions of the species are held for the memory; and therefore melancholy men are disposed to acquiring great learning through the multiplication of many species. Also they are suspicious by nature and they accept [things] boundlessly through the imagination, which has a greater concordance with melancholy than with any other complexion.

The relationship between mathematics and melancholy is also related to this earthy basis for the temperament:

> And the reason why melancholia has more proportion and concordance with the imagination than with other complexions, is that the imagination considers measures, lines, figures, and colors, which are best able to be imprinted in water and earth, because they have a denser material than fire or air.[83]

The connection of geometry to passions and negative emotions is perhaps nowhere clearer than in images of melancholia. The prominence of the mathematical instruments in Dürer's *Melencolia I* has been mentioned frequently. In *Saturn and Melancholy*, Erwin Panofsky, Raymond Klibansky, and

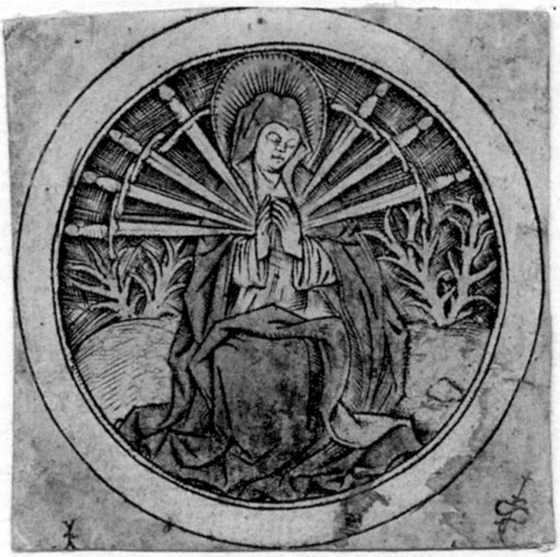

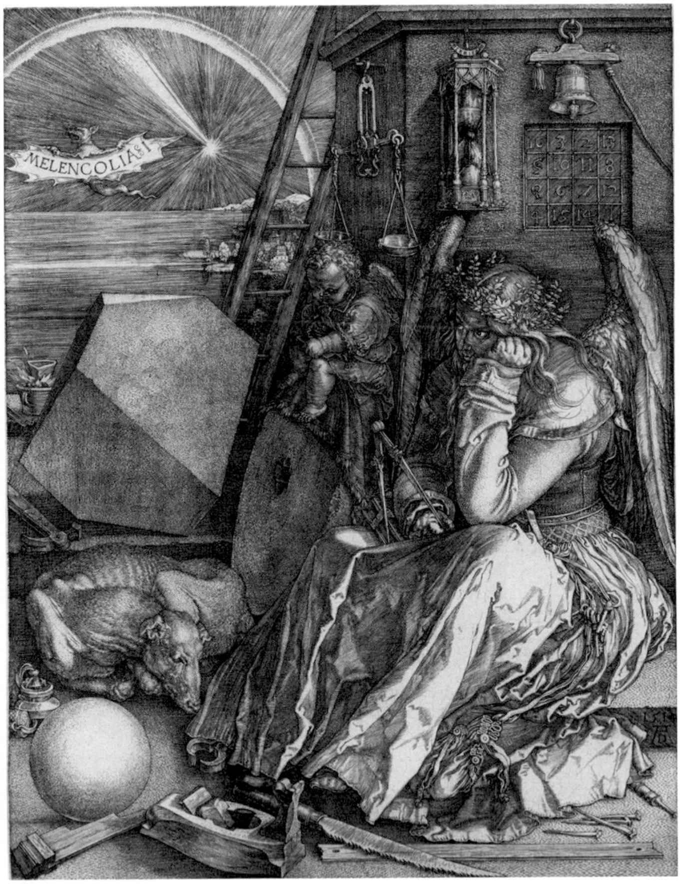

5.15 Anonymous Netherlandish (Monogrammist I.S.?). *The Virgin as Mater Dolorosa*, 1500–1525. Hand-colored engraving, 4 × 4.5 cm. © Trustees of the British Museum.

5.16 Albrecht Dürer, *Melencolia I*, 1514. Engraving, 23.9 × 18.7 cm. British Museum, London. © Trustees of the British Museum.

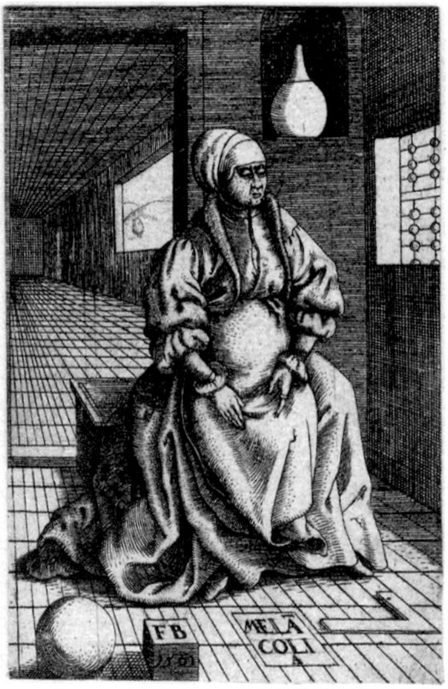

5.17 Monogrammist F.B. (Franz Isaac Brun?). *Melancolia*, 1560. Engraving, 7.1 (trimmed) × 4.7 cm. British Museum, London. © Trustees of the British Museum.

Fritz Saxl established a connection between Dürer's Melancholy and the iconography of the personified Geometria.[84] Dürer is not the only artist to connect melancholy with geometry. In a 1561 engraving by the Monogrammist F.B., Melancolia (fig. 5.17) sits—accompanied by a diagnostic urine flask, a carpenter's square (neatly aligned with the floor tiles), a sphere, and a cube—in front of a vertiginous spatial recession.[85] A gridded glass window presents no view, although the room in the background opens onto a landscape. The figure's heavy, seated body suggests a centered equilateral triangle in tension with the force of the perspectival diagonals. The image is thus relatively *un*balanced; mathematics seems to carry, here, the negative force of obsessional character.

As we saw, Bovelles places the wise man as an eye at the vertex of a triangle, looking down on the world. Yet as he lifts the mind with mathematics, mathematics also, just as insistently, attaches the human mind to the material world. Melancholia sits weighed down by the world, her imagination burdened with its own futility, her mathematical implements lying idle. Melancholy figures cradling head in hands project self-absorption and concentration. They are the despairing obverse of Bovelles's self-contemplating wise man, mired in eternal regress, like the ouroboros he also mobilizes as a figure of wisdom. As the obverse of spiritual ascent, Melancholy and the mourning Virgin are triangular for a reason.[86]

The Peasant's Geometry

The cowherd in the background of Bellini's painting might himself be sorrowfully meditating on Christ's sacrifice, in this field crisscrossed by property lines. The measuring of fields in the early sixteenth century is closely linked to the pan-European phenomenon of enclosure. Thomas More describes it in his Latin novel *Utopia*, associating it with the rapidly expanding wool trade: "And so that one glutton, a dire and insatiable plague to his native country, may join the fields together and enclose thousands of acres within one hedge, the farmers are thrown out: some are stripped of their possessions, circumvented by fraud or overcome by force; or worn out by injustices, they are forced to sell . . . they depart, I say, from house and home, all that was known and familiar to them, and they cannot find any place to go."[87] In Italy, enclosure was driven by the expansion of urban bourgeois interests into agriculture. The process involved rationalizing fields and integrating pasture and cropland in the name of individual property ownership. It thus reduced traditional common pastureland.[88] The land Bellini depicted was not timeless; it had a history. An index of interest in mainland property holdings (the *terraferma*) on the part of the Venetian upper classes is the early publication of Roman texts on villa life and agriculture. In 1472, Nicholas Jenson published, in Venice, a collection of the writings of Varro, Cato, Columella, and Palladius on rustic life. In 1508–10, Venice first lost its *terraferma* possessions to French and imperial armies in the wars of the League of Cambrai, and then regained them, giving these territories an especially important status.[89] Land like that depicted by Bellini was at issue, as territory and as property.

Such contexts raise the question: could geometry both measure fields *and* cultivate divine inspiration? The mathematical "middle" constituted by the human mind or soul was a kind of paradox, not a simple position to occupy. Humans remained irrational creatures, for the most part. If the triangle linked them to the dematerialized realm of God, it could also bring them painfully back to earth: the mind is pulled back by the body, by the world.

Artists found themselves caught in this paradox. Bellini's contemporary Albrecht Dürer is perhaps the best example of Renaissance mathematical aspirations to measure the human body; this appears clearly in his *Four Books on Human Proportion*. For many artists, mathematics was an empowering discourse that made it possible to claim an intellectual prerogative as

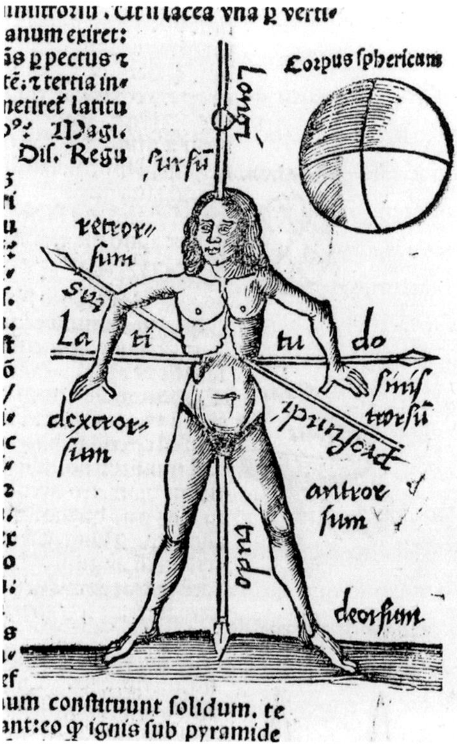

5.18 Anonymous. Speared Vitruvian man in Gregor Reisch, *Margarita philosophica noua . . .* (1508). University of Chicago Library, Special Collections Research Center. By permission.

practitioners of the liberal arts of the *quadrivium*. It enabled Dürer in particular to suggest that his work served as an especially significant site for the examination of the relationship between body and soul. Dürer's *Adam and Eve*, for instance, presents two bodies constructed via a very complicated system of human proportions that correlate to his prescriptions for the "mean" body (i.e., between the two extremes). The ideal body represented by Dürer's Adam is not just any body but a copy of a pagan idol, the god Apollo, as he himself acknowledges in a related engraving. Likewise, Eve is a figure of the *Venus pudica*. We might say Dürer has chastened these pagan idols by subjecting them to a system of numerical proportions.[90] An analogy for this might be found in Gregor Reisch's *Margarita philosophica*, which includes a sort of "Vitruvian man" (fig. 5.18) who illustrates the three-dimensionality of "bodies" in the chapter on geometry.[91] The woodcut does what Bovelles had cautioned against: it conflates the natural (human) and geometric body. The figure, speared by lines representing the three dimensions, length, width and depth, suggests that the body submitted to mathematical order is also the body wounded.[92]

An image by Dürer that traffics in mathematics, melancholy, and wounding is his the *Monument to Commemorate a Victory against Rebellious Peasants* (fig. 5.19) included in his artist's manual, the *Unterweysung der Messung.* As a triumphal monument commemorating aristocratic victory in the Peasants' Revolt of 1524–25, the woodcut has been the source of some puzzlement among scholars. It is hard for viewers, now, to accept it as a celebratory or lighthearted monument of victory by the powerful against the powerless. As Stephen Greenblatt and others have noted, many modern viewers expect the artist's sympathies to lie with the victims. Greenblatt wrote in "Murdering Peasants," the very first article of the first issue of the journal *Representations*, of the "perennial longing since Romanticism to discover that all great artists have allied themselves . . . with the oppressed and revolutionary masses."[93] Greenblatt suggests, following Panofsky, that Dürer's loyalty to Luther means he would not have supported the peasant revolt and would in fact have supported its brutal repression.[94] But the reading of sympathy for the peasant figure derives at least in part from Dürer's own visual choices. He depicts a seated peasant felled by a sword in the back—hardly a mark of noble conduct. And he uses a visual formula of melancholy—hunched over, head resting in hand—that he had also used for a Man of Sorrows figure (fig. 5.20), thus making the peasant Christlike. Dürer specifically uses the term *melancholy* to describe the peasant. The column on which the peasant is placed is constructed of the tools and products of his labors—given precise measurements by the artist in what seems like a parody of proportion. It suggests a type of Roman victory column laden with trophies, but to present upended jugs and wheat and eggs as trophies seems an ironic gesture at the very least: what have the conquerors conquered?

The measurements Dürer provides for the peasant column are a logical extension of his practice as an artist-measurer and chastener of idols. The primary function of the "figure on a column" type in Christian art is as pagan idol.[95] Roman monuments were meticulously measured by Renaissance archaeologists with a view to reconstructing proper architectural proportion—perhaps, too, as a way of restoring utopian peace and harmony. The column may indeed have been the architectural feature that received the most mensurational attention; Renaissance humanists painstakingly studied the proportions of its orders. And it was widely understood in architectural theory as the abstraction of a body—the Corinthian column, for instance, with its delicate proportions, was gendered female and thought to have been inspired by the tomb of a maiden topped by a flowering basket.[96]

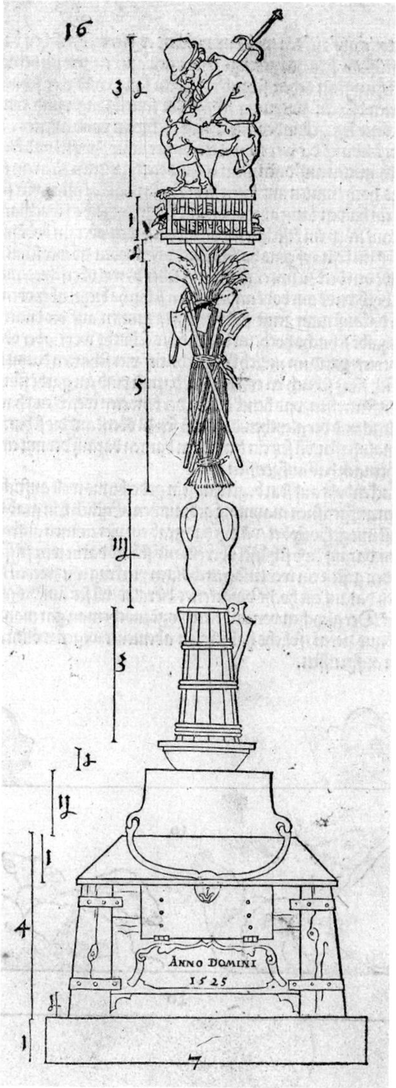

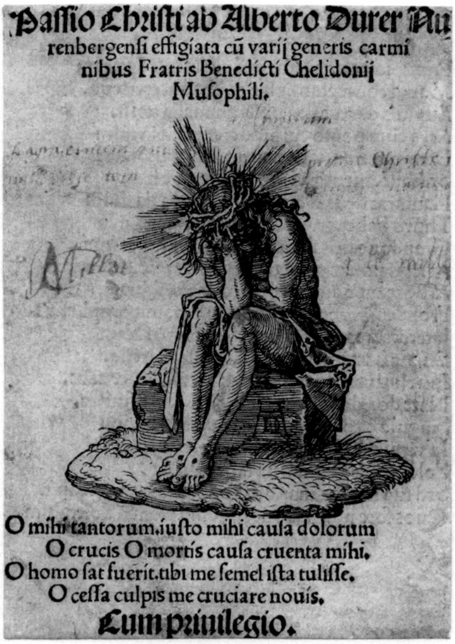

5.19 Albrecht Dürer. *Monument to Commemorate a Victory against Rebellious Peasants* in *Unterweysung der Messung* (1525). Woodcut. Base: 6.4 x 18.5 cm; column: 22.2 x 7.5 cm. Bibliothèque nationale de France, Paris.

5.20 Albrecht Dürer. *Christ as the Man of Sorrows*, title page of the *Small Passion*, ca. 1511. Woodcut and letterpress, 13.2 × 10.1 cm. British Museum, London. © Trustees of the British Museum.

The column—complicated figure of the Roman empire, of idolatry, of proportion, of the past—also figures prominently in the torments of Christ. The fallen peasant thus sends mixed messages. By his occupation and by the emotion of his pose, this figure belongs firmly planted in the earth. This abject figure has been both struck down and raised up in the position of the pagan idol. Dürer constructs a display of disproportion that echoes Vitruvius's writings on the grotesque:

> But those [rules] which were used by the ancients are now tastelessly laid aside: inasmuch as monsters are painted in the present day rather than objects whose prototype are to be observed in nature. For columns reeds are substituted; for pediments the stalks, leaves, and tendrils of plants; candelabra are made to support the representations of small buildings, from whose summits many stalks appear to spring with absurd figures thereon. Not less so are those stalks with figures rising from them, some with human heads, and others with the heads of beasts; because similar forms never did, do, nor can exist in nature.

Vitruvius is talking about ancient art, specifically Roman wall painting that employed grotesque and fanciful features that later inspired Renaissance copies. It is likely that Dürer knew this text, though it is not clear whether he would have known what type of images the Roman architectural theorist was describing.[97] Vitruvius continues:

> Now these things do not exist nor can they exist nor have they ever existed, and thus this new fashion has brought things to such a pass that bad judges have condemned the right practice of the arts as lack of skill. How, pray tell, can a reed really sustain a roof, or a candelabrum the decorations of a pediment, or an acanthus shoot, so soft and slender, loft a tiny statue perched upon it, or can flowers be produced from roots and shoots on the one hand and figurines on the other? Yet when they see these deceptions, people never criticize them, but rather take delight in them . . . unable to recognize what exists in accordance with authority and the principles of correctness.[98]

Perching a figure on a soft shoot, Dürer appears intentionally to present a scene that violates "authority and the principles of correctness," with its mathematical melancholia and its Vitruvian disproportion. The peasants' utopian demands may have seemed outrageous and their militancy frightening. But for Dürer and other members of his class—urban artisanal elites—perhaps the brutal repression of their revolt also seemed out of order. In defamiliarizing verticality, Dürer draws attention to social hierarchy.

In his *De sapiente*, Bovelles suggests that the wise man can distance himself from the world in order to achieve a point of view on it. The abject peasant, slain and placed incongruously atop a column, cannot avail himself of this position of epistemological privilege. Even for Bovelles's *sapiens* it might be a painful process: Bovelles uses a vocabulary of "purging" and "excreting" for the process of abstraction by which the image or species is rendered ever more ethereal as it progresses toward higher faculties of the mind.[99]

But can the regulation and promotion of fertility, the ordering of perception, and the measuring of the earth be connected in a strong sense? Is the triangle of perception to be associated with the pyramid of social hierarchy—or with that of the tomb? What might social class and the division of labor and land have to do with these mathematical questions? In the San Marco altarpiece, the elevated vertex of the triangle of figures expressed honor and majesty. But when understood as a triangle that links heaven and earth, spirit and body, the triangle becomes a melancholy one, aspiring to celestial realms but brought back painfully to earth. This melancholic triangle contains a tension: the elevated vertex is the point from which the wise man (in Bovelles's formulation) can look down upon the extended area of the world, yet the world itself, the base of the triangle, exercises a pull on the contemplative soul, dragging it down. Man is both at the vertex and at the ground, caught between two poles. The idea of the vertex of the triangle as both privileged point in a hierarchy and the point from which a view on the world can be constituted is indicated schematically in the triangular diagram in Lefèvre d'Étaples's 1533 edition of Boethius (fig. 4.6). The world at its most material forms the base and rises up through decreasing matter and increasing spirit, peaking in God. As Roussel's commentary states, "At a distance, the material at the base of the triangle of universals remains in its grossness [*crassitie*]; God as pure act is at the peak and cone."[100] This both *echoes* the perspectival pyramid and suggests a basis for understanding the assumptions of triangular composition as hierarchical and trinitarian. Bovelles wrote in his *Liber propriae rationis*: "Just as in the elevated cone of a pyramid, the points come together, which are numbered in the base at the bottom, so it is thus in the consubstantial mirror of the mind, or potential of the soul, in the apex and cone of the world, all earthly substances are called to coincide at a single unitary point."[101] Here the soul *rises* to the point at which it can liken itself to the divine. In a sense this triangle has collapsed two views that are differently collapsed in the theory of perspective. How to depict both viewer and viewed?

Bovelles's figure of fertile contemplation, in his triad of the first family, was Abel. But Abel's existence in the happy family structure is threatened: contemplation does not have long to live! And in this connection, it is worth noting that Abel is a shepherd—a herder existing outside the economy of landownership. His pastoral existence, though more pleasing to

God (that is, in its product—his offering) is cut short, as it were, by cultivation. Cain the cultivator/murderer might, too, be seen in biblical terms as inventor of the division of fields: a geometric sort of character.

Thus I conclude this chapter by gesturing toward the grazing animals in Bellini's painting and the contemplative herders in the background, meditating on Christ's sacrifice, prospectively and retrospectively. While the human soul may soar, here is where it is imprinted, drawn back inextricably to the body, the world, and the earth in its earthiness. The triangle carried with it the theological and intellectual ambitions of Bovelles and others—but also their disappointments, anxieties, and regrets.

6

The Third Place and the World

This chapter addresses more abstract triangles—those formed by an art object, a viewer, and a "third" position obliquely related to both. I've suggested that the triangle allows for reflection on relationality. If one can imagine arrangements of figures as triangular and one can imagine one's own implication in a picture, the viewer, too, might be a point in a triangle. At the edge of the scene, the third position might represent the connection between viewer and work, or yet another observer—a peripheral figure with an eccentric point of view, or a controlling figure (artist or sovereign) who sets images "in motion."

I will return to this point, but for now let's consider two very concrete triangles: diagrams that represent the practice of earth-measurement put to purposes of navigation and scientific knowledge, for in the encounter with other lands and cultures the notion of a different point of view was brought emphatically to the fore at the end of the fifteenth century. The first triangle is found in Amerigo Vespucci's *Mundus novus*, the second in a text from 1494, a few years before Vespucci's first voyage (though it already contains reactions to contemporary voyages to distant lands): Sebastian Brant's *Stultifera navis*, or *Ship of Fools.*

Vespucci's diagram of a right isosceles triangle maps "us" versus "them," Europeans versus South Americans, seen on the image (fig. 6.1) as "Nos" and "illi."[1] Thus it diagrams two points of view. This triangle, appearing in the 1504 edition of *Mundus novus*, was copied in a series of other tracts. Through the triangle, it links the geometry of mapping and navigation with the geometry of perspective. Many scholars indeed have seen perspective and cartography as related agents and reflections of a new worldview.[2] Triangulation was a basic tool for mapping the Earth and navigating via the stars and their geometric arrangements. If perspective and mathematically based navigation—and hence colonial endeavors—are linked through geometry as a technology of knowledge, this has also given scholars reason to equate the perspectival view with domination. In this chapter I will explore the relationship of power and perspectival geometry in other ways. I want to suggest some alternatives to the equation of mathematics with a scientific worldview that ensured or established the modern "subject" along with Western geopolitical dominance. The notion of the perspectival viewing position as central, powerful, and orthopsychic is complicated by perspective's own requirement (noted in Damisch's *Origin of Perspective*) of another point of view. The points of view produced by the triangles of perspective—specifically their relationship to power—are more complex than this simple equation proposes.

Vespucci's simple image suggests some ambiguities. The diagram constitutes an attempt to order complex new information as it dislodged and reconfigured European ways of understanding the ever-bigger and more complicated world. In Vespucci's text, the European position is described as the upright one. This is depicted as he describes it in the 1504 edition of his letter; but in at least two early versions of the woodcut, in 1507 and 1508 (fig. 6.2), the isosceles triangle rotates as if to sit on its vertex, making the two positions approximately equivalent. The triangle allows for a certain oscillation between positions and orientations: it diagrams two points of view, as it also demonstrates the curvature of the earth. By mapping the relationship schematically, the image in fact implies a kind of abstract equality between two points of view where the two legs of the triangle are equal. This abstract equality was, of course, to turn swiftly to painfully concrete inequality. How much did the acknowledgment of the possibility of another point of view affect Europeans' sense of themselves and the world?

As an abstract entity, the triangle was open to reconfiguration. This

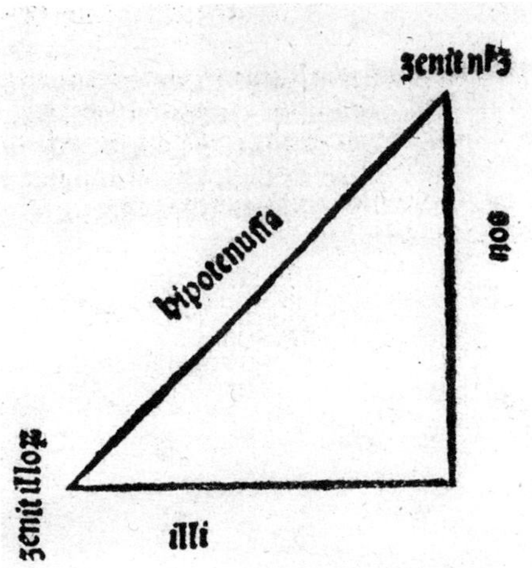

6.1 Anonymous. Triangular diagram in Amerigo Vespucci, *Mundus nouus . . .* (1508). Woodcut. Newberry Library, Chicago, Ayer 112.V5 1504b (Vault). Photo courtesy of The Newberry Library, Chicago.

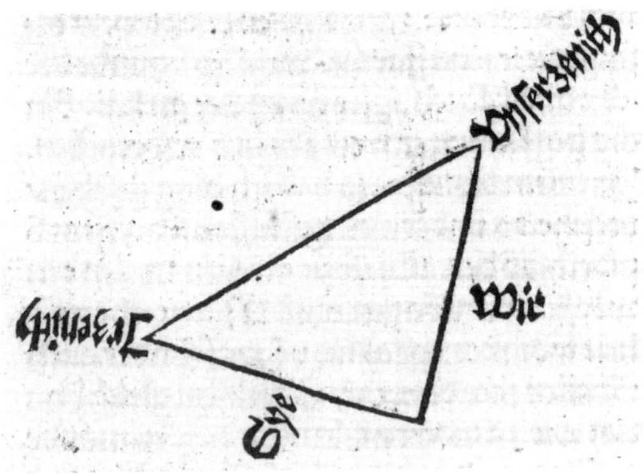

6.2 Anonymous. Triangular diagram in Fracanzano da Montalboddo and Amerigo Vespucci, *Newe vnbekanthe landt . . .* (1508), f. i5v. Woodcut. Houghton Library, Harvard University, US 2257.8.5*. By permission.

might be considered a kind of geographical relativism: difference understood as point of view rather than cultural or racial hierarchy. Whether or not there remains a possibility of oscillations between the two positions—Europeans here, South Americans there—this view of the world is still invested in the virtues, even wonders, of mathematical knowledge. But what if irrationality were such a basic human trait that any pursuit of worldly knowledge was itself folly?

For some observers an ever more energetic pursuit of worldly knowledge might not itself be worthwhile. Even in the realm of exploration, where geometry clearly made new knowledge possible, not all Renaissance observers heard word of distant lands with delight. As with esoteric inquiries in general, navigation had its critics, who felt that it was folly to delve too deeply into the secrets of the Earth. Perisauli's complaints about measure-

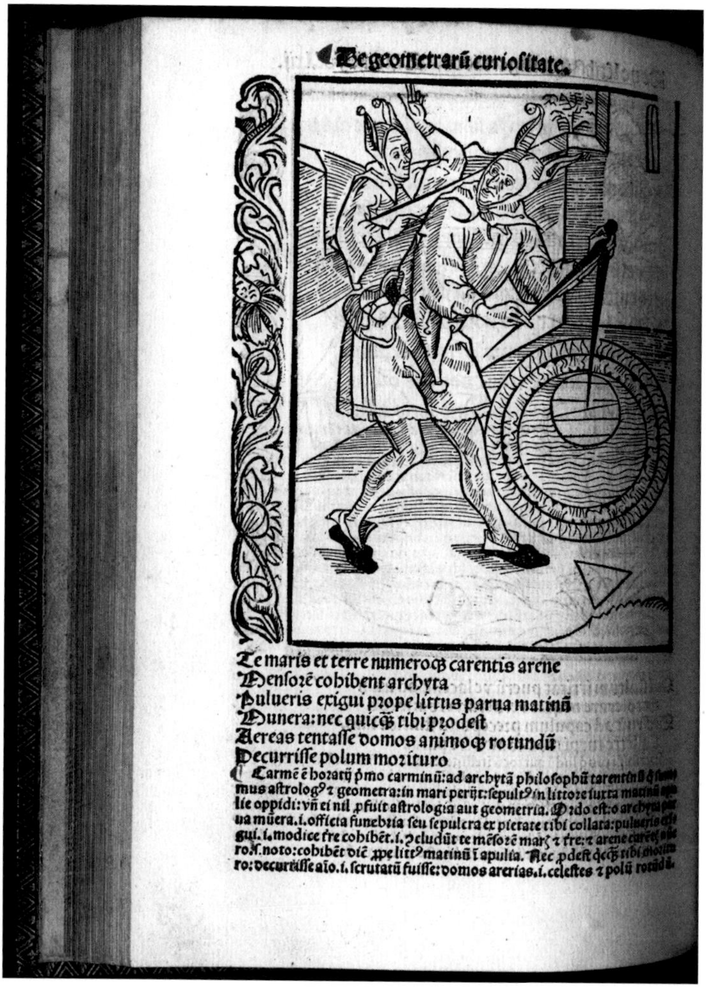
De geometrarū curiositate.

Te maris et terre numeroq; carentis arene
Mensorē cohibent archyta
Pulueris exigui prope littus parua matinū
Munera: nec quicq; tibi prodest
Aereas tentasse domos animoq; rotundū
Percurrisse polum morituro

6.3 Anonymous. Fool measuring the earth in Sebastian Brant, *Nauis Stultifera* (1506), f. 52v. Newberry Library, Chicago, Case Y 952.B7358. Photo courtesy of The Newberry Library, Chicago.

ment, which we saw in chapter 2, constitute a florid echo of those found in Sebastian Brant's Ship of Fools, the *Narrenschiff* or *Stultifera navis*. In the woodcut adorning the description of the follies of measurement, a fool attempts to measure the Earth (fig. 6.3). Philip Sohm has pointed to this image as a precursor to Dürer's *Melencolia I*, suggesting convincingly that Melancholia's dejected state, too, is the result of a failed attempt at earth-measuring.[3]

A different detail might catch our eye: on the ground near the foolish measurer lies a seemingly random triangle. From the account in the poem, we can recognize that it is not random at all: the triangle represents a figure traced on the ground by the Greek scientist Archimedes. Along with the knowledge they represent, Archimedes's figures are famous as indications of the great man's excessive attachment to geometry—to the exclusion of appropriate concern for his own life. When his home city of Siracusa was being invaded and a soldier entered his home, so the story goes, he told the soldier—indicating a mathematical diagram he had traced on the ground—"Fellow, don't touch my diagram."[4] According to the legend, the soldier then put him to death.

Any such diagram would, of course, have been much more complex than this simple equilateral triangle. Here, in the *Ship of Fools*, its simplicity renders the idea of excessive attachment all the more foolish. A running theme, indeed, in discussions of geometry is the excessive love felt for it by its practitioners. Archimedes provides a cautionary tale of attachment to knowledge in general, or abstraction in particular, to the exclusion of concern for well-being (and political prudence): the disconnect of the abstract from the concrete.

In practice, geometry did not always cohere with the human body or mind, which proved dangerously prone to error and resistant to measure. In his *Recette veritable*, Bernard Palissy, the sixteenth-century potter, surveyor, utopian architect, and Protestant, provides us with a sense of the problems inherent in measurement as applied to human beings. He tells the story of a dream in which his tools (level, square, ruler, compass, astrolabe) vie for the honor of preeminent implement. The tools ask him to adjudicate the quarrel, and he tells them that the honor rests with man, who made them all. But, they complain, man's untoollike, unmathematical, behavior makes him unfit to govern. They cry: "Man? Must we then obey and serve man, who is evil and full of folly? . . . Let us measure the head of a man . . . and you will see that he doesn't have a single straight line . . . in all his parts."[5] Palissy is skeptical but attempts the experiment, and finds, as he writes,

> I could not make a single definite measurement, because the folly that was in the head made its measurements keep changing. So I was confused, because I found the head to be first of one appearance, then of another, and even though at times there was some semblance of straight lines, as soon as I took out my tools to draw them, suddenly all the straight lines became oblique, at which I was very astonished, seeing that there was not a straight line in the head of man, because his folly made all the straight lines bend.[6]

He then produces an impressive catalogue of follies concocted by human brains. Human passions can't be contained or measured by the geometer's tools; they actively work to obstruct or destroy ordered geometries. One might see the oblique lines of which his tools complain as the orthogonals of a perspectival construction: a distorted deception taken for true representation. The human brain—bent and distorted as it is—can hardly serve as a paradigm of measured, rational space.

As the philosopher Karsten Harries has argued in *Infinity and Perspective*, even as the technique of perspective presented a convincing illusion of a privileged "point of view," it relativized that point of view by suggesting the possibility of infinite others. In his complicated and elusive book, Harries argues (as others have also done) for the importance of Nicholas of Cusa as the progenitor of philosophical modernity. He bases this on Nicholas's perspectivalism: the sense that any given point of view is only one of many possible points of view. According to Harries, this implies that once one has an understanding of a perspective *as* a perspective, one is already beyond it, capable of recognizing that appearances must be interpreted in order to achieve an objective understanding of reality.[7] For Panofsky, in *Perspective as Symbolic Form*, perspectival construction—that is, the viewing position constructed by Alberti—is the precondition for a notion of objective vision. For Harries, on the other hand, diversity of viewpoints is always already bound up in the possibility of objectivity. A more troubling possibility subsists, however: what if in fact there is no way to truth at all?

Punctum dolens

Returning to painting in the first century of Albertian geometric perspective, I want to explore what's at stake in the interpretive interventions made by two French scholars, Daniel Arasse and Hubert Damisch. For Arasse, a key moment that occurs in tandem with Albertian perspective is the development of an iconographic convention for the annunciation of the angel Gabriel to Mary. The iconography of the Annunciation presents a simple unit of narrative—what Arasse calls an "enunciation." For Arasse, this must be understood as the essence of making statements: enunciation grounds the possibility of narrative painting and the practice of the visual *historia*—which was, for Alberti and many who followed him, the highest aim of painting.

How does Annunciation represent enunciation? First, think in terms of the statement (subject + predicate) as the basic unit of narrative. In visual terms this would be structured to involve two partners, or even more abstractly, a movement from point A to point B—the angel announces to Mary. For Arasse, it is particularly significant that in Quattrocento annunciations the angel normally appears in profile: the event is depicted as a motion that is strictly perpendicular to the viewer's relationship to the picture.[8] The angel is the agent, with Mary as recipient. In conjunction with Alberti's theories about the veracity of the painted image and the fixed viewing point, Arasse argues, this positioning abolishes the participation of the viewer in the work, making the viewer the mere witness of an unfolding truthful event. The enunciative mode is defined in contrast with other predominant iconographic modes in the fifteenth century: in particular iconic, portrait-style depictions that in theory established only a relation between viewer and object, not a situation of witnessing a statement. One might liken this to the distinction made by Michael Fried between absorption and theatricality, except that in *this* absorptive act what we have is a dyad, not a single figure.[9] This is also the flip side of the power accorded to the viewer as privileged surveyor. Alberti's text, as I argued in chapter 2, actually represents viewer and object as equal. On the question of the closure of the viewer from the scene, I therefore disagree with Arasse; but I think he points us in the right direction.

If a statement is enunciated by the angel toward the Virgin, it is one that is often *doubled* by the presence of God the Father. In many depictions of the Annunciation, God the Father sends golden rays, and sometimes a dove representing the Holy Spirit, in Mary's direction. Piero della Francesca's famous Arezzo *Annunciation* (fig. 6.4), part of the *Legend of the True Cross*, is an example in which the painter explicitly plays geometrically with the arrangement of these three figures. Together, the figures form a right triangle on the painted surface. But God's delivery also closely follows the perspectival line of the architrave of Mary's nichelike space. We might read this presence as a concession to the iconography's legibility to human eyes—the event itself, as Arasse points out, is outside the realm of measurable reality. But in Piero's fresco we are forcefully made aware of the coincidence or enfolding of the event (the incarnation) and the statement of the event (the annunciation) in a single image. Another example we have already seen is Francesco del Cossa's Dresden *Annunciation* (fig. 3.1), which, as Arasse notes, is an exception to his rule of the profile angel.[10] But it is actually even more interesting: in it the two transactions—God and Mary and angel and Mary—seem to mirror one another geometrically through three dimensions.

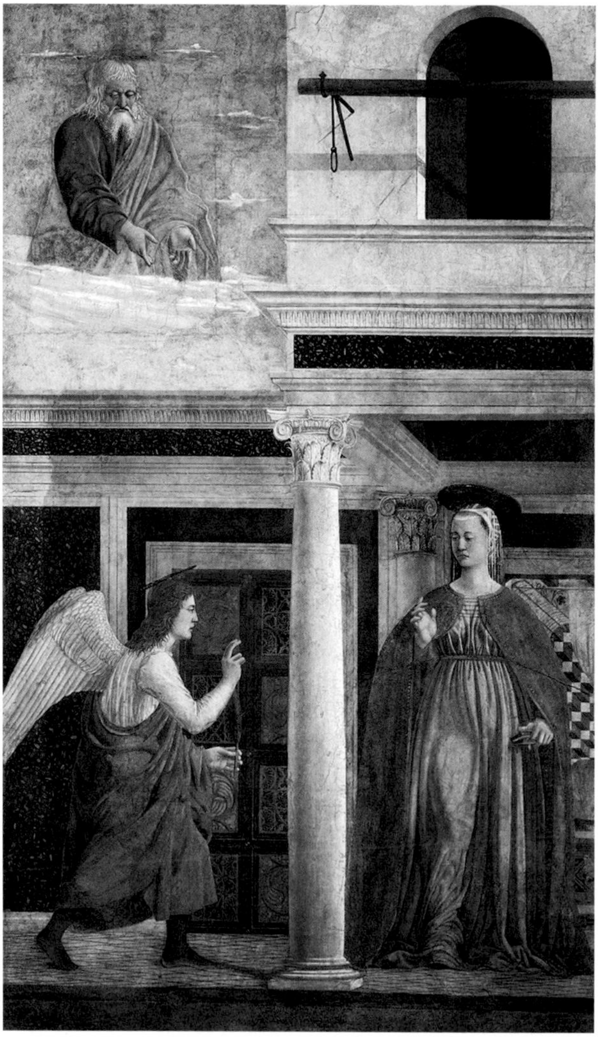

6.4 Piero della Francesca. *Legend of the True Cross: Annunciation*, ca. 1452–56. Fresco, 329 × 193 cm. S. Francesco, Arezzo, Italy. Scala/Ministero per i Beni e le Attività culturali / Art Resource, NY.

This doubling of event and representation in the same image resonates with diagrams that represent perspective using a distance point.[11] Earlier I mentioned Damisch's suggestion that perspective requires a "third party." On the one hand, he is making a point about figures that appear at the edges of paintings—figures who cannot be identified with Alberti's *festaiuolo* figure, who invites the viewer into the picture, but rather stand apart, as an observer embedded within the scene. Damisch writes, "The perspective paradigm effectively posits the other, in the face of the 'subject,' as always already there: it introduces a *third party*, previously excluded by an art (like that of the medieval period) that was essentially contemplative, that precluded all possibility of passing from one position to another as well as of entering into the painting as if it were a 'scene.' "[12]

Damisch calls this a "three-term structure," involving an I, a you, and a he or she. He plays this out in the structure of his own book, written as a dialogue between two voices for a reader. Worth noting as an aside is Damisch's odd equation of contemplative art with the art of the Middle Ages and narrative art with the Renaissance. Privileging narrative and the Renaissance, he associates them with one another far more strictly than the historical material can support, since narrative art predominated in the Middle Ages and the Renaissance was in fact the great period of proliferation of private devotional art. While their views on the function of such painting with respect to the viewer are diametrically opposed, both Damisch and Arasse privilege the form of painting that, as Damisch characterizes it, "not only shows but thinks."[13] I am tempted to suggest an analogy with Iamblichus's description of the dyad as formless: a simple devotional relationship between viewer and work, for Damisch, has no structure.[14]

But the argument that the construction of perspective assumes a third party gains strength from the practical construction of the perspectival image. Why are three positions implied? Perspective assumes the existence of a profile point of view. The profile position determines the correct distance from the picture plane of the frontal point of view, and hence the rate of diminution of the depicted scene in perspective. Perspective diagrams that use the system of the distance point set a point outside the frame of the picture, or at its edge, that represents the position of the viewer through a two-dimensional projection of the distance at which he or she stands (fig. 6.5). In figure 6.5, the ground plane is divided into equal units. From the points (C, D) that represent these divisions, lines (orthogonals) meet the vanishing or "centric" point. The viewer's distance from the panel is indicated by the line from the "distance point" (A) to a point on the panel (E). A diagonal drawn from the distance point to the ground plane below creates intersections with the orthogonals, establishing parallel lines that complete the grid of floor tiles and the panel's fictive space. This series of lines parallel to the ground plane (transversals) represents the floor's recession—at a rate keyed to the distance at which the viewer is ideally situated.

What's assumed by this construction is that the closer to the picture the viewer is held to stand, the more sharply the orthogonal lines (the ones per-

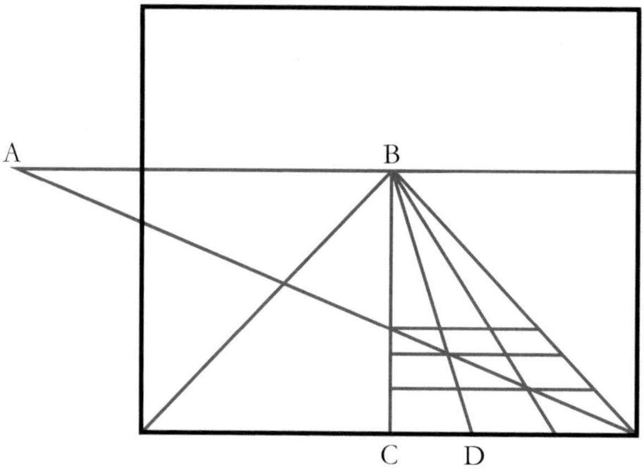

6.5 Diagram of perspectival construction using a distance point.

pendicular to the picture plane) should recede. They create a wide angle with the vanishing point, to mirror the wide angle the eye makes with the edges of the picture. Conversely, the farther away the view is held to be, the smaller the "incline" created by the pavement grid of floor tiles. Consider a point at a great distance from a small rectangle forming a pyramid with it: the acute angle formed at the vertex, the eye, gets smaller the farther away the rectangle is. Given an eye at the same level as the vanishing point, the angle of this view is the same angle at which recession of the floor tiles is constructed. (All these geometries are based, of course, on the notion of a visual triangle, or pyramid.)

Rather than entering into the intricacies of debates over the relationship between theory and practice in Quattrocento perspective—who knew what and when—I want to emphasize the function, in perspectival diagrams and practices, of the distance point or representation of the viewpoint of the onlooker. The distance point is an explicit way of articulating the viewer's position with respect to a perspectival image; when explicitly represented in a diagram it appears outside the frame of the "image," referring to the viewer's distance from the painted surface, structuring the angle of perspectival recession. It is a way of figuring the third dimension, collapsing it into two.

For Alberti, as I suggested in chapter 2, this "third" point (in a "little place"—*areola*—to the side of the picture plane) enabled the matching of viewer and work.[15] As in the first version of Vespucci's triangular diagram, the two points of view appear roughly equal. But as the practice took hold, artists took opportunities to use the position of the distance point—or an analogous position incorporated into the picture—to represent a site of authority, whether their own or the viewer's. Sometimes it was a figure of greater divine or political authority; in the case of God the Father in an Annunciation, the figure doubles, or certifies, the action of the narrative. Perspective's distance point resonates with this doubling of event and reception in the same image, as artists drew figurative inspiration from the technical process of constructing a perspective scene.

Lyle Massey has pointed out the oddness of the depiction of the distance point in technical diagrams of the practice of perspective: "In Vignola's second rule . . . [t]he viewer is asked both to imagine the viewpoint (which still appears as a figure) and to embody it at the same time. The demonstration becomes a kind of contorted folding of three-dimensional, geometric description (representation) onto a two-dimensional surface (object) in which the viewer is asked to both rationalize and inhabit the point of view simultaneously."[16] The ambiguity Massey sees "at the heart of perspective geometry itself" is inherent in the distance point, a depth folded ingeniously into width.[17] The line from the centric point to the distance point is a perpendicular line projected onto the two-dimensional field. Technically, to read such images properly we are to understand that this is not a doubling at all, simply a different representation of the position in front of the picture. But the point hints at a third place, distinct from both subject and object.

In his *L'annonciation italienne*, Arasse extends the question to consider the meaningful coherence, but also tension, between perspective and Annunciation. The key point is the moment of incarnation of God in human form as an unfigurable miracle. Arasse cites a passage in a printed sermon by Bernardino of Siena that emphasizes the status of the incarnation as a long string of miraculous entries: of the immeasurable within measure, the invisible within the visible, the unfigurable in the figure—and, notably, the *artisan in the work*. Arasse writes, "If, inasmuch as it implies the Incarnation, the Annunciation is the moment where the incommensurable comes into measure, how could perspective, 'symbolic form' of a commensurable world, have visualized this (latent) arrival of the unfigurable in figure? Must we think that this would have forbidden painters to depict the Incarnation, invisibly present at the heart of the Annunciation? Must we admit that painters could, at best, make an allusion to it by means of accessory motifs, secondary iconographic allusions?"[18] Perspective for Arasse was a tool that artists used to explore this notion of the representation of the invisible in the visible. Sometimes the point was to block the possibility of visibility, by placing an obstruction where the vanishing point should fall. As Arasse puts it, "Artists establish in the measured proportionality of the work an *interval* that visualizes the entry of incommensurable divinity in the human world of measure."[19] This could be an expedient solution to a system that resulted in many points of convergence or an acknowledgement of the very question of the unrepresentable.

Seeing Seeing

One of Arasse's primary examples is the Annunciation (fig. 6.6) found on the predella of the St. Lucy Altarpiece, whose main panel I discussed in chapter 4 (the pieces have been dispersed to several different collections). Domenico Veneziano's predella offers a classic formulation of both parts of Arasse's paradigm: it depicts the enunciation/Annunciation (without God the Father) and also forecloses the visibility of the vanishing point, insisting on the unrepresentability of incarnation. Only the "enunciation" is depicted, not the "event" of the incarnation itself. The background consists of an "enclosed garden" or *hortus conclusus*, a phrase from the Song of Songs commonly used to describe the Virgin. The vanishing point falls at the garden's closed gate. But in the altarpiece, when complete, what the angel announces—with finger pointing upward—would have been what appeared above, in the main panel. In a way, both images hide and reveal (by using perspective to *point*) at the same time. Together, the perspective lines of the main panel and the predella create mirrored triangles that do not touch but point toward one another, as if earth and sky struggled to be joined at a single point.

With the relationship between the two panels, the viewer is given privileged access to something that only the eyes of the spirit can see. The question of what one sees or doesn't see is also brought into play by the presence of St. Lucy herself, who functions as an artist in the work—or, at the very least, an observer. The early Christian saint was held to have torn her own eyes out to avoid being the object of attraction to suitors. Later she was martyred, an event depicted in another panel of the predella. Here is an image that calls into question Damisch's distinction between contemplation and enunciation. Lucy stands on the right edge of the painting, like Damisch's observers. She is above all a viewer, but a viewer who has eyes only for devotion. In picturing seeing itself, Domenico gives it a special divine privilege.

If Lucy is a figure for the painter, it is, however, a rather uncomfortable

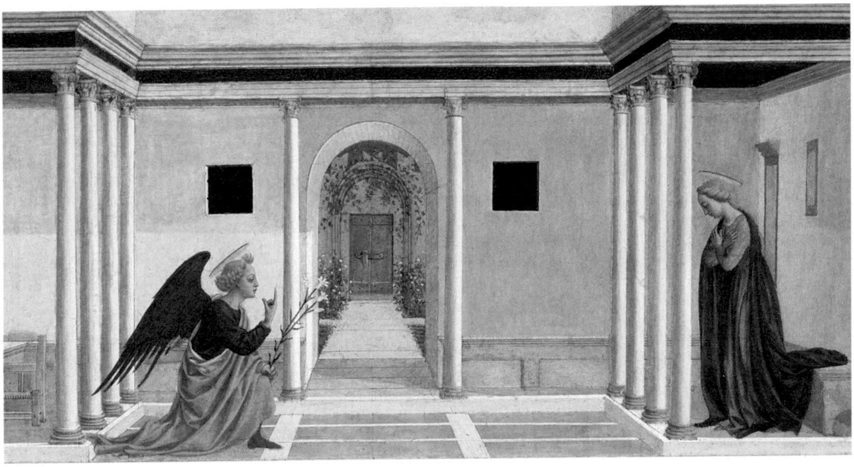

6.6 Domenico Veneziano. *The Annunciation* (predella panel from the *St. Lucy Altarpiece*), ca. 1445. Tempera on wood, 27 × 54 cm. Fitzwilliam Museum, University of Cambridge. Photo Fitzwilliam Museum / The Bridgeman Art Library.

place to be, given what occurs in one of the other predella panels: Lucy's martyrdom (fig. 6.7). The emperor-suitor stands above and to the right, ordering the beheading of the virgin saint. Here the vanishing point is the site of Lucy's own vanishing, the spot where the knife falls on her neck. The sovereign occupies a spot analogous to that often reserved for God the Father, except that he appears on the right of the scene and not on the left. If in the main panel the position "to one side" is occupied by a privileged and sympathetic observer, in the scene of martyrdom the figure who sets the scene in motion carries all the violence of authority with him.

The Sovereign View

Botticelli's *Primavera* (plate 7) also poses the question of the figure at the edge of the work: I will argue that it expresses surprisingly, in its very absence, the notion of the lateral viewing position that sets the image "in motion." Botticelli has merged the idea of the frieze, drawn from ancient sarcophagi, with a compositional schema like that of the highly symmetrical *sacra conversazione* (that is, a painting like the St. Lucy Altarpiece), peopled with figures from Greco-Roman mythology. We can trace a pronounced triangle here by following the sway of Chloris's body as she is grabbed by the

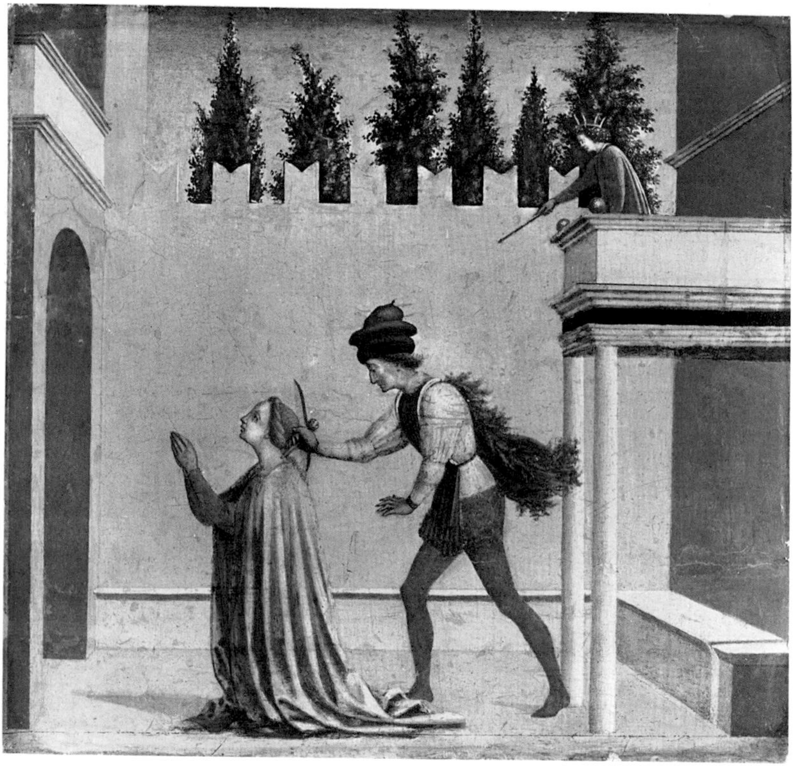

6.7 Domenico Veneziano. *Martyrdom of St. Lucy* (predella panel from the *St. Lucy Altarpiece*), ca. 1445. Oil on wood, 26.5 × 29.5 cm. Gemäldegalerie, Staatliche Museen zu Berlin, Inv. 64. Bildarchiv Preussischer Kulturbesitz / Art Resource, NY. Photo: Jörg P. Anders.

blue wind-god Zephyr at the right edge of the painting, moving up to Cupid above, and then tracing the implied course of his cocked arrow toward his target, the graces at the left.

Viewing the painting as triangular highlights its slight asymmetry. Having looked at paintings like Fra Angelico's and Domenico Veneziano's, we would expect Venus, like Mary, to stand at the actual center of the painting. But she is in fact slightly off center. To conceive her as the center of the panel would require adding approximately one figure's width to the right-hand side. Furthermore, she is immediately flanked by a triad of figures on either side, but on the left side only a major god also frames the scene. That is, Mercury anchors a triad of graces on the left, but on the right the Zephyr-Chloris-Flora triad—minor figures in comparison to Venus and Mercury—stands alone.

Edgar Wind claimed that the "crux of any interpretation of the *Prima-*

vera is to explain the part played by Mercury" and that the "composition of the painting is not fully understood, nor the role of Mercury quite comprehended, until he and Zephyr are seen as symmetrical figures."[20] Yet this is a strange symmetry. Zephyr is a kind of demigod, more like the nymphs or graces than like the stately Olympians Venus and Mercury. And Mercury's and Zephyr's relations to the female figures are quite different. Mercury rules over the three graces; Zephyr, a lesser figure in Greek myth, is the agent of Chloris's transformation into Flora. He also participates fully in the triadic group that seems to mirror the three graces on the other side, part of a triad rather than the ruler over one. The painting appears like a musical phrase missing its final note.

If a figure were missing from the scene, who would it be? What god might flank the scene on the right? I have argued elsewhere, in an article on which I substantially draw here, that it is Apollo.[21] I also suggested there that Wind came close to this notion without ever explicitly stating it. In his *Pagan Mysteries in the Renaissance*, he cites a text by Ficino that begins with the triad Mercury, Venus, and Apollo, and suggests a musical application to the *Primavera* involving Apollo.[22] Ficino was close to the Medici and served as an adviser to Lorenzo di Pierfrancesco de' Medici, ward of his older cousin Lorenzo, called the Magnificent; the painting was almost certainly made for the younger Lorenzo. In fact, the triad of Mercury, Venus, and Apollo (or the Sun), interpreted in many different ways, reappears insistently in the Ficinian corpus.[23] Ficino writes of them, "These stars are, among the celestial bodies, those which most affect by their favor the human genius."[24] In his *Three Books on Life*, he declares that the three heavenly guides of scholars are Mercury, Apollo (the sun), and Venus; a scholar should rise early in the morning when these three planets are in force.[25] In his commentary on Plotinus's *Ennead* 1.3, which deals with the means for the soul to ascend to the heavens through contemplation, Ficino takes the three gods as figures for the three types of humans who can follow paths of ascent: the philosopher (Mercury), the lover (Venus), and the musician (Apollo). From that starting point, Ficino works analogically and piles interpretation on interpretation. He equates the gods with the senses: Venus with sight, Apollo with hearing, and Mercury with the imagination. These are the three highest of the six senses of humans, which included the imagination or "common sense"; they contrast with smell, taste, and touch, the lower and more bodily senses. In his commentary on the *Symposium* Ficino states that these higher senses are the only ones required for Love.[26] Further, Mercury is the good, Venus the intellect, Apollo the soul. Mercury is truth, Venus beauty, Apollo concord. Whatever variety there may be in these interpretations, it is clear that the three gods persistently form a triad.

Were Apollo positioned to the right of Zephyr, he would produce a more satisfying triangular symmetry, according to typical Renaissance conventions such as those we have seen already: there would be two triads ruled over by Olympians flanking Venus (rather than a triad ruled by an Olympian corresponding to a couple of nymphs—really the same nymph—"moved" by a wind), a total of nine figures in three loosely triangular groups of three. Wind was fond of Neoplatonic triads and spent a chapter of *Pagan Mysteries* on the three graces as an emblem of Neoplatonic movement between God and the soul. The painting, in my view, contains two complete triads (the graces and Zephyr-Chloris-Flora) and one that lacks its third term (Mercury-Venus-Apollo).[27] Were Apollo to appear in the picture it would also make this painting a precursor, in reverse, of Mantegna's more complicatedly composed painting usually known as the "Parnassus" (fig. 6.8), where this triad appears very explicitly: Apollo in the left foreground, Venus in the top center (with Mars as her accessory), and Mercury in the right foreground.[28] In a more oblique way, Botticelli's *Adoration of the Magi* in the Uffizi (fig. 6.9) might also be seen as a formal parallel. Despite a difference in the proportions of the panel—which places the Virgin much higher than Venus—they are dead ringers for each other. A probable self-portrait of Botticelli appears in what would be the place of the missing third term in the *Primavera*; a young man with Medici features (sometimes identified with the younger Lorenzo, sometimes not) appears as an onlooker in the place of Mercury and in a similar (but reversed) pose. Even the retainers who stoop in reverence at the lower right of the Nativity seem to parallel the sway of Chloris in her flight.

Why would Apollo be absent from the picture? Perhaps because he was absent when a particular viewer was not in the room—that is, perhaps Apollo was intended to be embodied in a privileged viewer. A contemporary analogy is found in the absence, among the Virtues, of Justice—to be embodied by Borso d'Este—in the Sala dei Stucchi at the Palazzo Schifanoia in Ferrara.[29] Perhaps the "missing" god was to be figured by the elder Lorenzo

6.8 (opposite top) Andrea Mantegna. *Parnassus (Mars and Venus)*, 1496. Oil on canvas, 160 × 192 cm. Louvre, Paris. Erich Lessing / Art Resource, NY.

6.9 (opposite bottom) Sandro Botticelli. *Adoration of the Magi*, ca. 1475. Oil on wood, 111 × 134 cm. Uffizi, Florence. Erich Lessing / Art Resource, NY.

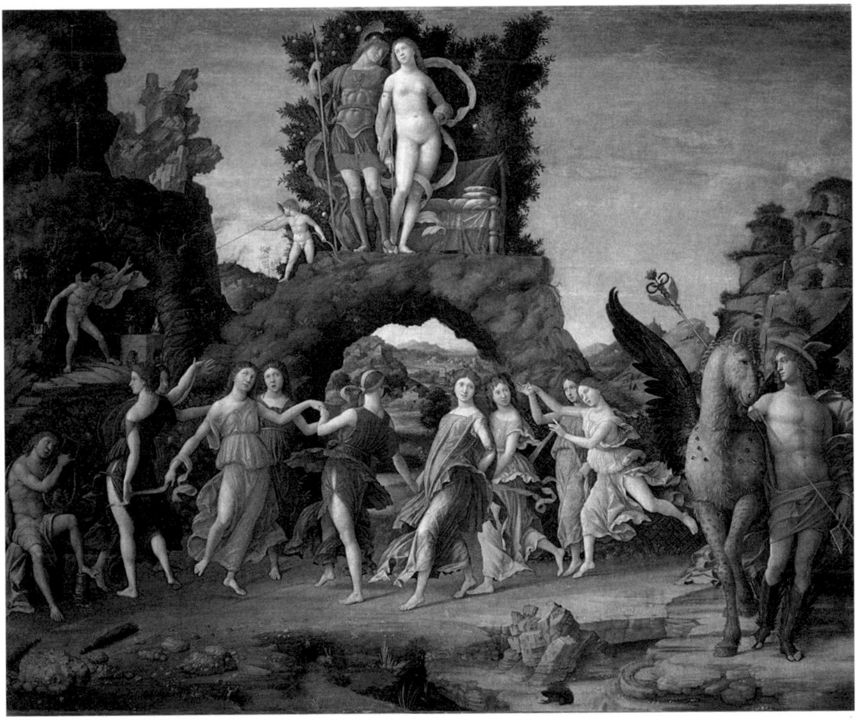

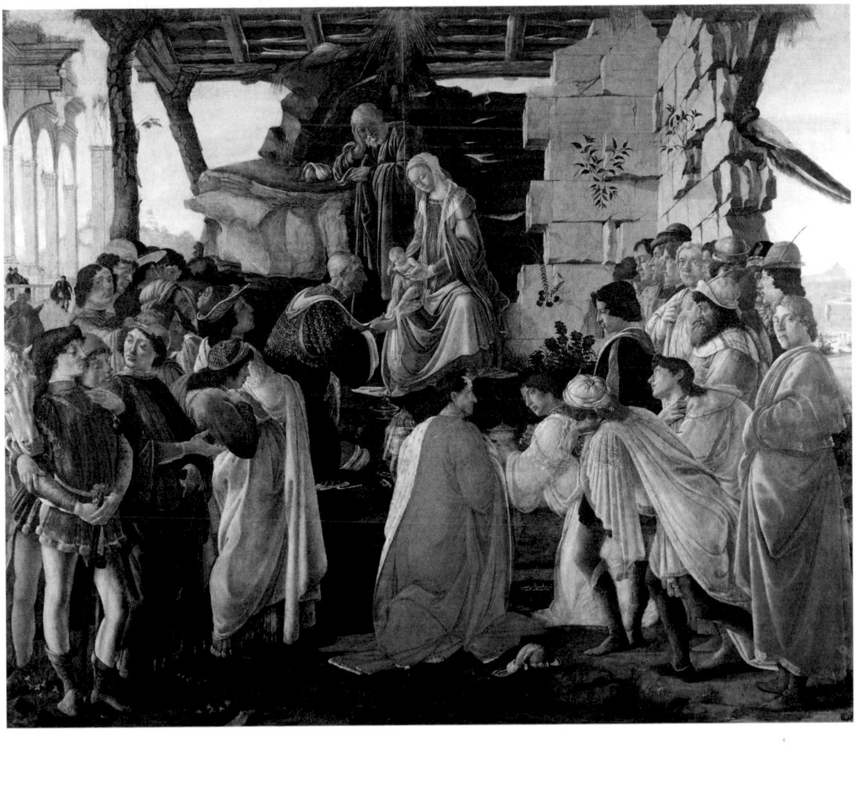

de' Medici—Lorenzo the Magnificent—himself. Agnolo Poliziano called Lorenzo "Phoebus" in his funeral ode and refers to him as Apollo's poet in his *Stanze*; indeed the grounds for such an identification are manifold. Like Apollo, Lorenzo was a musician and poet. Apollo is the inventor of the laurels of poets; a common pun, used by Poliziano, connects the name Lorenzo with laurel (Lauro). Lorenzo's family name might also be relevant; Apollo is also a patron of medicine (the art that gave the Medici their name). Any number of things might explain the absence of Apollo; and perhaps this was part of the courtly game the painting was designed to provoke. Perhaps ironically, one of the chapters of Wind's book is titled "The Concealed God": a reference to the god of Cusanian negative theology, perceptible everywhere and nowhere.

Apollo also has relevance to the group on the right. In another letter by Ficino, the sun god makes an appearance as the father of Flora. In the springtime she wanders into the realm of Venus, where she discovers bodily pleasures but also pains—from which she returns to her father, "author of life-giving medicine," to seek healing.[30] Apollo is also the rival of Zephyr for the attentions of another mortal-turned-flower, Hyacinth. As father of Flora, herself the namesake of the city of Florence, Apollo would appropriately govern this group.

If the picture had to be completed by a viewer or viewers, if in effect the viewer were part of the picture (and vice versa), this would confuse the categories of subject and object that we expect the Renaissance to clarify for us. While the basic hierarchical structure of Neoplatonism mirrored social hierarchies and can thus be read as partly an expression of sociopolitical ideologies, the relations it describes between appearance and essence, perception and its objects, body and soul, are complex and confusing—reflecting a confusion of boundaries between subjects and objects rather than an establishment of clarity.

To think of the "object"—the physical object—as a player in a complex theater of cultural exchange and performance, to think of it without objective boundaries, might require thinking about the "concealed god" in a different way. If my reading is correct, the concealed god alludes to a political figure. If we identify the missing figure with Lorenzo de' Medici, we might think of the painting as an expression of the "veiled signory" of the fifteenth century.[31] As the century progressed, the Florentine republic became more and more nominal, as the Medici exercised power covertly and, later, more overtly.[32] In the 1470s and 1480s, it was still part of the ruler's work to preserve the appearance of republican government.[33] Recent interpretations of archival documents suggest that the painting was made at Lorenzo's behest on the occasion of his young cousin's very political marriage, arranged by Lorenzo to his own ends. This cousin—Lorenzo di Pierfrancesco de' Medici—would later revolt against Lorenzo's son Piero, call in the French king to intervene in Florentine rule, and adopt the name Popolano (meaning "plebe"). These actions, politically expedient though they may have been, grew from a long-simmering conflict, both financial and political, between the branches of the family, a conflict that required legal arbitration in 1485.[34] The younger Lorenzo's at best ambivalent view of his older cousin and guardian is reflected in letters Ficino sent him. Lorenzo the younger, Ficino explains in one letter, traces his lineage to Phoebus. In the other, he warns him not to surround himself with bad influences. He compares the influence of bad advisers to an eclipse in which the moon suffers the absence of the sun because the Earth comes between them.[35]

Lorenzo the Magnificent played a very specific patriarchal role—which made him a hidden third party to the marriage and thus to the painting produced for it. This role also prompts us to consider the sexual ambiguities of male power in Florence. In the story of Apollo and Flora in Ficino's letter, Apollo abandons his daughter to the lustful realm of Venus. And yet in proximity to the figure of Zephyr, Apollo might also recall their shared male love object, Hyacinth. As Michael Rocke has shown, male culture, the political life of the republic, sodomy, and its prosecution were intertwined in Medici Florence. Male power depended in part on a vast age differential between husband and wife, and this produced a population of young men seeking nonmarital sexual activity and whose social and civic life took place largely in the context of all-male groups. Meanwhile, the Medici prosecuted sodomy at once comprehensively (in terms of arrests) and mildly (in terms of punishment) while allowing open celebration of it within their own circle. Along with the poetry of Poliziano and others, Neoplatonism played a role in this, authorizing male-male love while at the same time desexualizing it with varying degrees of seriousness.[36]

Among other aspects of the arguments about love transcribed by Ficino is the idea of reciprocal and hierarchical bonds produced by love between older and younger men. These individual bonds mirror the more general Neoplatonic structure of the cosmos, a ladder of levels from the most material to the most divine in which a direct relationship is established between each degree of hierarchy and the next (but not between those at more than one remove from each other). In this view, "Platonic love"—desexualized, hierarchical love between men—might also provide traction for an authoritarian romance, the love of the ruler.

Such a structure is served well by the distance point configuration. One is encouraged to identify with this originating point outside the picture that determines its structure. And the distance point is not just a neutral third position. It oscillates between independent existence as a force that calls the painting into being and a simple convention for representing the viewer's position. A viewer may be encouraged to identify with this originating point outside the picture—or, perhaps, to envy and compete with it. For the power relationship *between* the two positions is complex. Dürer's woodcut of a draftsman drawing a nude using Alberti's veil (fig. 6.10) shows the kind of "cheat" an artist might employ to bring the two positions in line.

In the woodcut, found in the second (posthumous) edition of Dürer's *Unterweysung der Messung*, the viewer's implied gaze has in a sense disordered the techniques of perspective, impeding a correctly foreshortened view. The viewer has to see what the draftsman sees, which is not what the viewer would see if the draftsman were seeing it. That is, in order to communicate the scene to the viewer of the print, Dürer has twisted the woman's body so that the viewer has a better view than the artist supposedly drawing her has. She is turned away from the artist to the extent that she is turned toward the viewer; the implication of the double view of perspective produces this angling of the body. One might say that the eye of the other casts doubt on the objective veracity of one's perceptions. But this is also a homosocial structure: (presumably) male competition over a woman's body.

The position outside the picture, with which a viewer uncomfortably identifies, might be aligned with Jacques Lacan's notion of the gaze; certainly Damisch had Lacan's psychoanalytic interpretation of perspective close at hand in formulating his own understanding of the technique. For Lacan, the image, especially in its perspectival aspect, presents us to ourselves as animal: caught, trapped, rendered wordless. As much as we might associate our power of sight with the position of the hunter—via metaphors of "capturing" a likeness—there is still the potential for the viewer to be the one captured, rendered animal. We might say that Lacanian psychoanalysis, in the broadest sense, deals with friction on the borders of what is human, the mapping of the human against other categories: woman, if the human is understood as male; the unconscious, if the human is understood as rational; animals, automata, and art. The "geom-

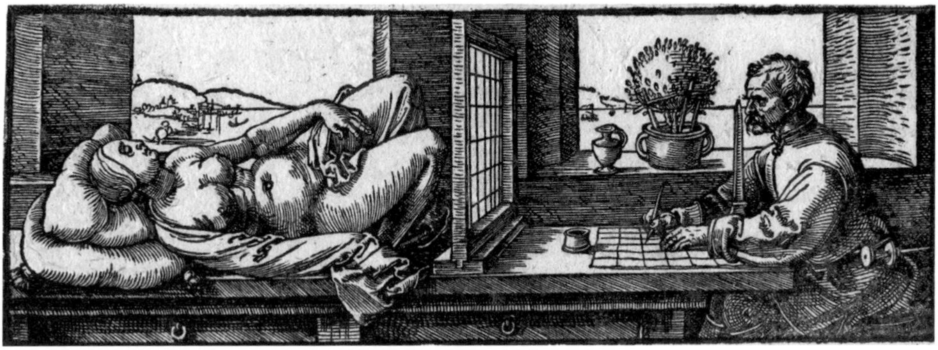

6.10 Albrecht Dürer. A draftsman drawing a nude model using Alberti's veil, in *Unterweysung der Messung*, 1538. Woodcut, 7.6 × 21.5 cm. British Museum, London. © Trustees of the British Museum.

etral dimension," the web of perspective, catches not only the represented object but also our gaze:

> For us, the geometral dimension enables us to glimpse how the subject who concerns us is caught, manipulated, captured, in the field of vision.
>
> . . . What is the desire that is caught, fixed in the picture, but which also urges the artist to put something into operation?[37]

More than Lacan, however, I find François Gantheret's notion of "wild perception," *perception sauvage*, useful. This is the "oblique look" that enables the viewer of Luca Signorelli's self-portrait (in one of his Orvieto frescoes, subject of a discussion by Freud) to see him seeing her. "It is laterally, obliquely [*de coin*] that the wild perception comes to us that puts us in the presence [*en présence*]. . . . But with Signorelli looking at us looking, we are—from the corner of our eye from which we see him seeing us—inside iconic, Byzantine representation, where he who is not the spectator, but in presence, finds himself close to the gods."[38]

What is this oblique look? A glance out of the corner of the eye bridges a distance with sudden immediacy. Gantheret's *perception sauvage* can be likened to Walter Benjamin's *mémoire involontaire*, which is not so much natural memory as its vestigial trace in urban modernity, a fleeting moment of receptivity to sensation that is contrasted with the instrumentalized use of one's archived memories—or, in this case, the instrumental use of vision.[39] Gantheret's view is a more transhistorical one—*literally* transhistorical, since the look that surprises us also bridges a historical gap. There might be no better way to connote psychological interiority in naturalistic painting than eyes turned to look, angled "sideways." This is an eminently legible look (the source, perhaps, of legends of roving-eyed portraits). It might also be a dangerous look; the sly, curious, or furtive look; the involuntary glance, the look out of the corner of the eye, the willfully ill-advised turn to look (Orpheus, Lot's wife). The bodily phenomenon of simply looking sideways, "out of the corner of one's eye," in a direction other than the one in which one is facing, slips easily into a cultural significance (however arbitrary) of doubleness, treachery, deceit—sly watchfulness, that of Georges de la Tour's card players and fortunetellers, or a dirty joke, as in Robert Doisneau's photograph titled *Un regard oblique*.

Signorelli's oblique look is that of a painter poised to inspect his own face in the mirror while painting his self-portrait. Yet it is *our* oblique look that Gantheret emphasizes. The perspectival space of the fresco is outside ours; it is as we move through our own space, he argues, that we allow ourselves to be caught unawares by the painted figure looking back. A depiction of illusionistic figures deployed in what is understood to be three-dimensional space opens up the possibility of a play of angled gazes, whether or not that space is presumed to be coextensive with that of the viewer. The movement of the spectator enlivens the image. The painted figure achieves presence through that glance that diminishes the "objectness" of the object.

The look that enables this animation of the image is oblique and fleeting. But it also depends on the potential for other kinds of encounters in the same space. This encounter does not take place in an enclosed psychological interiority but in a social space. While Gantheret calls this perception a "wild," or natural, perception, it is not distinct from social life. The sense of *someone there*, whether it is in a museum, a church, or one's own bedroom, relies on the potential for someone else actually *to be there*. This potential differs greatly according to the space in which the experience takes place, but it is implicit in all such encounters. It relies on an understanding of a social world iterated over time through diverse encounters, some more tightly scripted than others. While the tools of psychoanalysis may offer help for understanding space and sexuality as a web of trajectories of desire, shame, submission, and annihilation, this often appears to function solely on the level of an individual subject. The larger social world—what Lucien Goldmann, in a chapter on psychoanalysis in his 1967 *Marxisme et sciences humaines*, refers to as innumerable "transindividual subjects" in which the individual takes part—tends to drop out, ceding the primary role to the simple relation of subject and object.[40] What I draw from Lacan, then, is the notion that the gaze is not something that one can simply possess: it is not a masterful, surveying view. It is threat, risk, lack—surveying, constituting, and reducing the subject; a risk of social annihilation ("the moment at which we lost him who speaks"); the threat of capture within one's own trap. But the gaze is also a collective one—a social one.

When psychoanalysis is employed in the study of the visual arts, it often, like phenomenology, encourages us to imagine a solitary viewer who stands with a singular and totalizing attentiveness before the work of art, taking in a picture, and only a picture. The viewing situation assumes an idealized relation of subject and object, where viewing is an intrapsychic operation. This situation of solitary attentive viewing—whether or not we can say it exists now—was certainly a rare one in the early modern period, prior to the white walls of the modern museum. And in a situation of literal but complexly articulated (that is, not at all obfuscated) patriarchy, multiple authorities—multiple gazes—intersected overtly and covertly in extended households, courts, clans, citizenry.[41] How can we make sense of this social world?

For Lorenzo di Pierfrancesco de' Medici, there is a final twist. The younger Lorenzo was also the recipient of Amerigo Vespucci's letters that became the *Mundus novus.* Vespucci may have derived some of his interest in symbolic uses of mathematics from the Platonic movement in Florence; his uncle Giovanni Antonio Vespucci was also a close associate of Ficino. In Amerigo's triangle, as we saw, the abstraction of the mathematical knowledge that underlies discovery presents the theoretical possibility of equality—yet eventually serves to reinforce hierarchy.

These issues play out differently in another famous Renaissance image, Hans Holbein's *Ambassadors* (fig. 6.11). The instruments of the mathematical quadrivium on the shelves between the two men, emphasizing knowledge of the earth, surveying, navigation, and control, are undermined by the anamorphic skull that appears at their feet when the painting is viewed from the proper angle. We might say this painting provides the ultimate thematization of the position constituted by the distance point. It is only when a viewer takes up an oblique position with respect to the painting that its full impact makes itself felt. Thanks to a trick of our cognitive apparatus, we still see the ambassadors themselves in their finery and with all the accoutrements of earthly measurement.[42] But we are newly reminded of their, and our, mortality by the skull that takes shape before them. The painting

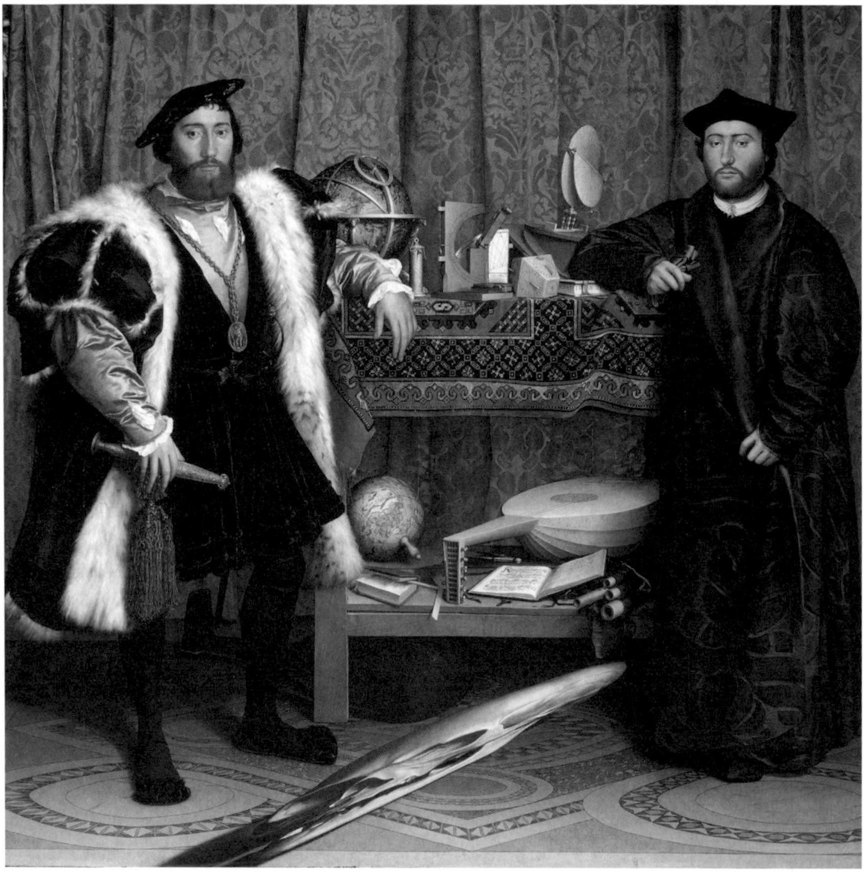

6.11 Hans Holbein the Younger. *The Ambassadors (Jean de Dinteville and Georges de Selve)*, 1533. Oil on wood, 207 × 209.5 cm. National Gallery, London. © National Gallery London/Art Resource, NY.

echoes the dark note on which Perisauli's satire of geometry ends: "Even if you know all of this [the measurements of all lands], you will never know the final limits [*ultima spatia*] of your life. And you will finally be an inhabitant of the grave, and surveyor of your tomb, and there will be nothing of you but fable and smoke."[43]

7

Jealous Geometries

In 1584, Giovanni Paolo Lomazzo published his *Trattato dell'arte de la pittura*. This text, coming at the end of the sixteenth century, provides a retrospective view on the art of the Renaissance and attempts to systematize it both chronologically and stylistically. In this text Lomazzo associates the triangle with Venus in a rather unusual way. He writes, "The Greeks, in imitation of those even more ancient than themselves, found the true venerable proportion in which nothing but extreme beauty and grace could be discerned, and they gave it, in the triangular mirror, to Venus the Goddess of celestial beauty from which all other beauty flows." Venus/Aphrodite is often associated with a mirror; often it is her attribute.[1] But Lomazzo's idea of a specifically *triangular* mirror of Venus appears nowhere else, at least as far I have been able to discern. A drawing by Fra Bartolomeo (fig. 7.1) shows the devotees of Venus seeking to catch her fecund image in their mirrors—the same subject as a famous painting by Titian. The group is composed in a clear triangle, but the mirrors themselves, as we would expect, are round. Lomazzo continues: "But we, leaving aside the mirror, will dem-

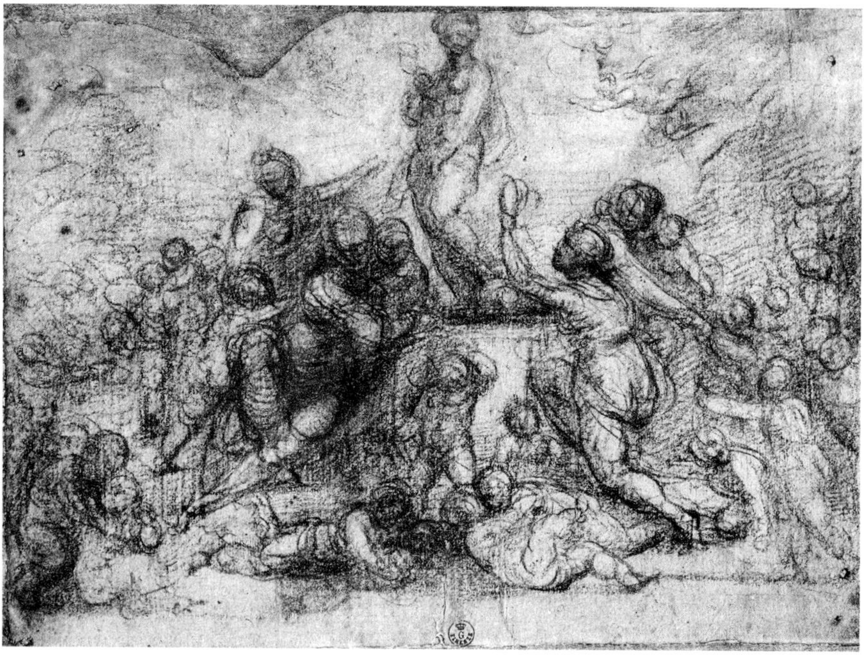

7.1 Fra Bartolomeo. Study for the Worship of Venus. Black chalk drawing, 20.5 × 28.8 cm. Uffizi, Florence. Finsiel / Alinari / Art Resource, NY.

onstrate it in the triangular figure, and in the isosceles triangle, which has two equal lines and one unequal, and a pyramidal form."[2] This is the pyramidal frame of the *figura serpentinata*, inspired by Michelangelo, which became a critical figure for reading the aesthetic conventions of Mannerism up through William Hogarth. Lomazzo next pursues the theme by examining the proportional divisions of the triangle that are appropriate to the depiction of differing body types.[3]

In leaving aside the mirror, one might say, he leaves aside self-reflection about his own stake in, and suppression of, the eroticism of the metaphor: its suggestion of the pubic triangle. This is a metaphor that indeed goes back to the Greeks: the triangular letter delta (in addition to referring to river deltas) was a term for the female genital region in ancient Greek, prominently mentioned in Aristophanes's *Lysistrata*.[4] Lomazzo's notion of the triangular mirror of Venus is suggestive and seems to have no precedent in ancient or Renaissance discussions of Venus or of the triangle.

The most direct and obvious source for Lomazzo is another Renaissance text, whose creative misreadings of ancient sources he then compounds.

Lomazzo's discussion of the representation of the gods in his treatise is selectively borrowed from Vincenzo Cartaro, author and compiler of the classic iconographic text *Le immagini dei dei de gli antichi*, first published in 1556.[5] In Cartaro's text, the triangle appears in association with Venus in two ways. First, he discusses the temple of Venus on the island of Paphos, drawing his account from Tacitus, who described the statue of Venus there as an aniconic coiled pyramid whose rationale was uncertain to the Roman historian.[6] Cartaro tells us that some associate it with the umbilicus, which he notes is the seat of female desire. Yet, he says, this is confusing, since the figure of Jupiter Ammon was depicted in the same way.[7] Cartaro falls back on the theory that this is one of those deep religious mysteries that the ancients did not want to reveal too publicly.[8]

But the pyramidal Venus makes another appearance in Cartaro's text, and this is the part that Lomazzo draws upon most directly. Cartaro describes a Roman coin of the reign of Numerianus with a figure of Venus Victrix, the victorious Venus, and next to her a triangle topped by a circle (fig. 7.2).[9] Some, Cartaro says, think this is an umbilicus—reminding us of the goddess of the Paphians. Others think it is a mirror, because of the account in Philostratus of the picture of the cupids in which, as Cartaro writes, "the nymphs set up a statue of Venus, because she was the mother of such beautiful offspring, and also dedicated [to her] a silver mirror."[10] In fact, Cartaro has misread the figure on the coin; it was just a letter *B* (fig. 7.3), and the female personage is most likely Minerva.[11] Though his reading of the coin is faulty, Cartaro's description happens to be reminiscent of Canaanite imagery of the goddess Tanit—often depicted as a triangle topped by a crescent and circle.[12]

Lomazzo lifts Cartaro's discussion of this figure but makes a few changes. He uses Cartaro's description of the coin but presents it as simply *the* way the Romans represented Venus. He entirely omits the umbilical reference while still maintaining the reference to the Paphians; he also, interestingly, corrects Cartaro's reference to Philostratus, stating that the nymphs dedicate the mirror to Venus because she has made *them* the mothers of beautiful children (the *amori*). He then repeats the speculation that the triangular object depicted next to the Roman image of Venus is a mirror.

What Lomazzo has done here is to clean up the story, omitting sexual desire, as he does when he describes Venus's attribute of the triangular mirror elsewhere as a mirror of Prudence, apt for self-examination.[13] But the coiling line of the "umbilicus"—and the flame with which Venus, according to Tacitus, was worshiped—reasserts itself in Lomazzo's description of

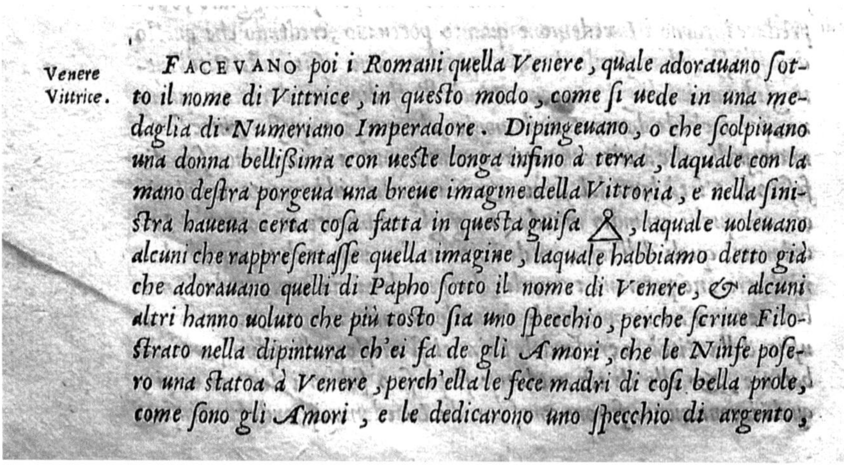

Venere Vittrice.

FACEVANO poi i Romani quella Venere, quale adorauano ſotto il nome di Vittrice, in queſto modo, come ſi uede in una medaglia di Numeriano Imperadore. Dipingeuano, o che ſcolpiuano una donna belliſsima con ueſte longa infino à terra, laquale con la mano deſtra porgeua una breue imagine della Vittoria, e nella ſiniſtra haueua certa coſa fatta in queſta guiſa △, laquale uoleuano alcuni che rappreſentaſſe quella imagine, laquale habbiamo detto già che adorauano quelli di Papho ſotto il nome di Venere, & alcuni altri hanno uoluto che più toſto ſia uno ſpecchio, perche ſcriue Filoſtrato nella dipintura ch'ei fà de gli Amori, che le Ninfe poſero una ſtatoa à Venere, perch'ella le fece madri di coſi bella prole, come ſono gli Amori, e le dedicarono uno ſpecchio di argento,

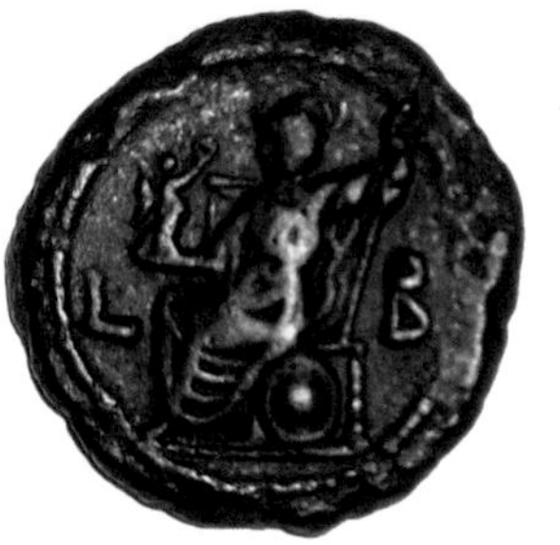

7.2 Vincenzo Cartaro, paragraph with typeset triangle and circle, in *Le Imagini con la spositione de i dei de gliantichi* (Venice: Marcolini, 1556), f. 115v. National Gallery of Art Library, Washington D.C. David K. E. Bruce Fund. By permission.

7.3 Alexandrian mint. Billon tetradrachm, reign of Numerianus, 3rd c. CE. Collection of the author.

the ideal shape constituted by the isosceles triangle framing a "serpentine" line. Elsewhere, to elaborate on this notion of the *figura serpentinata*, he cites a precept said to be from Michelangelo on ideal form. The figure should be made "pyramidal, serpentine, and multiplied by one, two, and three."[14] He enters into a lengthy discussion of the serpentine line as a flame, which allows the artist to achieve liveliness in the figure—in Lomazzo's phrase, artists call this *furia de la figura*, the "fury of the figure."[15] In Plato's account of the elemental structure of the universe in the *Timaeus*, the triangle was the elemental shape of fire, but also, in Tacitus, the Paphian Venus was worshiped with a pure flame rather than blood sacrifice.[16] The multiplication by one, two, and three suggests the *fertile* triangular number 6.[17]

What I would like to argue here is that the triangle had powers of attraction that drew these authors to it, despite the improbability of the source. Both Cartari and Lomazzo *want* to believe in the notion of a symbolic triangle associated with Venus. Perhaps we can say that this is part and parcel of an attraction that also manifested itself in the diagrammatic, compositional, and conceptual triangles we have seen scattered through European visual culture of the fifteenth and sixteenth centuries.

In mathematical texts, as we have seen, the triangle was associated with fecundity: for Charles de Bovelles, the triangle was *fecundissimus*, and this built on a long-running theme. We have seen the idea of fecundity in the triangle's compositional association with Mary and with other scenes of birth. On the one hand, the "fecundity" of the triangle seems to come through the association of the triangle with a particular bodily morphology: the mons pubis, or *mons veneris*, the "triangle" of Venus. In the series of erotic *Loves of the Gods* by the engraver Giulio Bonasone, the repetitious blankness of the triangular space enclosed by the tented bedchamber comes to take on somatic connotations. In modernism, the triangle and the female body both form basic units of representation; Tamar Garb has associated Paul Cézanne's interest in the form of Mont Sainte-Victoire with his particular obsessions with women as represented in, for instance, his *Eternal Feminine*.[18]

As we saw in chapter 6, the triangulation produced by perspective is tinged with the erotic, even erotic rivalry, in Dürer's woodcut of a draftsman drawing a nude woman. Like Uccello, Durer's draftsman (fig. 6.9) might have in mind "how sweet perspective is," as he screens himself from the naked woman lying before him. He peers across what Joseph Koerner calls his "rather phallic" pointing tool, obsessed, perhaps, with his own technical brilliance.[19] Dürer's draftsman reminds us that the grid of perspective does not simply map the visible world for a disembodied eye/I; it invites, seduces, affects, and sometimes confuses the body, which for Renaissance thinkers was closely bound up with the soul. As we saw, desire, not only the artist's but the presupposition of the viewer's, seems to subvert the proper workings of representational accuracy, twisting what should be a linear relation of artist to object into a triangular one of artist, object, and viewer.

This woodcut had a powerful influence on learned visual culture in the decades that followed (and perhaps ever since). Even such a famous and seemingly sui generis work as Titian's *Venus of Urbino* (plate 8) draws inspiration from Dürer's woodcut, echoing the female model in particular. Venus has been stripped of her minimal coverings and props herself up to look directly at the viewer. Not only does she recline on the same pillows as Dürer's artist's model, but the same potted plant sits on the back windowsill. We may not be inclined to expect that a print could have influenced a painter, however high the estimation in which the German artist was held in Italy.[20] Although this assumption has shifted slightly, thanks to recent scholarship, it might still come as a surprise that the champion of *colorito* would look to prints for inspiration. But at the time Titian painted his *Venus*, Dürer's *Unterweysung der Messung* was one of the few printed manuals for painters that existed (it was not until 1540 that Alberti's would be published, unillustrated). Titian was the master painter in a bookselling center Dürer had famously visited; he must have known Dürer's book. But if he is commenting on Dürer, what is he saying?

Titian also plays with perspective in the *Venus*, suggesting a more oblique reference to Dürer. Following interpretations by Rona Goffen and Daniel Arasse,[21] we might conceive of the more public space of the background as a space of exchange, marriage, and social hierarchy (represented by the dowry in the chest, the power relations between the women in the background) that contains a strong version of orderly geometric perspective in the receding lines of the floor tiles. The foreground, on the other hand, gives the viewer an intimate space of erotic presentation without socially intelligible markers (hence the confusion over identity: courtesan? bride? goddess?). Goffen and Arasse both argue that it is impossible, according to the rules of perspective, to make Venus, at the appropriate scale, occupy a space coherent with the perspectival inset. The inset might represent a kind of fantasy, like a thought bubble.[22] But is she dreaming this other, measurable space, or is it dreaming her?

Titian's perspective construction also, as Goffen and Arasse argue, connects to Venus's body symbolically. If one follows the lines of perspective in the inset space in the upper-right corner to where they would intersect behind the curtain, one arrives at what is also the point of intersection of a horizontal line leading from Venus's eyes with a vertical line leading from her crotch. The vanishing point thus seems to be doubly deflected from (and pointing to) the eyes and the genitals. Eyes, genitalia, vanishing point: these form another triangle. Venus's hand also plays a role: does it cover, touch, elicit? We play a game of touch and sight orchestrated by an artist who later became famous for painting directly with his fingers. And these deflections matter a great deal; image and desire stand in a supplemental relationship to one another, both complementing and supplanting each other. Titian's Venus might be Uccello's wife waiting for him to come to bed as he plays instead with perspective. But her teasingly covered sex is itself a "vanishing point."[23]

I have examined the triangle as divine order, political hierarchy, compositional strategy, and expression of the human relationship to the world. In this chapter I will address a type of triangle not so easily diagrammed: triangles of affect formed within networks of social relations, as they are mediated by objects and in particular by works of art. I am dealing here with love triangles. But I will also suggest, following up on the previous chapter, that the triangular structures of the perspectival image and the triangular structures of desire may map onto, facilitate, and even contaminate one another. How might an image like Titian's *Venus of Urbino* have served the relationship between patron and artist (as more than simply the carrying out of an instruction)? How might such an image have functioned for the artist in competition with other artists—or with the viewer? Triangles might appear as rivalry, or as a play of gazes or trajectories of desire implied by figures within a painting. They might invoke the structure of perspective, but not necessarily straightforwardly. Thus I return to the technique of perspective with a new context in mind. Perspectival space sets the stage for triangular relations in which artist and viewer struggle for possession of the object, but this is a perspective that is oblique and ungraspable, its relations mapped within a larger social sphere and aesthetic conventions.

On the one hand, perspective could be used to distort space, to play, to obstruct or deflect the gaze, or to direct it symbolically rather than objectively. On the other, orderly perspective may be seen as a defense against the frequently *dis*orderly jumble of experiences by which body, space, and vision relate to one another. Social conditions certainly changed during the early modern period—as did the technologies of representation. The visual image, with the circulation of prints, decorated objects, and small-scale paintings, came to occupy a new place in the formation of subjectivity and social life. One might argue that as "symbolic form," implying the illusion of entry into space, perspective might have been most appropriate to those classes of society that actually controlled space. (Hence the central role of "perspectives," for instance, in royal ceremonies of entry into cities, by which princes made ritual shows of entering and taking possession of them.)

In this context, can we still think of perspective as a culturally relevant "symbolic form"? Certainly, for such classes of society, perspective as a technology of visualization *was* related to architectural lines of sight (the siting of a villa, the deployment of windows on an urban facade). Architectural changes over the fifteenth and sixteenth centuries, for the wealthiest people, meant more spacious, "magnificent" palazzi organized around principles of classical reference, geometric harmony, and legibility. Other significant uses of perspective were found in mapping, surveillance, and military strategy, and in theatrical spaces where the view of the audience is directed toward a particular segment of space in which a performance takes place. (This might include actual performances using stage sets based on Vitruvius's comic, tragic, and satirical scenes; or it might include the theaters and ephemeral arches used in royal entries.) But other modes of visual experience also present in the Renaissance might imply other kinds of subjectivity.

Jealousy and the Spaces of Artistic Subjectivity

We have already seen the possibility of rivalrous gazes and oblique glances as the structuring "look" of a picture. Triangular relations of rivalry and competition might inspire the excessive and recursive emotion of jealousy or envy; an oblique look might be the evil eye. Like melancholy, these triangles challenge rational measure more than they support it. The iconography of jealousy takes shape in ambiguous groupings and images of unequal lovers. In the normative context of Raphael's workshop, it appears in the Vatican loggie as the triangular embrace of Isaac and Rebecca, spied upon by Abimelech. But it is in the story of Vulcan, Venus, and Mars that it finds its best instantiation. In the story, the lame (or perhaps impotent) blacksmith god, married to the resplendent goddess of beauty, must reckon with her trysts with the warrior god Mars. An initial example from among the many sixteenth-century painters and printmakers who produced versions of the story is Jacopo Tintoretto's (plate 1), painted around 1550, the work with which I opened this book. Tintoretto gives us a dramatic scene in which Vulcan's dominant, oversized form descends upon Venus's bed, tearing the sheets off to seek signs of her infidelity. Here as elsewhere, Tintoretto uses a sharply angled, disorienting perspective, an element of the nervous painterliness characteristic of his Venetian mannerist style. As was common in Renaissance painting, lines of the square floor tiles guide our understanding of the image's perspective system. In this case, they lead the eye inexorably toward the upper right of the picture, allowing us to make the discovery that Vulcan has yet to make: the helmeted figure of Mars, hiding under the table in the middle ground (only his face and the glint on his helmet emerge distinctly from the murky shadows). In the upper center of the picture—where in a more conventional construction we might expect to find the vanishing point—a mirror stares back at us instead, reflecting Vulcan's body from behind. The eccentric perspective in this painting certainly heightens the drama (and perhaps the comedy).

At the edge, where the perspectival orthogonals converge, we find faintly indicated Vulcan's forge, the work space from which he has undoubtedly emerged, as if sweeping with him the lines of perspective. Because Vulcan was an artificer, it is possible to view artists' exploration of the subject matter of Vulcan, Venus, and Mars as a kind of ongoing collective reflection on artistic identity. In the mid-sixteenth century, as the literary examination on visual arts became an established genre, this iconography came into its own as a self-conscious discourse on seeing and craft. It took shape in prints and drew on the print medium for some of its particularities. Making this imagery available in prints enabled other artists to copy it, and in copying it to reflect upon it, to identify or disidentify with it, and collectively to produce a dialogue on desire and erotic rivalry and their role in generating art and its audiences. Scholars have often studied the modes of self-presentation of Renaissance and Baroque artists—for instance in self-portraits in which they depicted themselves as humanists, antiquarians, and gifted creators. But we have not always considered the more oblique, complex, or private ways in which they visually theorized their craft or themselves.

Artists' fantasies, as expressed in printed images in particular, create a dialogue that runs alongside more public image-making. What do I mean by fantasy? Freud refers in his *Introductory Lectures on Psycho-analysis* to phantasy as a "half-way region," halfway between reality and dream. This makes sense from the point of view of Renaissance theories of the mind, in which the fantasy or imagination was often understood as, in fact, a mediating region, one responsible for transit between the senses and the higher faculties (reason, judgment, intellect). If we look for expressions of the fantasies of Renaissance artists we might think first about certain kinds of drawings; next we might think of prints, which mediate between private and public in quite interesting ways. They might be disseminated widely or more narrowly, signed or not, appreciated in private or more publicly displayed. Prints resonate with certain qualities of the *fantasia* as a mental faculty understood as physical space, a material site of inscription—that is, of sense perceptions and their recombination and reinvention.

I've titled this chapter "Jealous Geometries": how can geometries be jealous? My point is, following up on my initial arguments about Renaissance subjectivity, to emphasize what from a modern point of view might seem a blurring of the subjective and the objective: of feeling and things, passions and physical spaces. Objects and passions resonate with one another; the faculties of the mind appear as material, physical spaces. Castiglione describes a built environment that reads emotionally: "We find that a lover sometimes rejoices to look at a window, even though it is shuttered, because it was there that he once was favoured by the sight of his lady."[24] This is made vivid in a poem by Filenio Gallo: "Happy, adventurous and rich stones, where the fair breast of my lady rests, panes of glass blessed above all things, near where my longed-for flower stands; bitter jalousies [*gelosia*, meaning shutters], which let slip just a glimpse of my heart's aim, rich, precious, graceful window frame, which, adorned with such a goddess, surpasses all others."[25] In another poem the jalousies cast a frivolous shadow on the beloved's face and are later berated for hiding her light.[26] Might the geometry of perspective space, too, provoke jealousy? Perspective both assumes one point of view and, as we have seen, implies the potential for others. The perspectival space represented in a picture, then—or in some pictures, anyway—might be not a "real" space that would encourage a sense of dominance and control, but a fantasy space that brings us into contact (possibly contagiously) with the artist's *fantasia*.

Mythological prints and paintings enabled artists to produce a subjective space (like the imaginary spaces created in devotion through meditation) in which ideas might be worked out through narratives of personification. In the case of mythological painting, a further reference might be the theatrical conventions of personification. Personified concepts were common subjects of plays (and political ceremonies such as royal entries into cities) in late medieval and early modern Europe. In these contexts the gods were often difficult to distinguish from personified qualities we are accustomed to call "allegories."[27] The mind was a theater of conflicting passions and faculties; stories played out by the gods of Greece and Rome can be seen as a version of the medieval *psychomachia* (a battle among personified virtues and vices for control over the soul). In printed *istorie*, the spatial coordination of figures functions like what Victor Burgin calls "perverse space," in which vision is mobile and multiple, and no single point of view allows the spectator to relax into a safe position of dominance over the scene.[28] Rather, visual pleasure, passion, and attachments circulate among creatures and objects, glancing off one another, revealing various erotic possibilities. Burgin suggests we keep in mind that within Freud's notion of "drives," any drive—such as "objectification"—also implied its obverse ("identification").

For some, imagination and the associated passions are specifically linked to disproportion and disorder. Gianfrancesco Pico della Mirandola, in the orbit of the ferociously reformist friar Savonarola, writes that "he who obeys the dictates of the perverted [*incurvi*, literally "bent"] sense and deceitful imagination, at once loses his dignity, and degenerates to the brute."

The imagination also plays an active role in aesthetic creation. Vasari associated Dürer's prolific production of prints with his "extravagant" imagination. But this active role has risks, presenting the more likely potential for error. For Gianfrancesco Pico, the imagination is "a power of the soul which conceives and fashions likenesses of things." Just like mathematics, it exists "midway between incorporeal and corporeal nature, and is the medium by which they are joined."[29] While technically the imagination is subject to the higher faculties, it can, under certain circumstances, take its own initiative and fall into "error." Imagination is the mother of "ambition . . . cruelty, wrath and passion."[30] It is also the basis for desire, according to Aristotle; sixteenth-century French writer Symphorien Champier equates the imagination and desire even more directly.[31]

The imagination's disorder could, under certain circumstances, lead directly into envy and jealousy.[32] Among the various passions to which the soul was prone, jealousy was considered one of the most disordered and disordering emotions, the generator of irrational imaginings. Juan Luis Vives warned, "Strong envy spoils our judgment, much more than any other emotion."[33] Benedetto Varchi, a founder of the Florentine Academy, who closely followed Vives in his text on envy, wrote, "You alone, without any measure or moderation, even against any common usage or human reason, are impoverished by the wealth of others, and made rich by their poverty."[34]

In his book *Theater of Envy*, René Girard examines envy in the Renaissance through a concept he calls "mimetic desire"—desire aroused by envy, which provokes imitation and competition.[35] This implies a certain receptivity on the part of the mimetic desirer not only to the object of desire but to the desires of other subjects. Jealousy and envy were also particularly closely related to vision and to art. Envy, *invidia*, was associated with vision: its verbal form, *invideo*, contained *video*, the Latin verb "to see." The Greek word *zelotypia*, used in humanist Latin for "jealousy," was understood by both Vives and Varchi as "zeal for beauty" or "envy of form," because they construed *typia* as equivalent to the Latin *forma*.[36] Varchi calls jealousy "a fearful suspicion of the lover that the beloved thing, which he would not want to share with anyone, should make free with itself [*fare copia di se*] to another."[37] This suspicion, in addition to involving fear of improper coupling (the near homonym *coppia* and the sexual connotations of the phrase *fare copia di se*), suggests a resistance to *copia*, copying, or abundance. Indeed, in Gabriele Paleotti's discussion of portraiture in his *Discorso intorno alle immagini sacre e profane*, he uses the same phrase to describe the act of having a portrait made—a copy of oneself.[38]

But what, Varchi asks himself, of the "things" we don't want to enjoy ourselves but also do not want anyone else to enjoy in an improper fashion—daughters, sisters, mothers? At this point he decides that jealousy is "a fear or suspicion that anyone we don't want should enjoy any beauty; and this for two reasons, either in order to enjoy it ourselves, or so that only those we want enjoy it."[39] Varchi emphasizes beauty; he also opens the definition of jealousy to objects of either sex and material objects of beauty. Envy, of which jealousy is one species, is for Varchi explicitly a disturbance not only of estimation but also of vision ("seeing too much" or "not seeing" at all). He also identifies envy with imitation and competition, and seems almost to be speaking specifically of artists when he says that the envious are "great simulators and dissimulators."[40] As envy distorts one's sense of measure, it also confers skills in both positive and negative deception (i.e., creating appearances and hiding the truth). The envious are thus like artists, skilled in the manipulation of visible signs.[41]

And thus Vulcan is a natural avatar of the artist. In representing Vulcan menaced by Mars for possession of his wife, artists might be confronting tensions over the uncontrollable dissemination (and potentially copying) of their work, or their insecure status at courts such as that of Fontainebleau. Vulcan figures the manual labor of artistic production in Velázquez's much later painting in which Apollo visits Vulcan's forge; Apollo, as the "light of day," revealed Venus and Mars's affair to Vulcan, and for Velázquez he also represents learning enlightening craft. In Matteo Maria Boiardo's *Tarocchi*, the notion of Vulcan as artist is suggested by the "new forms" in which he caught Venus and Mars, though the ambiguity of the phrase means that the forms might also be the lovers' sexual positions: "Jealousy made Vulcan in new forms catch Venus and Mars in his nets" ("Gellosia fe' Vulcano in forme nove / Pigliar Venere e Marte entro le rethe").[42]

How, then, did Renaissance artists use the figure of Vulcan? In a drawing by Parmigianino (fig. 7.4) the artist displays his technical mastery in the confident description of the figures' limbs and in the deft handling of the surface differences between bodies and cloth. The torso of Vulcan emerges with firm definition from his drapery while the faces of the other two figures—Mars and Venus, apparently "caught in the act" by the jealous god—appear faintly, but convincingly, from beneath upswept cloth. Daniel Arasse interprets Parmigianino's drawing as a reflection on the artist's relation to his work, imagined as a specifically erotic relation: it represents the moment of creative inspiration—its momentary quality expressed in the billowing draperies.[43] Arasse equates this inspiration with the god's obvious sexual

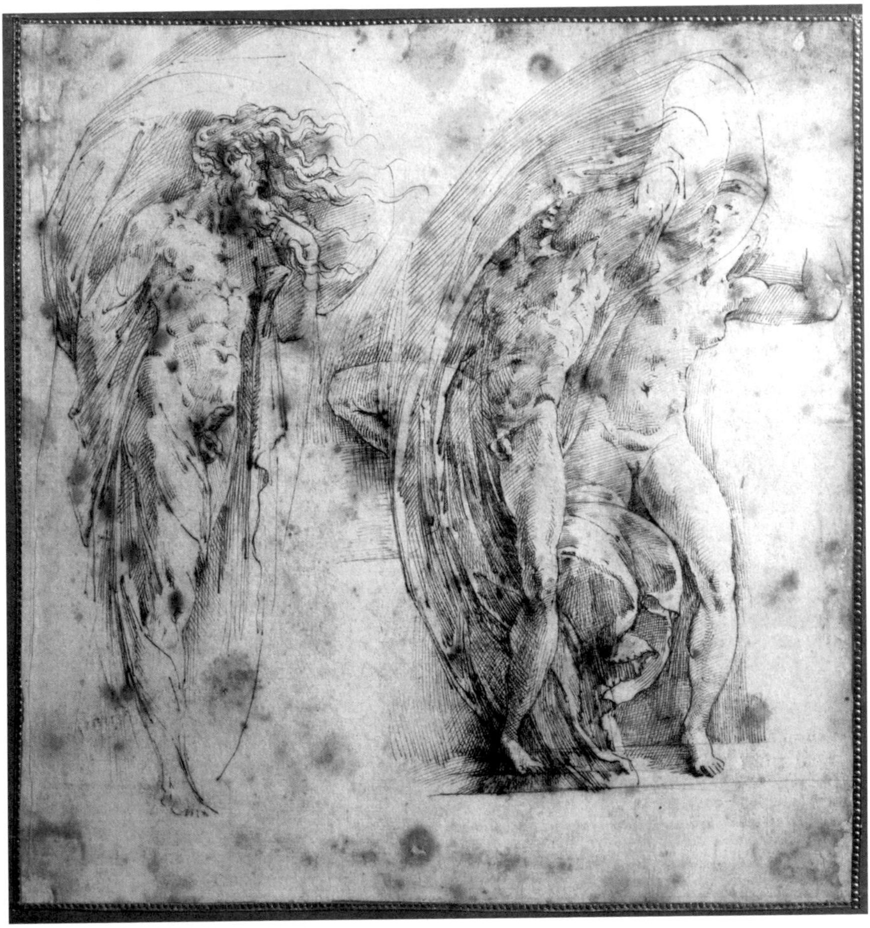

7.4 Parmigianino (Francesco Mazzola). *Vulcan, Venus, and Mars*, before 1540. Red chalk drawing, 20 × 19.2 cm. Galleria Nazionale, Parma. By permission of the Ministero per i Beni e le Attività culturai—Sopritendenza BSAE di Parma e Piacenza Archivio MIBAC.

excitation. Yet Vulcan's excitation is produced not only by desire for his wife but by a scene of blatant infidelity.

Jealousy displays itself not only in the mythological narrative but also in the further dissemination of the image that enables Arasse's own reading. To ground his reading of Vulcan as the artist, Arasse uses a retooling of Parmigianino's motif (fig. 7.5) in a fresco painted at the chateau of Fontainebleau by Francesco Primaticcio. In the fresco, from the Chamber of the Duchesse d'Étampes, Vulcan is transformed into Apelles painting Alexander the Great and his mistress Campaspe. According to the story, Alex-

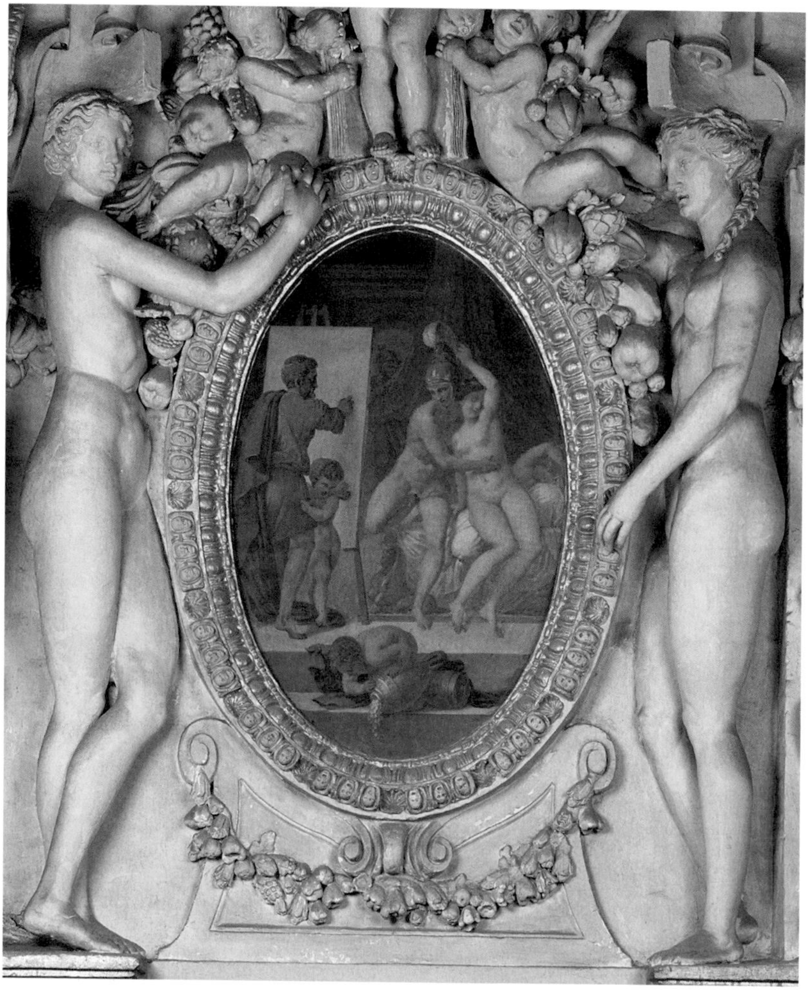

7.5 Francesco Primaticcio and Abel de Pujol Alexandre-Denis. *Apelles Painting Alexander and Campaspe*, ca. 1541–44, restored 1836. Medallion with stucco figures on the east wall of the Royal staircase, formerly the chamber of the Duchesse d'Étampes. Château de Fontainebleau, France. Réunion des Musées Nationaux / Art Resource, NY. Photo: Jean-Pierre Lagiewski.

ander gave Campaspe to the painter as a reward for his excellent service. Arasse reads Apelles convincingly back into the figure of Vulcan. But perhaps there is more to the story. How are we to think about the fresco's "derivative" character? If, as Arasse suggests, the drawing was a private and personal one, how did Primaticcio get his hands on it?

An etching related to both drawing and fresco was made (fig. 7.6) by Master L.D.—a Fontainebleau printmaker (so called for his monogram)

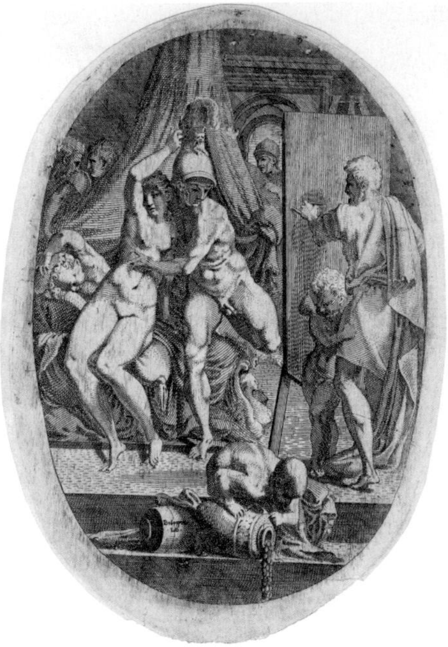

7.6 Master L.D. *Apelles Painting Alexander and Campaspe*, ca. 1541–44. Etching, 37.5 × 26 cm. British Museum, London. © Trustees of the British Museum.

who worked closely with Primaticcio.[44] As with most such prints, we assume it was made as a copy of the fresco. But this might not be so certain. According to Vasari, another printmaker also at Fontainebleau, Antonio Fantuzzi, was in fact the artist Antonio da Trento, who was also accused of stealing prints and drawings from Parmigianino.[45] Vasari's story may well be garbled, for other evidence militates against this identification.[46] But there may be a kernel of truth in it: perhaps Primaticcio or his assistants brought some of Parmigianino's designs to France, in printed if not in drawn form. Primaticcio's design could have been inspired by a now-lost image. There is further irony in the subject matter, which normalizes Primaticcio's appropriation of Parmigianino's composition by turning a story about infidelity and jealousy into a story about gift giving. The artist is no longer condemned to derive inspiration from his jealous gaze at his wife with her lover; instead he receives Alexander's favor, and his mistress. Situated as it is in the bedroom of King Francis I's own mistress, it also seems to suggest the particular favor in which the Duchesse d'Étampes held Primaticcio, hinting at female agency present in life (the duchess was an important policy adviser to the king) if not in art.

Thus we might see jealousy, or envy, as a productive force in Vulcan's—or perhaps Parmigianino's—erotically charged artistic creation. The theme of Vulcan was of continuing interest to the artist. Indeed, throughout his career Parmigianino used proxy figures—artists in their workshops, mythological creators, and philosophers—to reflect upon the artist's work. Vulcan was not the only one. One drawing in the Morgan Library shows the seated figure of Prometheus reaching to light a torch that he holds like a paintbrush, as he prepares to animate a figure that represents the first man. Private drawings of scholars and philosophers consumed in thought suggest that Parmigianino identified with the state of contemplation, perhaps including mathematical pursuits: on a sheet of studies by the artist, in the collection of the Uffizi, a fragmentary hand contorts to clutch the word *giometria* between thumb and middle finger. One odd drawing at the British Museum shows a philosopher, perhaps Archimedes, seated with an armillary sphere at his feet, holding an object (probably an instrument of some kind) so close to his face as to block the view of the window that stands before him.[47]

It is a similar oblivion produced by obsessive labor that we find depicted in Enea Vico's engraving after another design by Parmigianino that demonstrates the continuing import of Vulcan to the artist. Vulcan hammers away at his workbench while Venus sleeps in bed in the background—in one state of the print. The print exists in at least two others: the first, in which Venus appears with Mars (Vulcan is still absorbed in his work [fig. 7.7]) and the second, in which they have been rubbed out—the engraved figures literally burnished from the copperplate (fig. 7.8). It is only in the third state that Venus appears on her own (7.9). These three states exist in an uneasy narrative relation to one another—it is not immediately obvious to the viewer which came first. David Freedberg argues that the adulterous engraving was "censored," and Diane Russell expresses surprise that the state with the figures rubbed out exists in more than one (ostensibly trial) impression.[48]

What if the burnished version imaginatively suggests the angry inter-

7.7 (opposite, top) Enea Vico, after Parmigianino. *Vulcan, Venus, and Mars* (first state), ca. 1540. Engraving, 22.5 × 33.5 cm. British Museum, London. © Trustees of the British Museum.

7.8 (opposite, center) Enea Vico, after Parmigianino. *Vulcan, Venus, and Mars* (second state, with Mars and Venus erased), ca. 1540. Engraving, 22.8 × 34 cm. British Museum, London. © Trustees of the British Museum.

7.9 (opposite, bottom) Enea Vico, after Parmigianino. *Vulcan, Venus, and Mars* (third state, with Venus sleeping), ca. 1540. Engraving, 23 × 34.2 cm. Published by Antonio Salamanca. British Museum, London. © Trustees of the British Museum.

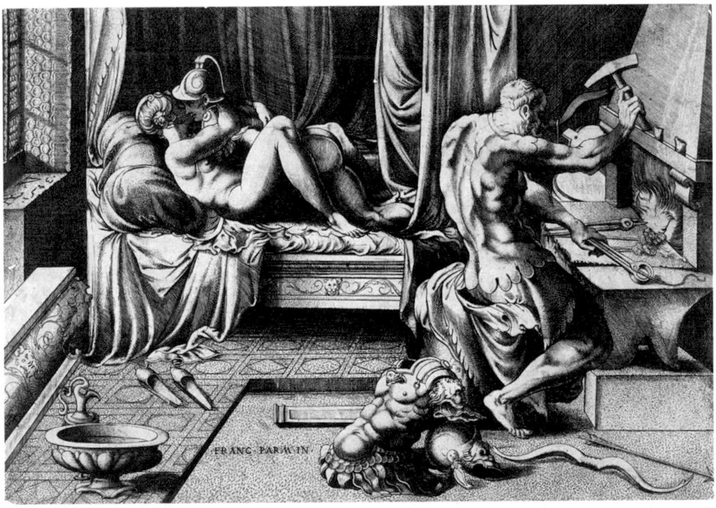
·FRANC·PARM·IN·

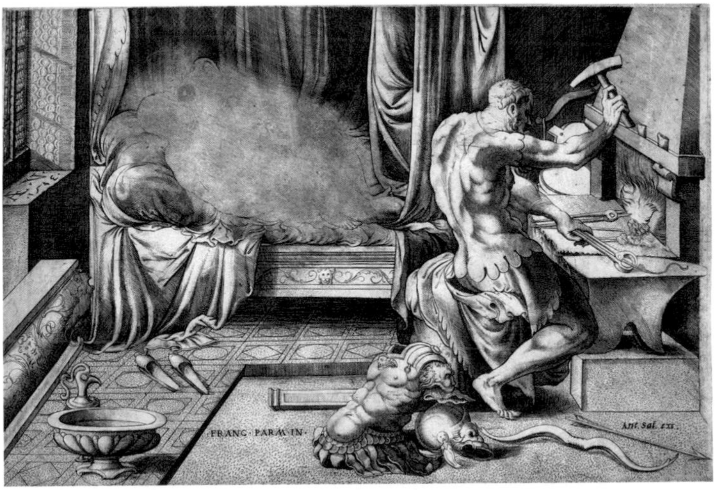
·FRANC·PARM·IN·
Ant. Sal. exc.

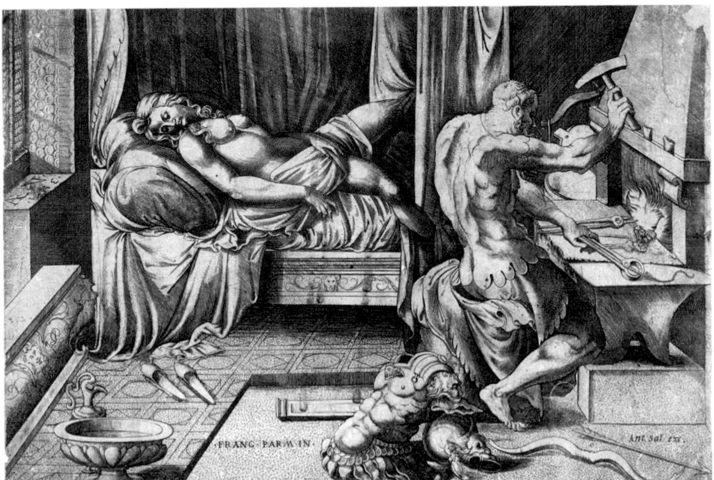
·FRANC·PARM·IN·
Ant. Sal. exc.

vention of the jealous god himself? In Parmigianino's drawing, the tryst of Mars and Venus serves as a spur to sexual arousal and, presumably, creativity. In a now lost drawing related to the series of prints by Vico and known through a copy, Vulcan crouches at his workbench in the same pose, but Venus reclines, somewhat awkwardly, directly above his anvil, holding the infant Cupid—as if to suggest that the fiery forge is in fact the site of sexual reproduction.[49] What if, in the series of prints, the bed of Mars and Venus in fact represents the thought bubble of Vulcan's imagination as he toils? The different states of the print seem to suggest a narrative of fantasy, desire, and jealous anger that animates them as a series and conflates the character Vulcan and the artist. But who is the artist? Many hands intervened in one way or another on this plate, and Vulcan is only a convenient placeholder. If we read artistic rivalry into the story, it might imply competition among the parties to the print's creation (Parmigianino? Vico? Salamanca?). Who is the true author?

A similar story might be told of Parmigianino's drawing in the British Museum, catalogued as "A group of three nude men" (fig. 7.10).[50] I read it as more ambiguous: yet another structural transformation of the three figures of the love triangle. The thighs, hair, and chest of the seated, backward-facing figure read feminine, even if her back and arms are muscular. One might additionally read the rope "she" holds as, again, a studio device to allow the model to hold the pose, were in for the fact that the end of the rope drops limply over the figure's fingers, and the hatching marks underneath her arm suggest at least the possibility that the object is actually a net. The two clearly male figures fit easily into the characters of Mars (the standing male) and Vulcan (the melancholic, half-seated male). Mars in a sense switches places with Vulcan, by comparison with the Parma red chalk drawing. But the part of the sheet he occupies has been cut off and reattached, as if registering tension over the figure's place in the narrative.[51]

In these images many issues might be at work—sexuality, artistic rivalry, the position in which the artist is placed by his service to a patron, his efforts at capturing nature. Mars might represent the patron, or the audience. Is Vulcan jealous of Mars or of Venus, or of both? Might envy and jealousy complicate clear distinctions between who may be an object of desire and who may be a rival? Is the artist jealous of women's procreative power, of the position of the other man, of the position of the woman? With imagination and desire closely linked, Venus might represent the work of art or beauty itself. An artist might be jealous of other artists or of the viewer's possession of his artifact. Is the excessive, obsessional character of jealousy one that

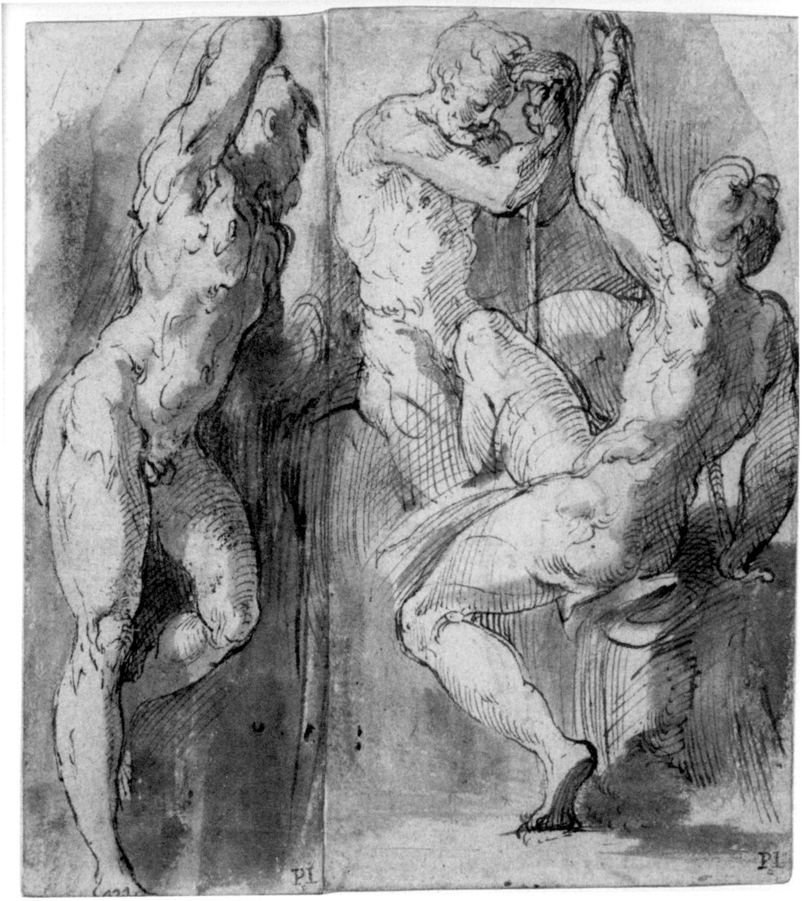

7.10 Parmigianino (Francesco Mazzola). *Vulcan, Venus, and Mars* (or *Three Nude Men*), 1518–1540. Pen and brown ink with brown wash over graphite on paper, 12.6 × 11.2 cm. British Museum, London. © Trustees of the British Museum.

necessarily works itself out through the imposition of order, of the socially sanctioned forms for intimate attachment? In relation to the works of other artists, might Vulcan's jealous passion produce a stylistic hyperbole, an individual *maniera* based on an intertwining of envy and imitation?

Print culture's metaphorical promiscuity may have contributed to such emotions. Viewers may have viewed prints in private—we might even say that some of them, such as erotic prints, constituted around themselves a penumbra of private space. But they were also viewed in social gatherings. Printmakers worked in workshops that involved a variety of roles; they made prints for the market, but they also, deliberately or not, made prints for other printmakers. The artist here is not the masterful subject we have come to associate with Renaissance artists and, especially, Renaissance perspective, but a fragile one, sustained by jealous passion and enmeshed in complex relations of cooperation and competition. Along with Vasari's story of Parmigianino's assistant, records show that Andrea Mantegna paid thugs to rough up rival printmakers who were counterfeiting his work—a more violent solution than that of Dürer, who sought judicial relief against Marcantonio Raimondi for similar actions.[52]

A court case that took place in Rome in 1577 involving the heirs and rivals of the print publisher Antonio Lafreri provides further insight into the machinations of a very competitive field. Testimony refers constantly to the envy felt by other printmakers for the Lafreri *bottega*.[53] After Lafreri's death, rivals apparently tried to steal from his stock and materials, and one of his workers, the young Girolamo of Modena, was found dead in the Tiber. Lafreri's heir Claudio Duchetti testified that he asked around to find out what sort of company the boy kept and found that he frequented only other printmakers. Duchetti "then turned it around in [his] mind [*fantasia*] and imagined that this was an assassination against the workshop, which is envied by many in the profession."[54] Personal and collective envy are intertwined: another printmaker states that "Messer Geronimo was murdered through envy of his being in the workshop of Messer Claudio, because the workshop is envied because it is the biggest in Rome and has the most beautiful works in this medium."[55] Something else that emerges in the testimony, however, is not so flattering to Lafreri and Co.: Lorenzo Vaccari testifies both that Lafreri counterfeited some of his prints—and that Lorenzo did not pay him back in kind. He states, perhaps disingenuously, that counterfeiting was not special cause for ire or vengefulness, for "it was Messer Antonio who counterfeited all the new prints of the rest of us shopmasters, because that was the kind of man he was and he had the means more than any of the rest of us did."[56] The intensity of competition in this nascent industry is startling. Prints did not by themselves bring rivalry into being, but they made images easier to copy, artistic thefts more tangible, and the topic of rivalry itself more visible.

In prints in particular, the different figures in the narrative of Vulcan, Venus, and Mars might have represented the different roles in the production process, which in addition to the basic tasks of engraving and printing could involve inspiration, invention, transfer, copying, theft. On the last of a series of twenty-six French prints illustrating the story of Jason and the Golden Fleece, a little vignette of Vulcan (with Cupid) and Venus appears that serves as the artists' signature (fig. 7.11). The two artists are the engraver

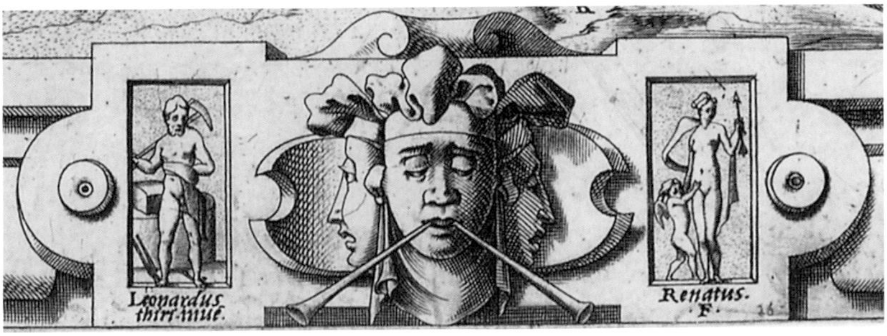

7.11 René Boyvin, after Léonard Thiry. Vulcan and Venus (detail) from *Medea Fleeing in her Chariot* in Jean de Mauregard, *Livre de la Conqueste de la Toison d'Or*, 1563. Engraving. Print Collection, Miriam and Ira D. Wallach Division of Art, Prints and Photographs, The New York Public Library, Astor, Lenox and Tilden Foundations. By permission.

René Boyvin and the painter and draftsman Léonard Thiry, who worked with Rosso Fiorentino on the Galerie François I at Fontainebleau and made the drawings that served as the design for Boyvin's engravings. The signature delineates their roles: "Leonardus Thiri inue" (*inuenit*, or "invented [it]"); "Renatus F." (for *fecit*, "made [it]").

One thing that's odd here is that, as we have seen, the more common coupling of Venus is with Mars; Cupid is usually understood to be *his* child, not Vulcan's. So the pairing masks the trouble in the Olympian household, trouble that Thiry displayed fully in a drawing (fig. 7.12). Here Vulcan traps the guilty couple in a finely woven net of threads that spring from his hand like the rays of vision or the lines of perspective. Hand substitutes for eye, and it produces a flexible net that wraps itself around the two bodies, clinging to them in voyeurism and capture. Here the geometric form on which perspective is based becomes enfleshed as love triangle, with Vulcan presented as "master" of the situation both as artist and as husband. Are the triangles of perspective necessarily love triangles? Vulcan is the figure of the artist as embodied and desiring viewer. This view is further bolstered by a very similar engraving by Georges Reverdy (fig. 7.13) that emphasizes the gridlike quality of the net—which could hardly look more like the draftsman's grid in Dürer's woodcut illustration.[57]

The perspective grid has often been viewed as a technique of capture. And on some level it surely is. But Dürer's draftsman has also shown us what can't be captured: a 90° view of the same figure; the subjective experience of his model—the actual physical control that might underlie visual representation but for which visual representation is not a sufficient sub-

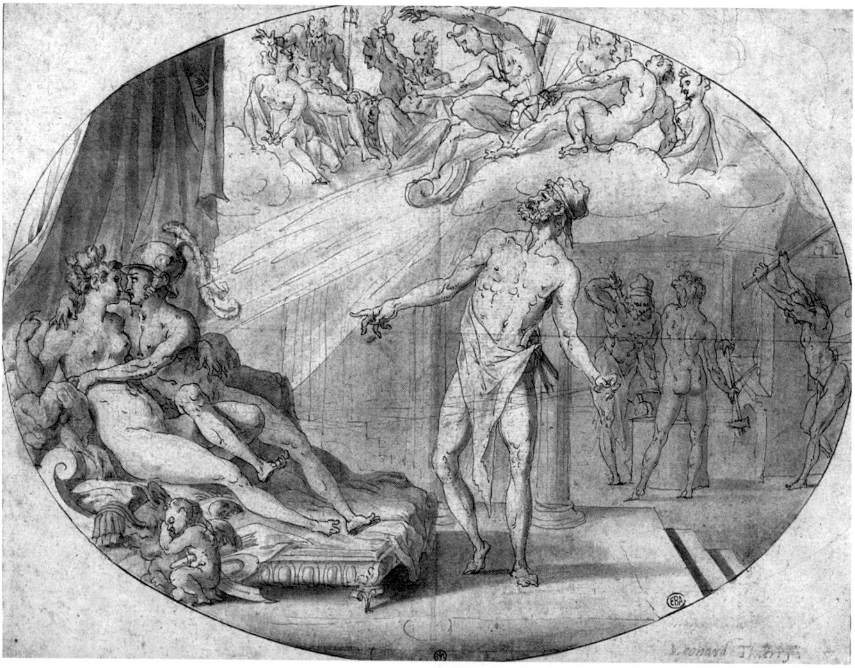

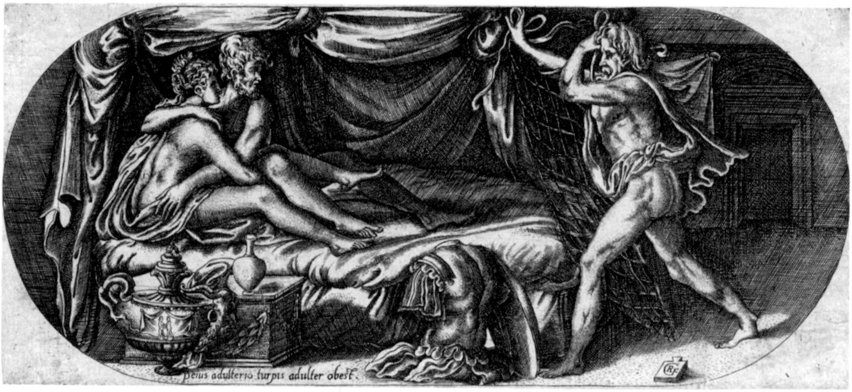

7.12 Léonard Thiry. *Venus and Mars Surprised by Vulcan*, ca. 1550. Ink and wash drawing. École nationale supérieure des beaux-arts, Paris. Photo: Jean-Michel Lapelerie.

7.13 Georges Reverdy, *Vulcan Trapping Mars and Venus in his Net*, between 1529 and 1565. Engraving, 12 (trimmed) × 27.3 cm. British Museum, London. © Trustees of the British Museum.

stitute. Owning a picture of a woman is not the same as owning a woman, even if both could in some sense, in Renaissance culture, be "owned." And with prints, any number of other viewers might own the same image. In another print (fig. 7.14), probably based on a drawing by Thiry, possibly a copy

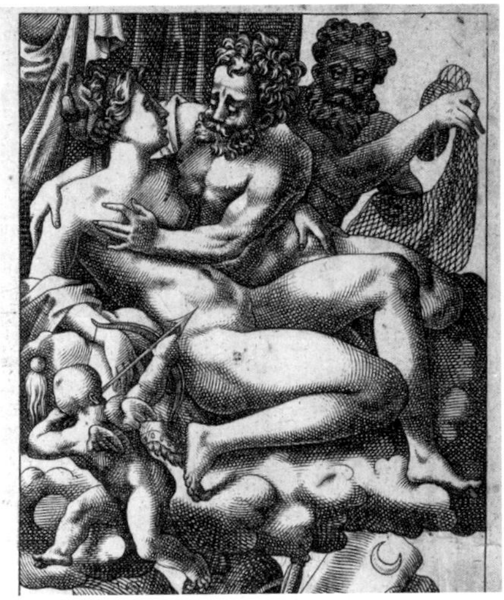

7.14 Anonymous, after René Boyvin. *Vulcan Trapping Mars and Venus in his Net* (detail from a mythological festoon design), ca. 1560. Engraving. British Museum, London. © Trustees of the British Museum.

of an engraving by Boyvin but almost certainly not by Boyvin's own hand, Mars and Vulcan are identical, so Vulcan is *further* enmeshed in the scene of scandal: caught in his own trap. Their resemblance might also be seen as an oblique expression of the identical copies produced by printmaking.

If we return, now, to Tintoretto's *Vulcan, Venus, and Mars*, we begin to see the painting as a link in a chain of prints that includes not only those I have mentioned but also others, for instance, by Caraglio and Antonio Tempesta. And this might make us see it differently. The lines of perspective lead our eye inexorably to the helmeted figure of Mars, who hides under the table in the middle ground.[58] I suggest we see there an inversion, with Mars representing the artist and Vulcan looking quite a bit like portraits of Titian (fig. 7.15). We might call this scene a drama of competition for the possession of beauty itself, with Vulcan as Titian, Mars as Tintoretto, and Venus as Titian's own Venus of Urbino. While this Venus is more animated, her body, facial type, and hairstyle are essentially the same, as is the small dog that has wakened from its slumber. While the perspectival lines draw us to Mars, we also notice the mirror that stares back at us. It recalls the mirror in Jan van Eyck's Arnolfini portrait, but in this case it displays Vulcan unceremoniously from the rear, making him the "butt" of the joke, the potential object of sexual violence.[59] Vulcan's powers—mastery, surveillance, unveiling—are themselves stripped bare by the mirror, the artist in the image, the image that looks back via a homoerotic gaze. Rather than imagining a viewer who can keep himself distanced from the world through his objective view on it,

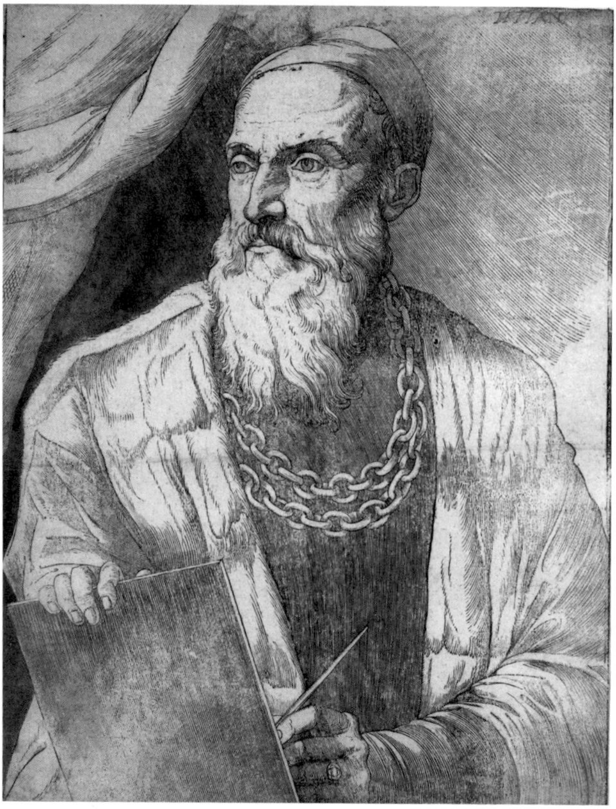

7.15 Giovanni Britto, after Titian. *Portrait of Titian*, ca. 1550. Engraving with brown wash added. British Museum, London. © Trustees of the British Museum.

Tintoretto explicitly implicates both artist and viewer in the scene. Perspective is not thereby abolished, but it serves to disorient us more than to orient us; the sharp angle suggests the disproportion inherent in jealous rage and artistic envy. While other artists might have seen themselves as Vulcan, Tintoretto presents himself as the bellicose younger god—hiding under the table.

How might Mars take proper possession of Venus? Michel de Montaigne, in his essay "On Some Verses of Virgil," compares two amorous scenes, one between Vulcan and Venus and the other between Mars and Venus. In the essay he connects the policing of sexuality with rhetoric, epistemology, and imagination. Sexual desire is natural, he argues, and the attempt to repress raw natural desire or knowledge of sexual practice by linguistic regulation can only fail, *impressing* knowledge through its very repression. He considers this specifically with respect to the education of girls. Interdiction

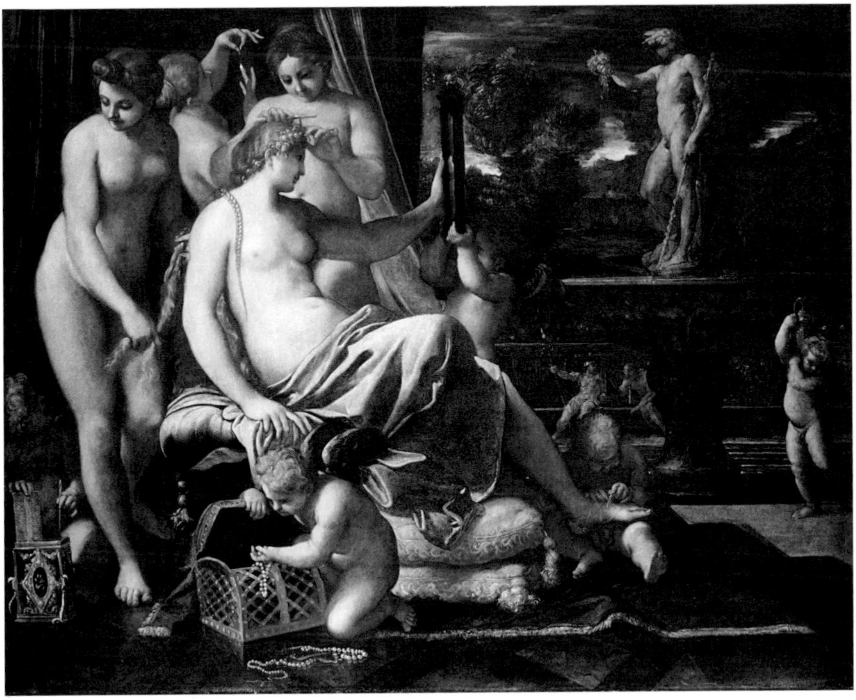

7.16 Annibale Carracci. *Venus Adorned by the Graces*, 1590–95. Oil on canvas, 133 × 170.5 cm. Samuel H. Kress Collection, National Gallery of Art, Washington. Image courtesy of the Board of Trustees, National Gallery of Art, Washington.

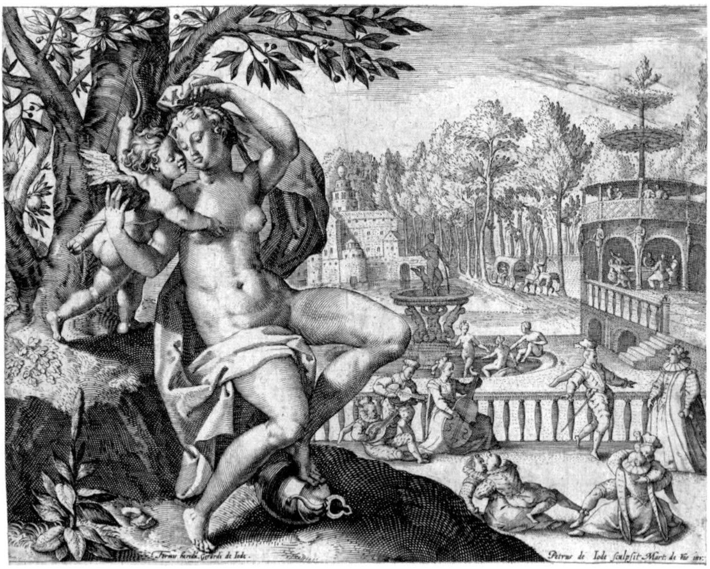

7.17 Pieter de Jode, after Maarten de Vos. *Spring*, 1590s. Engraving, 17.7 × 22.7 cm (sheet). Staatsgalerie Stuttgart.

imprints the thing in the fantasy; constraint, like clothing, can only increase desire. Montaigne compares Virgil's verses on Vulcan and Venus to Lucretius's description of the tryst of Venus and Mars, and finds Virgil's exuberance forced, artificial, unnatural; the words are too beautiful for the thing. Lucretius's extramarital eloquence, on the other hand, is "not soft" but "pregnant," "solid," "vigorous." It "does not please so much as it fills and ravishes." The things themselves, Montaigne says, are firmly imprinted in the poet's soul, and "it is the lustiness of the imagination that elevates and swells the words." The ostensible naturalness of this eloquence, however, can exist only within the artificial (we might say cultural) context produced by the husband's interdiction. That is, an extramarital affair, as "natural" as its passions might be, requires the constraints of marriage to exist.[60]

A final Renaissance scene of this triangular story unfurls in the background of Annibale Carracci's *Venus Adorned by the Graces* from the 1590s (fig. 7.16). One must look closely to find Mars and Vulcan, but they are there, almost indiscernible, seated in the background behind and underneath a fountain statue of Bacchus modeled on Cellini's *Perseus*. Vulcan exacts an adulterer's fee—we might call it a bride-price—under a statue that dissimulates the vanquished female gaze of Medusa. Mars's payment to Vulcan restores sexual normativity and regularizes the relationship between the two male rivals. And here nature, in the person of Venus, appears in the foreground, distracting us from the shady transaction between the two male gods. Venus dominates the image, but she is not the one who gets to tell the story. Collaborators rather than rivals (or perhaps, like the Carracci themselves, collaborators *and* rivals), they engage in an act of commerce.

Like its predecessor the *Venus of Urbino*, this has been viewed as a marriage picture, and it may well be. But we may also see Venus as Nature adorned by Art, as an element of the identity of this artist thought of as a "reformer of painting." He and his brothers and cousin also designed and made prints and borrowed widely from them (as well as from sculpture, as this painting demonstrates). In this case the painting's composition bears a strong resemblance to an undated scene of *Spring* after Maarten de Vos, in which Venus is paired only with her son and lovers cavort energetically in the background around a more distant fountain, its crowning figure posed nearly identically to Carracci's Bacchus (fig. 7.17). The two images are close contemporaries, but their exact relationship may be undecidable, and thus whether it is the two gods who have been transformed into multiple pairs of lovers or the reverse. The transformation could, it seems, have operated in either direction. But even as he leans toward Mars with coiled intensity,

Vulcan's jealous passion seems to have subsided, in the brushed surfaces and calm of Carracci's classicizing pastiche. If Vulcan's creations are mere artifice, and if his physical deformity is code for impotence, perhaps he has good reason to be jealous of the reality he has tried to capture, the "nature" embodied in the adulterous sexual act. Perhaps what Art envies—and desires—most is Nature. With Venus/Nature adorned by the (artful) graces, Vulcan claims his due.

Epilogue in the Female Eye

The images we have seen present depictions by men of intimate interior spaces that mobilize women's bodies but have little to do with women's experience of them. If we look to the Renaissance for the development of a new form of subjectivity, it might be one of susceptibility, subjection, and restriction rather than the confident assertion of control that we have come to expect from modern ideas about perspective. But what does this subject have to do with women—more often objects than subjects, even when the passions are triangular?

Perspective may present an ideal form of viewing suited to particular spaces, events, and social classes. But sources that could help us get to a social and visual organization of space that would apply more broadly even among these same classes of people—say, for women—are scant; records do not often spell out the specific everyday uses of space. Literary works written by and for upper-class women can provide some alternative understandings of space and vision, at once more restricted (and potentially frightening) and more intimate. In the *Heptaméron*, Marguerite de Navarre writes with great concreteness of the spaces in which her tales take place. These spaces form not a broad vista but a kind of affective itinerary (or, to borrow from a slightly later female author, Mlle. de Scudéry, a *carte du tendre*)—one of winding stairs and galleries, alcoves and corners, glimpses of windows, secret doors, darkness, and disguise. Within a semi-private space, two people might envelop themselves in privacy by sitting at a remove from others, pretending to read.[1] The contours of space are not geometrically defined: smaller, dimly lit rooms offer no broad perspectival vistas. Depending on location, style, and class, they might in fact be defined more by cloth—often the largest consumer expenditure—than by architecturally representable lines and angles. Draped in curtains, hung with tapestries, covered with mats or rugs, rooms themselves rhymed with clothed bodies, in a way distinct from the geometric correspondences suggested by the Vitruvian Man or Francesco di Giorgio's mapping of a man's body on a church plan. Correspondingly, in courtly literature vision is integrated with the other senses, especially touch; Castiglione talks of the sight of gloves being taken off and put on, or a glimpse of a beribboned leg as a lady lifts her skirts slightly while walking to church.[2]

The ideal palazzo as expressed by writers like Alberti may not have reflected the more chaotic and communal reality of Quattrocento life.[3] One thing that seems certain is that social hierarchy was reflected in spatial segregation.[4] While many working-class women were required to be in public as a matter of course for purposes of laboring, buying, and selling, upper-class women were confined indoors, praying, reading, spinning and weaving, and engaging in conversation.[5] At the same time, the social surveillance of women's virtue meant that privacy of any kind was the exception, and public places were viewed as sites of danger. According to Juan Luis Vives, in his *Education of a Christian Woman*, the more a young woman is seen in public the more potential faults she has to answer for; it is as if being seen outside at all is itself a form of stigma. For young women it was a necessary evil: in the portion of his text on married women, Vives suggests they stay inside even more than unmarried women, who must be seen in public in order to find husbands. Any place where men and women mix, especially under relaxed (or, worse, festive) circumstances, is especially hazardous to young women's virtue. And the eyes are particularly dangerous. Vives writes that the young girl "should not desire to see or be seen, or cast her glance this way and that"; "she will keep her eyes cast down, and will raise them but rarely and with modesty and decorum."[6] Dangers of sight include not only the potential to fall in love, to be "corrupted by the contagion"[7] of the vices she sees around her, but even to be infected by the evil magic of paid go-betweens: "the young girl should not even tolerate the sight of them, for they are basilisks or *katoblepae* that transmit a deadly poison from their eyes and annihilate you with a single glance. Let no one think that I am exaggerating. Some of these women are so skillful that they can often captivate you with their eyes with no words spoken. Others make use of spells and incantations, and would to God examples of these black arts were more rare. Why, by a mere greeting or smile or even a look, this serpent can defile a girl upon whom it has gazed."[8]

Even church might not be a safe space for the young woman. They were public spaces in which unscripted activities might take place: in Counter-Reformation churches an attendee might find herself the object of a surreptitious portrait commissioned by an admirer.[9] Young women were expected to regulate their bodies, their facial expressions, and their gaze, but also to pay particular attention to place. Many writers point to location (whether country and climate or specific types of local places) as sources of love and jealousy. "It is not to be permitted," Vives writes, "that a young woman and a man should converse alone anywhere for any length of time, not even if they are brother and sister. . . . Nor would I have them retreat into a corner of a crowded hall or room."[10] Elsewhere Vives writes about jealousy that "circumstances of place are also relevant: whether the place has no access and is closed and shut off; whether it is a sacred or a crowded place; whether it is exposed and open to the eyes of all, to friends and people loyal to us, or to enemies, busybodies, or gossipers; whether the place is entrusted to the care of a prudent and watchful custodian or not."[11] According to Vives, climate and complexion also play a role—some peoples are more prone to jealousy than others, but the physical and social configuration of specific public and private places also causes jealousy to increase or decrease.

While love affairs are condemned by Vives in the strongest possible terms, they formed the basis of the literary tradition of courtly love, still powerful in the Renaissance. In this context, the lady's look is often associated with wounding and with the petrifying gaze of Medusa or the monstrous basilisk that kills with a lethal look.[12] In Maurice Scève's *Délie*, an emblem of a basilisk gazing in a mirror is labeled "Mon regard par toy me tue": literally, the basilisk's eye kills the basilisk by means of the mirror, but figuratively, the lover kills himself by looking at his beloved. The emblem is paired with a poem in which the lover must bow his head to avoid the direct gaze of the lady's eye.[13] This eye that is both "active and passive," wounded and wounding, troubles distinctions of agency: whose fault is it when love is borne by a look? Poetic lovers parsed the meaning of small gestures via a well-developed and admittedly unreliable hermeneutics of the look.

In the 1538 narrative *Les angoysses douleureuses qui procedent d'amours* (*The Painful Anguish That Proceeds from Love*), written by "Hélisenne de Crenne" (a probable pseudonym for Marguerite Briet), the narrator suggests that it is women and not men who are the victims of love. She tells the story of a love triangle from the point of view of the young wife, victim of the violent throes of love and of a jealous husband. Her love affair is conducted entirely without touch; she casts "artificial and immodest glances" at her lover, but there are few opportunities for them to speak, much less to spend time together in a private space. Early in Hélisenne's story, she arises one morning. "Getting dressed, I went to open the window, and looking to the other side of the street, I saw a young man, also looking out his window, whom I began watching attentively . . ." She describes his beauty, then continues: "After looking too long I withdrew my gaze; but I was forced to turn my eyes back, toward him, he watched me as well, which made me very happy; but I was surprised to find myself so drawn to looking at this young man. . . . I was accustomed to taking and captivating men, and only laughing about it, but myself, miserably, I was captured. I could not take my eyes away, and desired no other pleasure than this one."[14]

Hélisenne's narrative, largely a psychological one with little action, develops a grammar of looks, facial expressions, and thoughts. While they relate to the courtly topos of the wound borne by a look, Hélisenne's looks have little to do with the frequent literary manifestation of the look as the instrument of a powerful lady. Looks are immodest, fixed, and immobile, dissolute, "artificial," and seductive.[15] Hélisenne hopes the flames of love will be cooled ("refrigerated") by opportunities for gazing on the young man's beauty, but instead she finds herself all the more heated.[16] Looks are able to communicate attraction but also indifference and complaint; when the young man seems to be pointing her out to his friends, which displeases her greatly, she responds by casting displeased looks. Her husband and the lover control space and other forms of power; her husband moves the household when he begins to become jealous, and her lover counters by hiring musicians to play under her window.

In Ficino's theory of love, for which he relies on Plato, vaporous spirits are emitted from and received by the eyes; these spirits are the source of love, tears, sparkles, fire, and agitated blood. The eye is a major focus of the courtly poetry of the later Middle Ages and Renaissance, which feeds the extensive love discourse summed up in Robert Burton's *Anatomy of Melancholy*. The eye is the source of love: it is the glance, the flash, the flame, of a lady's eyes that "send[s] the arrows" of the god of Love. The eyes are both aggressive and receptive. According to Burton the eye is "both Active and Passive"; it "betrays the soul"; it "wounds and is wounded, is an especial cause and instrument, both in the subject and in the object. As tears, it begins in the eyes, descends to the breast, it conveyes these beauteous rays, as I have said, to the heart."[17] Eyes are mobile and dangerous: "we speak of wandering, wanton, adulterous eyes, which, as he saith, lie still in wait, as so many soldiers, and when they spy an innocent spectator fixed on them, shoot him through and presently bewitch him."[18] Danger lurks in this mobile, rolling, flickering gaze.

For Hélisenne, the look is the privileged mode of communication because of the restricted set of spaces in which she spends her days. Most of the time she is in her own room—in bed, seated, or, as often as she can, at the window, where she experiences her confinement but also the chance to exchange glances with her beloved. Adrian Randolph has suggested that the Renaissance window be understood as a feminine space: the gaze is spatialized, but not in the commanding or distancing sense of traditional understandings of perspective, rather in a way that bridges distance.[19] But even more often, Hélisenne dwells in her *ymaginacion*, in which "the semblance, effigy or similitude of the young man was painted and described."[20] She goes to church, but there is little thought of religion; numerous times she makes or attempts contact with her lover by remaining in the church until the crowd has left. At home, her husband discovers her looks at the window and becomes angry. He moves the household when he recognizes that the man occupies a house with a neighboring window. Later, in a public place with her husband, she sees her lover.

> My husband, who saw things clearly, incessantly watched my movements and expressions, and in observing continually where I was sending my artificial looks, he saw my lover, and so he began to say, pointing him out to me and pretending not to have noticed me looking at him: My dear, look at that villain who has caused so many disagreements to bubble up between us, and as far as I can tell he does not wish to desist from such thoughts nor put an end to his outrageous folly. . . . And so I wish and command you not to use your habitual looks, and if you do not stop, you will enrage me so, that the impetuosity of my anger may make me exceed the bounds of reason.[21]

The eye in *Les angoysses*, finally, is not only a weapon or communicative medium but also a profoundly productive fountain of tears. Hélisenne's lover describes her eyes in conventional terms as "green and radiant" (using alternatively the terms *vair* and *vert*, gray and green), and she later takes up this same phrase to describe her own tearful eyes: "Through incredible sadness I distilled abundant tears from my radiant green eyes."[22] The eye and its watery emissions melt the distinction between inside and outside. Fully inside her sadness, she nevertheless conjures a vision of the tearful brilliance of her own eyes.

This eye is vulnerable rather than masterful, and space is tactile as much as visual. If things seen can have "masterful" effects on viewers, as I have suggested before, mimetic desire might involve not only the imitation of another desiring subject but the desire to imitate the object, to be shaped by it. Artist and spectator each have a thoroughly imperfect relationship to the object, one whose incompleteness is guaranteed by the inevitable presence of the third party in the triangle. In some works of art, the artist seems to have consciously set himself up in a rivalry with the viewer, as with the Alexander imagery at Fontainebleau. In a triumph of virtuosity and representational equivalence, the artist exchanges an image of the king's lover for the lover herself. Vulcan is not so lucky; he may control representation but not Eros (who may not in fact be his son). This occurs in other ways too; for instance, we might look to the famous *Susanna and the Elders* by Jacopo Tintoretto (fig. 8.1). The two elders anchor the trellis or screen, sustain it spatially and perspectivally, by marking its endpoints; their bodies create the perspective system of the image. They also create a triangle with the figure of Susanna. Perhaps they are Dürer's artist, peeking around the screen for a better view—what the viewer can freely see. They hold up a theatrical curtain in front of which Susanna is displayed for our own guilty view. The trajectories of the multiple gazes embedded in the image are oblique. And it matters that the Elders are two: in the story, told in the biblical Apocrypha, it is when Daniel separates the two men that they are found to be guilty—after they have accused Susanna of seduction. Their twoness is their downfall: the single viewer divided against himself, or two viewers whose difference seals their guilt. In this image, seeing may be a figure for knowledge, but it is also a figure for desire and its chaotic effects, for the elusiveness of its objects. The viewer is doubled too, into the one who desires and the one who condemns the Elders' desire.

In these visual possibilities of *in*direction in desire and the look, the pre-

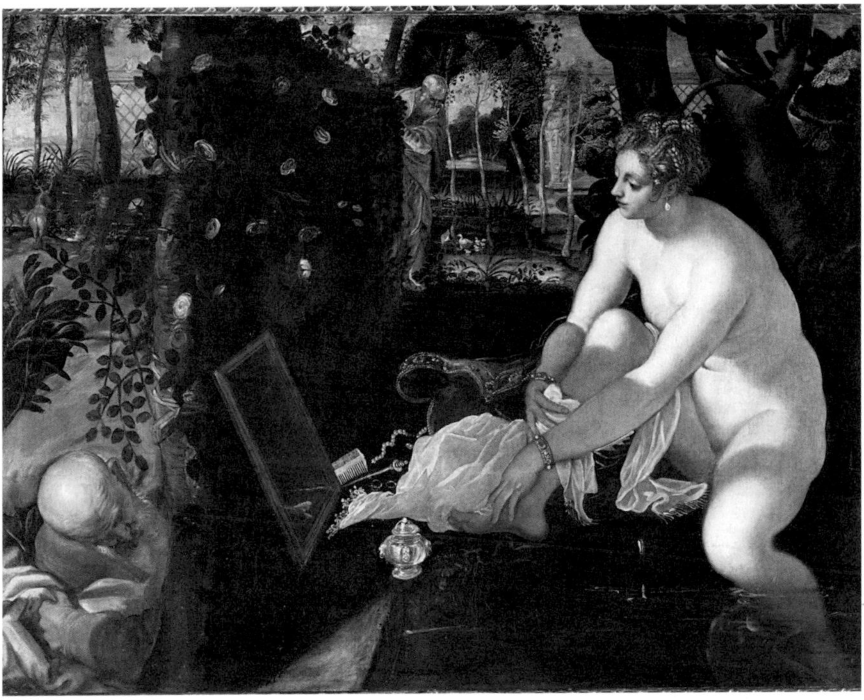

8.1 Jacopo Robusti Tintoretto. *Susanna and the Elders*, ca. 1560–62. Oil on canvas, 146 × 193.6 cm. Kunsthistorisches Museum, Vienna. Erich Lessing / Art Resource, NY.

tense of mathematical rigor in perspective is subverted by voyeurism, as the object of the gaze implicates and disconcerts the viewer. A print by Georges Reverdy (fig. 8.2) can illustrate the strange and recursive properties of the visuality of Renaissance desire. Two reclining female nudes find themselves linked by visual rays to a disembodied eye held aloft by a winged figure—forming a triangle. The appended text reads, "Avert your eyes from adorned woman; on account of the sight of woman many have perished."[23] This is a standard case, it appears, of lustful viewing masquerading as moralism, reveling in the very thing it seems to prohibit. We might be perplexed (as other interpreters of this image have been) by a male figure in a winged cap, looking very much like traditional representations of Mercury.[24] He averts his own eyes (as the caption tells him to) while holding up the single eye. On close inspection, we can see in the Reverdy print a dreamlike condensation of motifs from the myth of Perseus. When Perseus (wearing Mercury's winged sandals) meets the Graiae, sisters who shared but one eye amongst

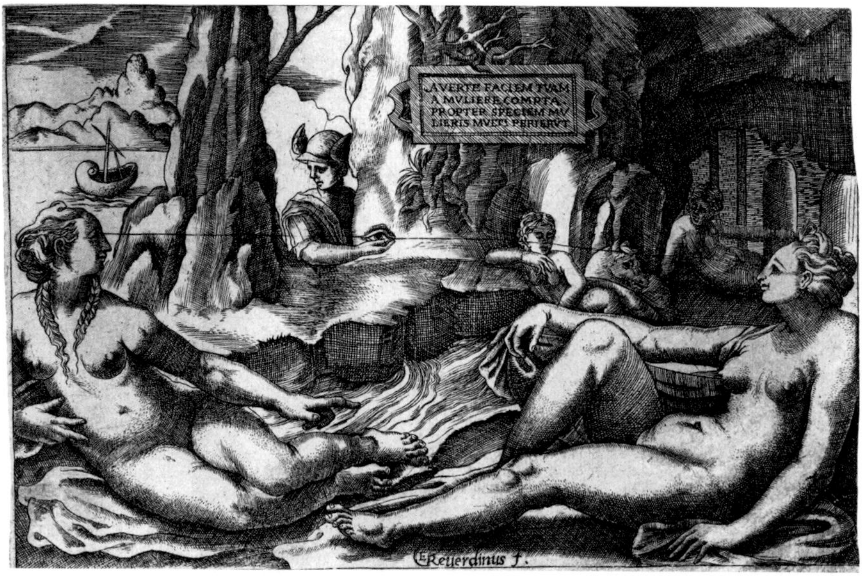

8.2 Georges Reverdy, *Perseus and the Eye of the Graiae (The Dangers of Concupiscence)*, ca. 1560. Engraving/etching, 13 (trimmed) × 19.9 cm. British Museum, London. © Trustees of the British Museum.

them, he steals their single eye to force them to reveal the location of the cap that will render him invisible. The cap makes it possible for him to defeat the horrible female gaze—that is, Medusa's—to whose serpent locks the nude women's braids seem to allude. In this scene, Perseus already wears the cap and brandishes the eye as if it is itself Medusa's head, ready to entrap the viewer—or the two women, frozen into representation. Should the viewer identify with the eye? Should we gaze, or not, at the reclining nudes? Does the eye look back?

Medusa, in all her terrifying, petrifying power, can be apprehended only in oblique representation, by the polished mirror surface of Perseus's bronze shield. To kill the Medusa, to avoid being transfixed by her petrifying stare, one must look at her image in a mirror angled just so, to see her without her being able to see you, without her having the chance to return the look and make you its object. Is this picture itself the polished shield—enabling the viewer to look at the women without being subject to their charms? This single eye, a vanishing point of its own, peers out, too, from Titian's painting *Venus with a Mirror* (fig. 8.3). Perseus, too, is caught in the picture—relentlessly drawn by the inextricable connectedness of the body to the imagi-

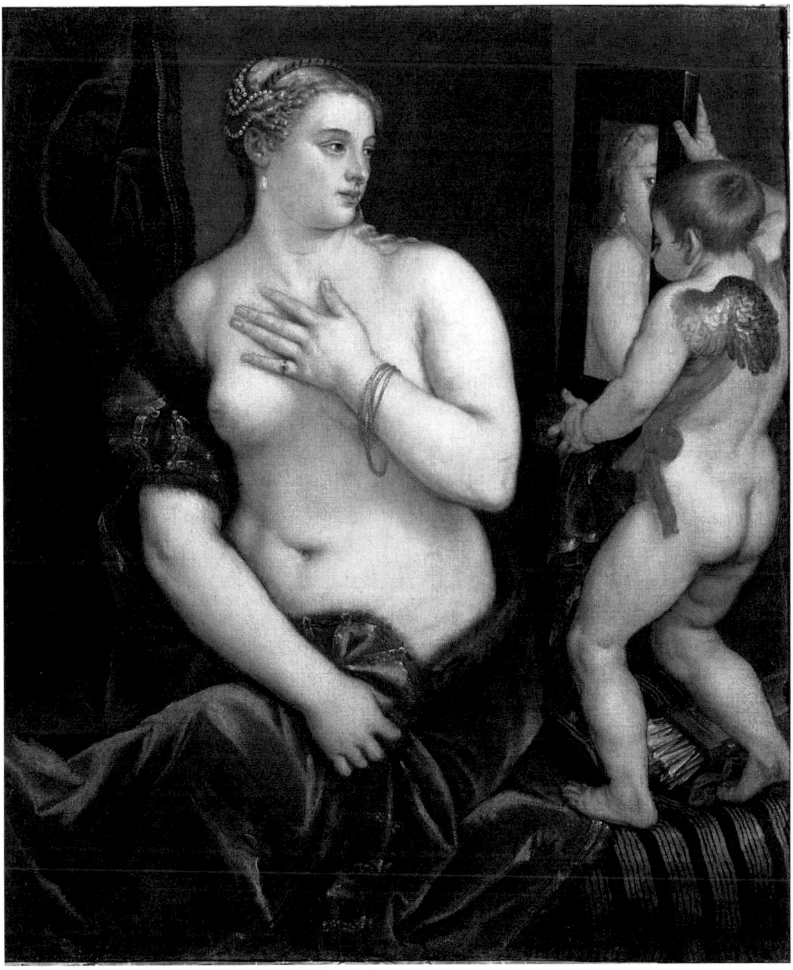

8.3 Titian (Tiziano Vecellio). *Venus with a Mirror*, ca. 1555. Oil on canvas. Gemäldegalerie Alte Meister, Staatliche Kunstsammlungen, Dresden. Bildarchiv Preussischer Kulturbesitz / Art Resource, NY. Photo: Elke Estel / Hans-Peter Klut.

nation, that most corporeal, imagistic, and Saturnian of the faculties. His disembodied eye is a far cry from command. Are we in fact to feel implicated in the gaze of this lone eye, to which lines like the triangular lines of perspective seem to run? As Hélisenne puts it, "I could not take my eyes away, and desired no other pleasure than this one."[25]

Notes

A list of source abbreviations can be found on p. 231.

CHAPTER ONE

1. Translations of postclassical texts (and this vocalization) are mine unless otherwise noted.

2. Ernst Gombrich, "A Plea for Pluralism," *American Art Journal* 3, no. 1 (Spring 1971): 85.

3. Harald Keller, *The Renaissance in Italy: Painting, Sculpture, Architecture* (New York: H. N. Abrams, 1969), 111.

4. Ben Shahn, arguing against the Greenbergian position, cites the triad and triangle as prime example of an apparently pure "form" that inevitably carried significant content. Shahn, *The Shape of Content* (Cambridge, MA: Harvard University Press, 1957), 53–55.

5. Heinrich Wölfflin and Wassily Kandinsky, both living and working in Munich, both emphasized the triangle. For Kandinsky, the triangular form was key to spirituality in art. See Kandinsky, "On the Spiritual in Art," in Kenneth C. Lindsay and Peter Vergo, eds. Kandinsky, *Complete Writings on Art* (New York: Da Capo, 1982, 1994), and Daniel Adler, "Painterly Politics: Wölfflin, Formalism, and German Academic Culture, 1885–1915," *AH* 27, no. 3 (June 2004): 431–56. The two may have met (see Peg Weiss, *Kandinsky in Munich: The Formative Jugendstil Years* [Princeton, NJ: Princeton University Press, 1979], 82); Kandinsky cited Wölfflin in unpublished lecture notes (Wassily Kandinsky papers, 1921–37, Getty Research Institute, Research Library, accession no. 850910, series I.A. Notes for Semester I, 1925–1931, box 1, folder 16).

6. James Elkins, "The Case against Surface Geometry," *AH* 14, no. 2 (June 1991): 143–74. See also reply by Franz Gnädinger, "The Case against Surface Geometry," *AH* 15, no. 3 (September 1992): 399–402. Daniel Arasse wrote bluntly, "I do not believe in the secret geometry of painters," in "Le regard de l'escargot," in *On n'y voit rien* (Paris: Denoël, 2000), 38. He refers to Charles Bouleau, *The Painter's Secret Geometry: A Study of Composition in Art* (New York: Harcourt, Brace and World, 1963). See also Thomas Puttfarken, *The Discovery of Pictorial Composition: Theories of Visual Order in Painting, 1400–1800* (New Haven, CT: Yale University Press, 2000), 10–11. Puttfarken points out that by reading individual panels of Giotto's Arena Chapel frescoes on the same terms as a framed quadrilateral, early twentieth-century scholars constructed these panels as autonomous objects, subject to diagrammable compositional unity.

7. See, e.g., the *Apotheosis of St. Thomas* by a follower of Francesco Traini at S. Caterina in Pisa, or the anonymous birth tray with the *Triumph of Venus*, ca. 1400 (Louvre, Paris) with famous lovers attached by threads to her genitals.

8. It is reminiscent of the triangle made by the wings of two angels and a dove in Filippo Lippi's *Adoration* from the hermitage at Camaldoli, now at the Uffizi.

9. The lines mark the location of the lance and ladder in the scene, forming a triangle with the ground. Warren Sanderson, "Geometry on a Carolingian Wall," in Nancy Wu, ed., *Ad Quadratum: The Practical Application of Geometry in Medieval Architecture* (Aldershot, UK: Ashgate, 2002), 16, with several examples of medieval geometric composition. Thanks to Ben Tilghman for the reference.

10. Ibid., 18.

11. John Canaday, *What Is Art? An Introduction to Painting, Sculpture, and Architecture* (New York: Alfred A. Knopf, 1980), 175.

12. Hubert Damisch, *The Origin of Perspective* (Cambridge, MA: MIT Press, 1994), 363.

13. Canaday, *What Is Art?* 172.

14. See John E. Murdoch, *Album of Science: Antiquity and the Middle Ages* (New York: Charles Scribner's Sons, 1984).

15. Michael Evans, "Geometry of the Mind," *Architectural Association Quarterly* 12, no. 4 (1980): 32, 47–49. See also Aden Kumler, *Translating Truth: Ambitious Images and Religious Knowledge in Late Medieval France and England* (New Haven, CT: Yale University Press, 2011), on the theological ambitions of manuscript images, including diagrammatic images.

16. See Renzo Baldasso, "La stampa dell'editio princeps degli Elementi di Euclide (Venezia, Erhard Ratdolt, 1482)," in Lisa Pon and Craig Kallendorf, eds., *The Books of Venice/Il libro veneziano* (Venice: La Musa Talìa and Biblioteca Nazionale Marciana; New Castle, DE: Oak Knoll, 2009), 61–100, and Baldasso, "Illustrating the Book of Nature in the Renaissance: Drawing, Painting, and Printing Geometric Diagrams and Scientific Figures," PhD diss., Columbia University, 2007, 235–37. Baldasso extends a brief comment made by Hyatt Mayor in *Prints and People: A Social History of Printed Pictures* (New York: Metropolitan Museum of Art, 1971), unpag. text opposite fig. 74. In the Venetian context (and elsewhere), humanist printing also expressed its fascination for geometry with triangular endpieces and tapering texts at the ends of chapters and books; the fashion for shaped text that diminished to a point in a triangular form may have drawn inspiration from Greek magical spells for luring a beloved or calming a fever. See Christopher A. Faraone, *Disappearing Acts: The Text, Image, and Medium of Ancient Greek Curative Amulets* (forthcoming).

17. Luca Pacioli, *Summa de arithmetica, geometria . . .* (Toscalano, Italy: Paganino Paganini, 1523), 4r: "sancta theologia nostro intellectui non bene iungi sine matheseos cognitione potest."

18. The triangle is of course connected to the number three, another popular topic. See Emory B. Lease, "The Number Three, Mysterious, Mystic, Magic," *Classical Philology* 14, no. 1 (January 1919): 56–73.

19. Plato, *Timaeus*, in *Collected Dialogues of Plato, including the Letters*, ed. Edith Hamilton (New York: Pantheon, 1961), 53C–55C: "In the first place, then, as is evident to all, fire and earth and water and air are bodies. And every sort of body possesses solidity, and every solid must necessarily be contained in planes; and every plane rectilinear figure is composed of triangles . . ." See also John Hendrix, *Platonic Architectonics: Platonic Philosophies and the Visual Arts* (New York: Peter Lang, 2004), 75.

20. Nicomachus of Gerasa, *Introduction to Arithmetic*, trans. Martin d'Ooge (New York: Macmillan, 1926), 240–41 (2.7). See also Boethius *Arithmetica* 2.19: "the principle [or origin] and element of all forms is the triangle" ("omnium formarum principium elementumque, est triangulum"; Boethius, *De institutione arithmetica libri duo . . .*, ed. Gottfried Friedlein (Leipzig: B. G. Teubner, 1867), 104). Gérard Roussel elaborates upon this notion in his commentary (*Girardi Ruffi, in dvos Arithmeticae Boetii libros, Commentarius*) in Boethius, *Arithmetica, duobus discreta libris: Adjecto commentario . . .* (Paris: Simon de Colines, 1521), 84v, stating that the triangle is not resolved into an alien nature but into its own; also that it is the "root and principle" (93r) and "essence, idea, and exemplar" of all planes (91v).

21. Nicholas of Cusa, *De possest*, in Jasper Hopkins, *A Concise Instroduction to the Philosophy of Nicholas of Cusa* (Minneapolis: Arthur J. Banning, 1986), 937 (¶44). See also Bovelles, *Ma. Corp.* (*De mathematicis corporibus*), in *Que hoc volumine . . .* (Paris: Henri Estienne, 1510[=1511]), 193r.

22. Georg Stuhlfauth, *Das Dreieck: Die Geschichte eines religiösen Symbols* (Stuttgart: Kohlhammer, 1937), catalogues early examples of the triangle used in mystical ways.

23. Nicholas of Cusa, *De beryllo*, in *Metaphysical Speculations: Six Latin Texts*, ed. Jasper Hopkins (Minneapolis: Arthur J. Banning, 1998), 820 (¶60).

24. Bovelles, *Liure singulier et vtile touchant l'art de geometrie* (Paris: Simon de Colines, 1542), 55r: "la premiere, & la plus mystique de toutes les figures angulaires." Gérard Roussel describes the triangle as a symbol of the Trinity: "just as the triangle determines every figure and is the origin of all figures, so truly the ultimate triangle and divine triad determines all things and is the origin of them all" ("quemadmodum trigonus omnem discernit figuram & est omnium figurarum principium: ita sane summus trigonus, diuinaq[ue] trias, omnia discernit & est omnium principium"). Roussel, commentary in Boethius, *Arithmetica*, 94r.

25. Bovelles, *Liber de duodecim numeris*, in *Que hoc volumine*, 156r.

26. On triangles in seventeenth-century philosophy, see Charles McCracken, "Knowledge of the Soul," in Daniel Garber and Michael Ayers, eds., *The Cambridge History of Seventeenth-Century Philosophy* (New York: Cambridge University Press, 1998), 1:796–832.

27. Roussel, commentary in Boethius, *Arithmetica*, 43v: "Q[uare] autem numerus perfectus omnis, trigonus triu[m] [a]equalitatem lateru[m] obtinens: trinitatis quoddam est signaculu[m]. . . . Ita assurgere pro[m]ptum: ad summam vnitatem, aequalitatem & vtriusq[ue] summam connexionem . . ."

28. Nicholas of Cusa, *De theologicis complementis*, in *Metaphysical Speculations*, 747–788, 747 (¶1).

29. Hugh of St. Victor, *Geometriae practica*, ed. Roger Baron (Notre Dame, IN: University of Notre Dame Press, 1966), no. 3, Praenotanda, p. 19: "Nunc vero considerandum est quomodo in hac trianguli figura omnium dimensionum ratio consistat." Frederick Homann translates this sentence as "We now examine how the right triangle scheme can be found in geometric configurations" (leaving out "omnium" and taking "dimensionum" as "geometric configurations"; I think it is better translated "measurements"). Frederick A. Homann, *Practical Geometry [Practica Geometria] attributed to Hugh of St. Victor* (Milwaukee: Marquette University Press, 1991), 35.

30. See Paul Frankl, "The Secret of the Medieval Masons," *AB* 21, no. 1 (March 1945): 46–60; James S. Ackerman, " 'Ars Sine Scientia Nihil Est': Gothic Theory of Architecture at the Cathedral of Milan," *AB* 31, no. 2 (June 1949): 84–111. For Cesariano's diagram of the building, see Vitruvius, *De architectura libri decem,* trans. Cesare Cesariano (Como, Italy: Gottardo da Ponte, 1521), 15v; on his anachronism, see Alexander Nagel and Christopher S. Wood, *Anachronic Renaissance* (Brooklyn, NY: Zone, 2010), 163–64.

31. "Geometras facit admirandos, astronomos, abaciq[ue] peritos. . . . Nimis aut[em] su[n]t magne excellentie ad pastores [?] com[m]endacione pictores, sculptores, imaginu[m]q[ue] auctores atq[ue] architecti." Jacques Lefèvre d'Étaples, *De magia naturali*, Biblioteca Apostolica Vaticana MS. Vat. Regin. Lat. 1115, 60v–61r. See also Eugene F. Rice Jr., "The *De magia naturali* of Jacques Lefèvre d'Étaples," in Edward P. Mahoney, ed., *Philosophy and Humanism: Renaissance Essays in Honor of Paul Oskar Kristeller* (New York: Columbia University Press, 1976), 19–29. As Rice describes it, the triangle "inspires love of justice and equity, for the equilateral triangle is the figure of *aequalitas*. By analogy with the creative power of the Trinity, it stimulates artistic creation in humans" (24). See Letizia Pierozzi and Jean-Marc Mandosio, "L'interprétation alchimique de deux travaux d'Hercule dans le *De Magia naturali* par Lefèvre d'Étaples," in *Chrysopoeia* (Paris: Société d'Etude de l'histoire de l'alchimie, 1992–1996), 5:191–264.

32. Noted in Luca Pacioli, *De divina proportione* (Venice: Paganino Paganini, 1509), 3r: "Per la qual cosa lantico e diuin ph[ilosoph]o Platone non immeritamente ladito del suo celeberrimo Gymnasio ali de geometria inexp[er]ti denegaua quando vn breue al sommo de la sua principal porta a lettere magne intelligibili pose de queste formale parolle. videlicet. Nemo huc geometrie expers ingrediatur. Cioe chi non era buon geometra li non intrasse."

33. Roussel, commentary in Boethius, *Arithmetica,* 92r: "numeri diuinae menti immanentes: rerum sunt exemplaria. Atque hae sunt diuinae mentis id[a]e[a]e:quas fouent Platonici sed non satis apposite."

34. In the *Metaphysics* 986a, Aristotle said that the Pythagoreans believed that numbers were the "principles of things." Aristotle, *Metaphysics*, ed. and trans. Hippocrates G. Apostle (Grinnell, IA: Peripatetic, 1979), 21; see W. A. Heidel, "The Pythagoreans and Greek Mathematics," *American Journal of Philology* 61, no. 1 (1940): 1–33, 12, and Christopher S. Celenza, "Pythagoras in the Renaissance: The Case of Marsilio Ficino," *Renaissance Quarterly* 52, no. 3 (Autumn 1999): 703. See also Michael J. B. Allen, *Nuptial Arithmetic: Marsilio Ficino's Commentary on the Fatal Number in Book VIII of Plato's "Republic"* (Berkeley: University of California Press, 1994), and "Marsilio Ficino: Daemonic Mathematics and the Hypotenuse of the Spirit," in Anthony Grafton and Nancy Siraisi, eds., *Natural Particulars: Nature and the Disciplines in Renaissance Europe* (Cambridge, MA: MIT

Press, 1999), 122. Nicomachus of Gerasa wrote in his *Introduction to Arithmetic I* that "the pattern [of all things] was fixed, like a preliminary sketch, by the domination of number preexistent in the mind of the world-creating God, number conceptual only and immaterial in every way, but at the same time the true and the eternal essence, so that with reference to it, as to an artistic plan, should be created all these things" (Nicomachus, *Introduction*, 189 [1.6]). Boethius picks up on this idea in his adaptation of Nicomachus: "From the beginning, all things whatever which have been created may be seen by the nature of things to have been formed by numbers. Number was the principal examplar in the mind of the creator." *Boethian Number Theory*, ed. and trans. Michael Masi (Amsterdam: Rodopi, 1983), 75–76. Gérard Roussel's commentary specifically states that the "species" in God's mind at creation were numbers (Roussel, commentary in Boethius, *Arithmetica*, 48r). "Quo fit, vt se habeat ad deum primus suae me[n]tis numerus, primaq[ue] species."

35. Marsilio Ficino, *Commentary on Plato's "Symposium on Love,"* trans. Sears Jayne (Dallas: Spring, 2000), 45 (speech 2, chap. 1).

36. See my discussion in Rebecca Zorach, "Renaissance Theory: A Selective Introduction," in James Elkins and Robert Williams, eds., *Renaissance Theory* (New York: Routledge, 2008), 18–19, 21.

37. Nicole Oresme, *Nicole Oresme and the Medieval Geometry of Qualities and Motions . . . "Tractatus de configurationibus qualitatum et motuum,"* ed. and trans. Marshall Clagett (Madison: University of Wisconsin Press, 1968).

38. Basilios Bessarion, *In calumniatorem Platonis libri IV*, in *Kardinal Bessarion als Theologe, Humanist, und Staatsmann: Funde und Forschungen*, vol. 2, ed. Ludwig Mohler, Quellen und Forschungen aus dem Gebiete der Geschichte 22 (orig. Paderborn: F. Schöningh, 1927; Aalen, Germany: Scientia-Verlag, 1967), 2.9.21.

39. Bovelles *Liber propriae rationis* 48r: "Mathematica entia, inter physicas & metaphysicas substantias, medium locum tenent: re quidem, subiecto h[a]ere[n]tia & alligata:ratione vero & cogitatione secreta, abiuncta, separata. Physica co[n]sideratio: rerum naturalium tam materia[m] q[uam] formas inuoluit. Metaphysica vero scientia, materiae respuens considerationem: formas absolutas versat. At mathematic[a]e disciplinae velut medi[a]e ambarum: ea quide[m] tractant, quae cum re ipsa, subiectis h[a]erere & inesse probentur: ratione tamen & cogitatione, ab iis ope mentis abiunguntur."

40. Nicholas of Cusa, *De theologicis complementis*, 748 (¶2).

41. Proclus, *A Commentary on the First Book of Euclid's "Elements,"* trans. Glenn Morrow (Princeton, NJ: Princeton University Press, 1970), 24.

42. Vasari, *Le vite dei più eccellenti pittori, scultori e architetti*, ed. Licia Ragghianti Collobi (orig. 1568; Milan: Rizzoli, 1942), 1:543. The statement is equivocal: it might alternatively be read as "how sweet is that prospect," i.e., the prospect of going to bed with his wife.

43. Boethius, *Commentaria in Porphyrium* (Paris: Migne, 1847), 64, 0099B.

44. Georg Rheticus, the first follower of Nicolaus Copernicus, displayed a veritable devotion to triangles, writing several books about them. See Dennis Danielson, *The First Copernican: Georg Joachim Rheticus and the Rise of the Copernican Revolution* (New York: Walker, 2006), especially 94–95.

45. "[T]he triangle is the first finite and terminated magnitude; in it the infinite angle shines forth as a whole" (Nicholas of Cusa, *De beryllo*, in *Metaphysical Speculations*, 819 [¶58]). Within each triangle (however different they are from one another) the "triune beginning, which is both one and three, shines forth" (820 [¶60]).

46. Ibid., 819 (¶58).

47. "Relinquit[ur] igit[ur] vt substa[n]tia sensibilis mu[n]di p[er]fecta sit in tria[n]gulo." Bovelles, *Physicorum elementorum libri decem* ([Paris:] Josse Bade, 1512), f. XIXv.

48. Iamblichus (attributed), *The Theology of Arithmetic*, trans. Robin Waterfield (Grand Rapids, MI: Phanes, 1988), 45.

49. Ibid., 41.

50. Ibid., 49. The author also cites Anatolius: "Triangles both reflect and are the first substantiation of being plane" (51).

51. Plutarch, *De E apud Delphos* 388a–c; quoted in Walter Burkert, *Lore and Science in Ancient Pythagoreanism*, trans. Edwin L. Minar Jr. (Cambridge, MA: Harvard University Press, 1972), 34.

52. Giovanni Pico della Mirandola, *Conclusiones sive Theses DCCCC*, ed. Bohdan Kieszowski, Travaux d'Humanisme et Renaissance 131 (Geneva: Droz, 1973), 67–68, under "1. Per numeros triplares, quas a Platone in Timeo ponuntur in triangulo animam significante, admonemur quousque in formis numerandis sit progrediendum per naturam illius, quod est prima forma formans." This view is echoed in Antoine Favre, *Les entretiens spirituels*, ed. Lance K. Donaldson-Evans (orig. Paris, 1602; Paris: Société des textes français modernes, 2002), 271, "Le monde et rond, l'ame triangulaire."

53. Ernst Cassirer, *The Individual and the Cosmos in Renaissance Philosophy* (New York: Harper and Row, 1963), 123.

54. Ibid., 145.

55. Ibid., 152.

56. Cardinal Gabriele Paleotti, in his Counter-Reformation *Discorso intorno alle immagini sacre e profane* (orig. Bologna, 1582; Bologna: Forni, 1990), 76v, 78r, discusses the power of images to transmute the senses literally into the images' own shape; he describes this as a penetrative violence.

57. David C. Lindberg, *Theories of Vision from al-Kindi to Kepler* (Chicago: University of Chicago Press, 1976), 108–10.

58. Roussel, commentary in Boethius, *Arithmetica* (1521), 91v–92r: "ipse trigonus, non tam numerus q[uam] numeri principium, non tam figura q[uam] eiusce principium, ante omnem numerum pariter & figuram. Sed quisnam isce trigonus omnem excutiens planum & formans, omnem nostrae mentis praeueniens numerum: nisi nostrae menti immanens trigonus idemque substantialis animae vnitas? Nam haec, prorsus numero necnon & figura prior est: atque ab hac, est omnis numerus pariter & figura. . . . Neque nostrae mentis numeri aliud sunt: praeterqua[m] eiusce imagines & contractiones. . . . Quare quemadmodum humanae mentis trigonus, in vnitate, & vnitas in trigono omne[m] creat numerum, format & terminat: ita propemodum diuinae mentis superimmensus trigonus in summa vnitate, & summa vnitas in superimme[n]so trigono, omnem creaturam ad esse deducit, format & definit. Q[uare] si numeri nostrae menti immanentes animae coincidere[n]t essetq[ue] ide[m] quod anima: co[n]spicuu[m] foret quenq[ue] numerum contractu[m] & eu[n]dem extra anima[m] subsistentem, in anima coincidere."

59. Charles de Bovelles, *Liber de sapiente*, in *Que hoc volumine* (Paris: Henri Estienne, 1510[1511]), 126r. For a modern facing-page edition with French translation, see Charles de Bovelles, *Le livre du sage*, ed. and trans. Pierre Magnard (Paris: J. Vrin, 1982), which contains references to the folio numbers in the 1511 Latin edition. "Liquet etiam Sapientis Animam in sui principii, id est Trinitatis, numero feliciter consummari."

60. Bovelles, *Liber de sapiente*, 131v. Elsewhere the sharp intellect is a triangle (or acute angle) while the soft, receptive memory is a circle (or obtuse angle). *Liber de intellectu*, in *Que hoc volumine*, 18v–19r.

61. Marsilio Ficino, *Commentary on Plato's Symposium on Love*, trans. Sears Jayne (Dallas: Spring, 2000); Robert Burton, *The Anatomy of Melancholy*, ed. Floyd Dell and Paul Jordan-Smith (New York: Tudor, 1927), 617–23, summarizing prior Neoplatonic authors.

62. Francesco Patrizi, "De intellectu," in *Nova de universis philosophia libris quinquaginta comprehensa* (Venice: Roberto Meietti, 1593), 31v (bk. 15), "Atque ita sit, vt cognitio, nihil sit aliud, quam Coitio quaedam, cum suo cognibili." Cited in Cassirer, *Individual*, 134 and 134n11. "Appetit quidem intellectus ens ipsum, & ad id redire nititur, vt cum illo coeat, & illud coeat secum: vnde coitio fiat, & cognitio" (ibid.).

63. Bovelles, *Physicorum elementorum libri decem*, f. XXXIIv, "spiritale[m] vero actione[m]: esse me[n]tis a me[n]te eruditione[m] spiritalisq[ue] luminis ac totius scientiae: ex vna in aliam commeatum. . . . Quod e[ni]m i[n] naturali actio[n]e est gigne[n]s: id i[n] spiritali e[st] erudiens."

CHAPTER TWO

1. Carol Strickland and John Boswell, *The Annotated Mona Lisa: A Crash Course in Art History from Prehistoric to Post-modern* (Kansas City: Andrews and McMeel, 1992), 34.

2. Thomas Puttfarken, *The Discovery of Pictorial Composition: Theories of Visual Order in Painting, 1400–1800* (New Haven, CT: Yale University Press, 2000), 3–8.

3. See Arnold Hauser, *Social History of Art: Renaissance, Mannerism, Baroque* (London: Routledge, 1951), 2:1–28 (chap. 1, "The Concept of the Renaissance").

4. "[L]itterarum nonnihil doctus, geometriae praesertim et earum artium quae ad picturae ornamentum accederent": Elizabeth Gilmore Holt, *A Documentary History of Art*, vol. 1, *The Middle Ages and the Renaissance* (New York: Doubleday, 1957), 200n8. Wolfgang Stechow also footnotes geometry as "perspective" in *Northern Renaissance Art, 1400–1600: Sources and Documents* (Evanston, IL: Northwestern University Press, 1999), 4.

5. See chapter 1, n6.

6. Michael Baxandall, *Painting and Experience in Fifteenth Century Italy: A Primer in the Social History of Pictorial Style* (orig. 1972; New York: Oxford University Press, 1988), 86–94.

7. This is also noted by Thomas Golsenne and Bertrand Prévost, introduction to Leon Battista Alberti, *La peinture*, ed. and trans. Golsenne and Prévost (Paris: Seuil, 2004), 23.

8. On Alberti, see *inter alia* Anthony Grafton, *Leon Battista Alberti: Master Builder of the Italian Renaissance* (New York: Hill and Wang, 2000), and Joan Kelly Gadol, *Leon Battista Alberti: Universal Man of the Early Renaissance* (Chicago: University of Chicago Press, 1969).

9. Margaret D. Carroll, *Painting and Politics in Northern Europe: Van Eyck, Bruegel, Rubens, and Their Contemporaries* (University Park: Penn State University Press, 2008), 26.

10. Alberti's contemporaries emphasized his mathematical expertise, classing him among the mathematicians: Cristoforo Landino (in "Fiorentini eccellenti in dottrina," part of the proem to his commentary on Dante) considered him a mathematician, though his learning stretched to many fields. Landino, *Proemio: Commento di Christophoro Landino fiorentino sopra la comedia di Dante Alighieri poeta fiorentino* (Venice: Bernardino Benali &

Matthio di Parma, 1491). "Ma dove lascio Battista Alberti o in che generazione di dotti lo ripongo? Dirai tra' fisici. Certo, affermo lui esser nato solo per investigare e' secreti della natura. Ma quale spezie di matematica gli fu incognita? Lui geometra, lui aritmetico, lui astrologo, lui musico e nella prospettiva maraviglioso più che uomo di molti secoli." Filarete wrote that he was "uomo dottissimo in più facultà è in questa molto perito, massime nel disegno, il quale è fondamento e via d'ogni arte che di mano si faccia, e questo lui intende ottimamente, e in geometria e d'altre scienzie è intendentissimo; lui ancora ha fatto in latino opera elegantissima." Filarete (Andrea Averlino), *Trattato di architettura*, ed. Anna Maria Finoli and Liliana Grassi (Milan: Il Polifilo, 1972), 10–11 (bk. I).

11. Leon Battista Alberti, *De pictura praestantissima, et numquam satis laudata arte libri tres absolutissimi* . . . (Basel: Westheimer, 1540), 1.12: "Ex quo illud dici solitum est visum per triangulum fieri cuius basis via quantitas cuiusve latera sunt iidem ipsi radii qui a punctis quantitatis ad oculum protenduntur." This sentence renders a repeated axiom of studies of vision and perspective; for Alhazen and Albertus Magnus, see Peter Theiss and Otto-Joachim Grüsser, "Vision and Cognition in the Natural Philosophy of Albert the Great (Albertus Magnus)," *Documenta Ophthalmologica* 86 (1994): 130. See also Witelo: "Omnium formarum uisibilium distincta uisio fit secondum pyramide[m], cuius uertex est in centro oculi, basis uero in superficie rei uisae, ex quo patet, omne quod uidetur sub angulo uideri." Witelo, *Vitellionis mathematici doctissimi (Peri optikes)* (Nuremberg: Hans Peterlein, 1535), 59 (3.18). In the published English translation, "pyramidem" is rendered as "cone": "Distinct vision of all visible forms takes place according to a cone, the vertex of which is in the center of the eye, while [its] base [is] in the surface of the visible thing." Witelo, *Witelonis perspectivae Liber secundus et Liber tertius*, ed. and trans. Sabetai Ungaru, Studia Copernicana 28 (Wrocław, Poland: Ossolineum, 1991), 3.18.125.

12. Plato, *Timaeus* 53C53c–55C55c.

13. Leon Battista Alberti, *Ludi matematici* (Milan: Guanda, 1980), dedicated to Meliaduse d'Este (1406–52), illegitimate son of Niccolò III d'Este. Pierre Souffiin makes the point that Alberti's presentation was neither consistently correct nor atypical of his time: Pierre Souffiin, "La *Geometria practica* dans les *Ludi rerum mathematicarum*," *Albertiana* I (1998): 87–104. On the basis of the unillustrated *De pictura* and *De architectura*, scholars assume Alberti was opposed to illustration—but manuscripts of the *Ludi* in fact contain numerous drawings. See Mario Carpo, *Architecture in the Age of Printing: Orality, Writing, Typography, and Printed Images in the History of Architectural Theory* (Cambridge, MA: MIT Press, 2001).

14. Jehane R. Kuhn, "Measured Appearances: Documentation and Design in Early Perspective Drawing," *JWCI* 53 (1990): 116. On surveying techniques as the basis for perspective, see also Martin Kemp, "Science, Non-science, and Nonsense: The Interpretation of Brunelleschi's Perspective," *AH* 1, no. 2 (1978): 134–61.

15. Sebastian Münster, *Rudimenta mathematica* (Basel: Petri, 1551), 53: "Per triangulum deprehenduntur locorum distantie."

16. Faustino Perisauli, *De triumpho stultitiae*, ed. A. Viviani and G. Fabbri (Florence: Il Fauno Editore, 1963), 74–77.

17. Bovelles, *Liure singulier*, 3v.

18. Evgeny A. Zaitsev, "The Meaning of Early Medieval Geometry: From Euclid and Surveyors' Manuals to Christian Philosophy," *Isis* 90, no. 3 (September 1999): 523. Zaitsev cites Plato's *Republic* 6 and Proclus's *Euclid's "Elements"* 11: "We must therefore posit

the soul as the generatrix of mathematical forms and ideas." But in the same text Proclus argued that mathematicals occupied a "middle ground." In medieval manuals, the distinction was often presented as one between practical geometry (measurement) and theoretical, or speculative, geometry. See, e.g., Thomas Bradwardine, *Geometria speculativa*, ed. and trans. George Molland (Stuttgart: Franz Steiner, 1989), 20–21. Bradwardine's text was published in Paris by Guy Marchant in 1495. Also Nicolaus Burtius, *Musices opusculum* . . . (orig. Bologna: Ugo Ruggeri, 1487; Bologna: Forni, 1969), h2r: "duplex geometria . . . scilicet theoricalis et practicabilis." A similar view is expressed in Hugh of St. Victor's *Practica geometriae*: "His breviter prelibatis, deinceps considerandum est quod omnis geometrica disciplina aut theorica est, id est speculativa, aut practica, id est activa." Hugh of St. Victor, *Geometriae practica*, no. 3, Praenotanda, 16–18, 39.

19. Plato, *Republic*, in *Collected Dialogues*, 6.510d; 7.526c–527c.

20. Leon Battista Alberti, *On Painting*, trans. John Spencer (New Haven, CT: Yale University Press, 1966), 43; trans. Cecil Grayson (orig. 1972; London: Penguin, 1991), 37.

21. Piero della Francesca echoes this in his *De Prospectiva Pingendi*, ed. G. Nicco-Fasola (Florence: Le Lettere, 2005), 63.

22. Twentieth-century art historians have overemphasized Alberti's metaphor of the picture as a view through a window into real space. See Puttfarken's account, *Discovery of Pictorial Composition*, 20–25, including discussion of Schapiro on frame as focusing device. As Stephen Campbell says in "The Art Seminar," the transcribed round-table discussion in James Elkins and Robert Williams, eds., *Renaissance Theory* (New York: Routledge, 2008): "Many of us find it exasperating to hear that Western representation is about the Albertian window, which came into being in the fifteenth century" (242).

23. Other theorists depend similarly on concrete metaphors. In explaining the visual pyramid, Filarete even refers to the account by Diodorus Siculus of the pyramids of Egypt (Filarete, *Trattato*, 649 [bk. 22]). The way they come together into the viewer's eye is like a "gabbia d'uccegli fatta di vermene, sottilissime, o vero uno cappello fatto di giunchi" (649).

24. Alberti, *De pictura* 2.44.

25. The image as proportional visual representation was already a key concept in Thomas Aquinas's discussion of the Trinity. To comment on Augustine's view that the *imago* of the Trinity was found in the human soul, Aquinas required a definition of *image*, and he defined it in terms of composition: "An image . . . represents a thing in a more well-defined manner [than a vestige], according to all its parts and the arrangement of its parts, from which it is possible to perceive something of the inner characteristics of a thing [*de interioribus rei*]." Thomas Aquinas, *Scriptum super libros Sententiarum Magistri Petri Lombardi*, ed. Pierre Mandonnet and Maria Fabianus Moos (Paris: P. Lethielleux, 1929–47), vol. 1, 1 *Sent.* D3.q3.a1.sol, cited in (with Merriell's translation) D. Juvenal Merriell, *To the Image of the Trinity: A Study in the Development of Aquinas' Teaching* (Toronto: Pontifical Institute of Medieval Studies, 1990), 43n9. Merriell also cites (44–45n13) Hilary of Poitiers's definition of the image, also used by Aquinas, by which the image is "rei ad rem coaequandae imaginata et indiscreta similitudo" (the imagined and undifferentiated similitude of a thing that co-adequates to the thing; Hilarius Pictaviensis, *Liber de synodis seu fide orientalium*, PL 10:490b-c). Elsewhere Aquinas specifies the necessity of proportion in the relationship between image and object as in "a small image, in which the proportion of the parts to one another is equal to that of the large object of which it is

the image." Cited in Merriell, 46 and 46n18: "ut patet quod in imagine parva, equalis est proportio partium ad invicem sicut in re magna cujus est imago" (1 *Sent.* D28.q2.a1.sol).

26. Alberti may have drawn inspiration from the similar triangles often illustrated diagrammatically in manuscripts of Aristotle's treatise *On Memory and Reminiscence.* Aristotle shows how memory stores information proportionally by using the principle of similar triangles. As with the theory of perspective, mental images are *small versions* of the objects in the world, proportional to them in the same way that similar triangles are, with time and distance used as examples. See John E. Sisko, "Space, Time and Phantasms in Aristotle, *De memoria* 2, 452B7–25," *Classical Quarterly* 47, no. 1 (1997): 167–75. In his 1556 commentary on Aristotle, Claude Champier explains that although mental images are much smaller than the things they represent, they correspond to them proportionally; he illustrates this with an equilateral triangle. Champier, *In librum Aristotelis de Memoria & Recordatione commentarii* (Paris: Sébastien Nivelle, 1556), 33–34.

27. Filippo Camerota makes this point in *La prospettiva del Rinascimento* (Milan: Mondadori, 2006), 80.

28. Alberti, *De pictura* 1.40. "Haec omnis diuidendi pauimenti ratio, maximè ad eam picturae partem pertinet, quam nos compositionem, suo loco, nominabimus."

29. In the ancient world space "was conceived as an aggregate or composite of solids and voids, both finite, and not as a homogeneous system." Erwin Panofsky, *Renaissance and Renascences in Western Art* (New York: Harper and Row, 1960), 122.

30. Gadol, *Leon Battista Alberti*, 75. Alberti's *Descriptio urbis Romae* also illustrates this orientation.

31. Piero defines design as "profili e contorni," in Piero della Francesca, *De prospectiva*, 63. For a returned exile like Alberti, this notion of things in their proper places might have had a special appeal.

32. Giovanni Paolo Lomazzo, *Trattato dell'arte de la pittura, scultura et architettura* (Milan: 1584), 22: "Oltre'di ciò ha anco d'usar il pittore queste linee proportionate con certo modo, & regola, laquale non è altro che quella che vsa, & con che procede l'istessa Natura in fare vn suo composito; doue prima presuppone la materia, che è vna cosa senza forma senza bellezza, & senza termine, e poi ne la materia introduce la forma, che è vna cosa bella, & terminata. Cosi fà il pittore ilqual piglia vna tauola, che ne la faccia non hà se non vna superficie o vn piano senza bellezza, le cui parti non hanno finì ne termini, & egli l'abbellisce, & termina delineando, & disegnando in lei vn'huomo, vn cauallo, vna colonna; & formando, o polindo tutti i suoi contorni: & in somma imitando con le linee la natura de la cosa che dipinge, cosi ne la larghezza, come ne la longhezza corpulenza, & grossezza." See also Frances Gage, "Giulio Mancini and Artist-Amateur Relations in Seventeenth-Century Roman Academies," in *The Accademia Seminars: The Accademia di San Luca in Rome, c. 1590–1635*, ed. Peter M. Lukehart (Washington, DC: National Gallery of Art, 2009). 247–87, on academic debates over the relevance of mathematics to painting in the late sixteenth–early seventeenth century.

33. Nicholas of Cusa, *Idiota de mente*, in *Nicholas of Cusa on Wisdom and Knowledge*, trans. Jasper Hopkins (Minneapolis: Arthur J. Banning, 1996), 571 (¶127–28); *De beryllo*, 827 (¶71), "to know is to measure." See also Bovelles, *Physicorum elementorum*, f. XLIIIIr-v.

34. Cusa, *De venatione sapientiae*, in *Metaphysical Speculations*, 1329 (¶80).

35. See also Luca Pacioli, *Summa de arithmetica . . .*, 3r: "E in q[ue]sto (commo dice il

seuerino Boetio) gra[n]dissimo [con]uenientia ue similitudine li numeri ha[n]no con li corpi creati eda natura p[ro]ducti . . ."

36. Ibid., 1r, introduction: "La quantita Magnanimo Duca: e si nobile & ecellente cosa:che molti philosophi per questo giudicata lanno alla substantia para: e concessa coeterna. Peroche hanno cognosciuto per ver[o] modo alchuna cosa in rerum natura senza lei no[n] potere existere." See also Gérard Roussel's dedication in Boethius, *Arithmetica*, 2v: "indeed, there is nothing divine or human, small or large, below or above, finally nothing so arduous or difficult that it will not have been anticipated by the clarity of numbers" ("adeo, nihil diuinu[m] siue humanum, nihil paruu[m] siue magnu[m], nihil inferius siue superius, nihil deniq[ue] tam arduu[m], tamq[ue] difficile: quod non sibi hoc numerorum ca[n]dore praesumperit").

37. Pacioli, *De divina proportione*, 3r: "alnumero naturale seco[n]do ogni sua diffinitione e ala mesura dela linea visuale."

38. See William Ivins, ed., *On the Rationalization of Sight, with an Examination of Three Renaissance Texts on Perspective* (New York: Da Capo, 1973).

39. Jean Pèlerin (Viator), *De artificiali perspectiva*, 2nd ed. (Toul: 1509), A1v: "Ou Nom du souuerain ouurier. Requis damis et studieux de choses subtiles et optables Le menre des viateurs plussierieusement et ouuertement coucher le tractie de perspectiue positiue naguieres mis sus: Considerant le prouerbe meritement trit: ung bien de quant est plus commun de tant est meilleur: et ung dit de quant est plus cler de tant est plus agreable: Et desirant faire plaisir et pareillement proffiter a tous & promouuoir de plus en plus beneditcions et prieres A propose resumer leuure et en latin et vulgar consecutiuement parcourir la matiere et a chcauns chapitres adapter propres et familieres figures faisants a lordre requis et plusample declaration dicelle Subiointes autres exemplaires auec briefz inductifz par chacun."

40. Ibid., A10v. "A la louenge du souureain artific dieu pere et saint esprit. Qui a perspicer & contempler la cite roiale de sa souuerainne maieste vueille conduire touz pacificz viateurs de la terre en laquelle comme noz peres sommes estraingiers et Pelerins. . . . Cellui qui a ce liure fait Prie pour touz de cueur parfait Et supplie treshumblement Prier pour lui pareillement." In the 1505 edition (unpaginated): "omnia ad laudem summi artificis dei patris et filii et spiritus sancti. Qui ad perspiciendam regiam gloriosissim[a]e majestatis su[a]e perducat omnes pacificos viatores terr[a]e in qua sicut patres nostri adven[a]e sumus et peregrini."

41. Ibid., B1v. "Signe reboutant lennemi Garde du corps de lame ami."

42. Ibid., C8v. "Et qui ne puet faire la voie / Son cueur par desir y envoie."

43. "Su[m]me faber reru[m] qui perspicis o[mn]ia solus/Ad te directo calle Viator eat. Amen." Ibid., C9r.

44. Viator is careful to state, "He who wishes to diminish [delimit the distances of, *limiter les distances* or, in Latin, *spacia diminuere*] a country landscape [*espace champestre*, Latin *planicies campestres*], by drawing the straight lines of a fictive underlying pavement, will accomplish his intention [*en pourtraiant les droites lignes de pauement sobzfaint, aura son intencion*] if he has not succeeded in doing it with some other geometral industry." Ibid., A6r.

45. See Alberti, *Peinture*, 234.

46. Leon Battista Alberti, *On the Art of Building in Ten Books*, trans. Joseph Rykwert, Neil Leach, and Robert Tavernor (Cambridge, MA: MIT Press, 1988), 241–42.

47. See Anne Dunlop, *Painted Palaces: The Rise of Secular Art in Early Renaissance Italy* (University Park: Penn State University Press, 2009), esp. chap. 2, "Art, Artifice, and the Rise of the Vernacular," 43–87.

48. Antoninus, bishop of Florence, *Summa theologica* (Verona: Augustinum Carattonium, 1740), col. 1285 (4.14.1.7). The sense of the passage is that the various forms of worldly knowledge (i.e., the seven liberal arts) can serve to make us understand the difference that exists between God and the world. Samuel Edgerton uses the phrase as the title of an essay, "*Mensurare temporalia facit geometria spiritualis*: Some 15th Century Italian Notions about When and Where the Annunciation Happened," in Irving Lavin and John Plummer, eds., *Studies in Late Medieval and Renaissance Painting in Honor of Millard Meiss* (New York: NYU Press, 1977), 115–30. He translates it incorrectly (in my view) as "spiritual geometry works to measure temporal things" (116), apparently assuming that spiritual geometry measures the temporal in a direct and practical sense. Antoninus instead argues that spiritual geometry demonstrates the limited nature and therefore inferiority of temporal things.

49. The confrontation between naturalism and the spiritual demands of art appears later in the fifteenth century, in Savonarola's critique of painters who give holy figures the faces and bodies of contemporary Florentines. See Girolamo Savonarola, *Prediche italiane ai Fiorentini*, ed. Francesco Cognasso (Perugia: La Nuova Italia, 1930), 3.1.391.

50. It is unclear why the comparison required the mirror; Kirsti Anderson suggests that the verification took place through a comparison of the mirror image of the painting with the mirror image of the building, thus relying on the authority of the mirror as an objective reference point. Anderson, *The Geometry of an Art: The History of the Mathematical Theory of Perspective from Alberti to Monge* (New York: Springer, 2007), 11.

51. "Egli è vero, niente di meno in disegno è vera, perché il disegno ancora lui non è cosa vera, anz'è una dimostrazione di quella cosa che tu ritrai o che tu vuoi dimostrare" (Filarete, *Trattato*, 657 [bk. 23]). I translate "vero" and "vera" as "correct" and "real" respectively to clarify the distinction in referent indicated by the gender of each adjective.

52. Maria Laura Cristiani Testi, "La Trinità di Masaccio, Unità di Cosmo," *Critica d'Arte*, ser. 4, 50, no. 4(1985), 110, citing Piero Sanpaolesi, *Brunelleschi* (Milan: Club del libro, 1962).

53. See Leon Battista Alberti, *Gli elementi di pittura,* ed. Girolamo Mancini (Cortona, Italy: Bimbi, 1864).

54. See for instance Ethan Matt Kavaler, "Renaissance Gothic: Pictures of Geometry and Narratives of Ornament," *AH* 29 (2006): 1–46.

55. Hellmut Wohl, *The Paintings of Domenico Veneziano, ca. 1410–1461: A Study in Florentine Art of the Early Renaissance* (New York: New York University Press, 1980), 37, citing Richard Offner, "The Barberini Panels and Their Painter," in *Medieval Studies in Honor of A. Kingsley Porter* (Cambridge, MA: Harvard University Press, 1939), 1:212–15, on the mid-fifteenth-century revival of a "Gothic mode of maintaining a decorative unity in the altarpiece by imposing on the composition a visible means of organization." For Offner, this implies an "intellectualization of form" (212), which accords well with the post-Council influence of Neoplatonic thinking.

56. Cennino D'Andrea Cennini, *The Craftsman's Handbook: The Italian "Il libro dell'arte,"* trans. Daniel V. Thompson Jr. (New York: Dover / Yale University Press, 1933), 43.

57. Symmetry as we understand it was practiced regularly, though when Renaissance

writers use the term they usually mean harmony of measurement rather than mirror-image symmetry. Jill Burke (*Changing Patrons: Social Identity and the Visual Arts in Renaissance Florence* [University Park: Penn State University Press, 2004], 134) notes the symmetrical order stipulated by patron Piero del Pugliese for the predella of the Innocenti *Adoration of the Magi*.

58. Pacioli, *De divina proportione*, 3r: "Se quella [music] recrea lanimo per larmonia. E questa [painting] per debita distantia e varieta de colori molto delecta. Se q[ue]lla suoi armoniche p[ro]portioni considera. E questa le arithmetici e geometrici."

59. John White, *The Birth and Rebirth of Pictorial Space* (London: Faber and Faber, 1957), 146.

60. Filarete, "guarda bene con l'occhio e con la mente," 644 (bk. 22); "gli occhi dell'intelletto," 650 (bk. 23).

61. On Protagoras, see Charles Trinkaus, "Protagoras in the Renaissance: An Exploration," in Edward P. Mahoney, ed., *Philosophy and Humanism: Renaissance Essays in Honor of Paul Oskar Kristeller* (New York: Columbia University Press, 1976), 190–213.

62. Martin Kemp, "Geometrical Perspective from Brunelleschi to Desargues: A Pictorial Means or an Intellectual End?" *Proceedings of the British Academy* 70 (1984): 89–132.

63. Ibid., 91–92.

64. Alberti, *De pictura* 1.32 (*recte* 35): "Condecens huius centrici puncti positio est non altius a iacenti linea quam sit illius pingendi hominis longitudo, nam hoc pacto aequali in solo et spectantes et pictae res adesse videntur."

65. Golsenne and Prévost, introduction and dossier in Alberti, *Peinture*, 34–37, 308–10.

66. Alberti, *On the Art of Building*, 29; Alberti, *De re aedificatoria opus elegantissimum et quam maxime utile* (Florence: Nicolaus Laurentii, 1485), c3r-v (1.12). Robert Tavernor discusses Alberti's use of *concinnitas* in terms of its basis in *numerus* in his *On Alberti and the Art of Building* (New Haven, CT: Yale University Press, 1998), 43–46.

67. See Leon Battista Alberti, *The Use and Abuse of Books* (*De commodis litterarum atque incommodis*), trans. Renée Neu Watkins (Prospect Heights, IL: Waveland, 1999).

CHAPTER THREE

1. Daniel Arasse, "Le regard de l'escargot," in *On n'y voit rien* (Paris: Denoël, 2000), 32–33.

2. God the Father has a triangular halo in the Bible of Borso d'Este, in the ca. 1505 *Coronation of the Virgin* at Santa Maria della Consolazione in Ferrara, and in a fifteenth-century Ferrarese manuscript of Petrarch's *Trionfi* (Bodleian MS. Canon. Ital. 83), as well as dal Cossa's *Annunciation*. Other examples include Antoniazzo Romano's altarpiece of the confraternity of the Annunciation, ca. 1489–90, Santa Maria sopra Minerva, Rome (where Mary also reads at a distinctly triangular lectern); a drawing by the Veronese artist Girolamo dai Libri, from before 1555, in the National Gallery in Washington; an *Antiphonary* by the Bolognese Master of the Barbazza Hours, Biblioteca Estense (Modena), Lat. 1010, f, 1r; an *Annunciation* by Giovanni Battista Cungi (from 1547) in the Museo Civico, Borgo Sansepolcro; and a sixteenth-century book of hours (use of Rome) written by Vincent Raimondi (Bodleian MS. Douce 19). This partial list provides some sense of the diversity of media (and concentration in central-northern Italy). See also an engrav-

ing by Wierix after Stradanus (published by Philips Galle), *The Holy Trinity* (before 1612), British Museum 1870,0625.732; an engraving of the *Coronation of the Virgin* by the Master of the Die, BM 1884,0412.13; an anonymous Italian drawing, BM 1946,0713.1444. In sculpture, the high altar of San Giorgio Maggiore in Venice, produced by Girolamo Campagna in 1593, represents God the Father with a triangular halo. The triangle appears as a stand-alone symbol for God in certain sixteenth–century images, such as an "all-seeing eye" figure that appears atop Pontormo's *Supper at Emmaus* and became popular in the seventeenth century. Scholars have questioned whether the triangle in Pontormo's painting is original, because of the eye in it; this *occhio divino* became a standard mode of representation for God only in the Counter-Reformation. Might the eccentric Pontormo have imagined the emblem to address questions of visual epistemology in the Emmaus story? John Shearman mentions a late sixteenth-century drawing after Pontormo's painting at the Hermitage showing a three-headed figure in place of the *occhio divino*, in his review of *Pontormo's Drawings* by Janet Cox-Rearick, *AB* 154, no. 2 (June 1972): 209–12. Note also a late sixteenth/early seventeenth-century Aegidius Sadeler (publ. Ph. Galle) *Annunciation*, and a print called *Godly Meditation* from England (ca. 1600) with God the Father replaced with a triangle labeled "Eternitie" (British Museum 1932,0709.4). On triangles and Padovanino, see Maria Loh, *Titian Remade: Repetition and the Transformation of Early Modern Italian Art* (Los Angeles: Getty Research Institute, 2007), 103–5. See also the alchemical-Kabbalistic images from ca. 1600 in Urszula Szulakowska, *The Alchemy of Light: Geometry and Optics in Late Renaissance Alchemical Illustration* (Leiden: Brill, 2000). A striking (later) architectural example is the pilgrimage church in Stadl Paura near Lambach, Austria.

3. James H. Beck (*Jacopo della Quercia* [New York: Columbia University Press, 1991], 90) is one of the few scholars to mention the triangular haloes in della Quercia's work. When the nave was completed at a height well above the original plans in the late sixteenth century, the new proportions (designed by Francesco Morandi, called Terribilia) were based on an equilateral triangle. See G. Dehio and G. von Bezold, *Die kirchliche Baukunst des Abendlandes* (Stuttgart: Arnold Bergsträsser, 1901), 2:528–33.

4. In the Vulgate, "Faciamus hominem ad imaginem et similitudinem nostram." Merriell, *To the Image of the Trinity*, 198–99.

5. Rupert of Deutz, *De trinitate et operibus eius*, PL 167 (Paris: Migne, 1854), Liber Primus (Genesis 1:1–23), 0199: "Quod extra se sancta trinitas non habuerit exemplar, ut conderet haec tria, coelum, terram et omnem ornatum eorum."

6. I thank Stephen Campbell for the suggestion that San Petronio was the key source for the use of the iconography in Ferrara.

7. Wolfgang Braunfels argues that the Throne of Grace came about as an illustration of the "Te Igitur" in the Canon of the Mass, a Eucharistic image that put God the Father in a priestlike position, sacrificing his Son. Braunfels, *Die Heilige Dreifaltigkeit* (Düsseldorf: L. Schwann, 1954), xxxv–xliii.

8. Otto Zöckler refers to the triangle as the oldest symbol of the Trinity, in "Religiöse Sinnbilder aus vorchristlicher und christlicher Zeit," pt. 3, "Trinitäts-Symbole," *Der Beweis des Glaubens*, n.f. 2 (1881): 297. He argues that its continuing use reflects the difficulty of representing the Trinity (300).

9. Augustine, letter 120, 2.7, in *The Works of Saint Augustine: A Translation for the 21st Century*, 2/2, *Letters 100–155*, trans. Roland Teske, ed. Boniface Ramsey (Hyde Park, NY: New City, 2003), 133.

10. Bernard McGinn, "Theologians as Trinitarian Iconographers," in Jeffrey F. Hamburger and Anne-Marie Bouché, eds., *The Mind's Eye: Art and Theological Argument in the Middle Ages* (Princeton, NJ: Princeton University Press, 2006), 186.

11. John of Damascus, *Three Treatises on the Divine Images*, ed. Andrew Louth (Crestwood, NY: St. Vladimir's Seminary Press, 2003), 29.

12. Ibid., 96–101.

13. Wyclif and the Lollards opposed the anthropomorphic depiction of the Trinity. W. R. Jones, "Lollards and Images: The Defense of Religious Art in Later Medieval England," *Journal of the History of Ideas* 34, no. 1 (January–March 1973), 30, 32.

14. McGinn, "Theologians," 195.

15. Ibid., citing Heinrich Seuse, *Leben Seuses*, chap. 53, in *Deutsche Schriften*, ed. K. Bihlmeyer (Stuttgart: W. Kohlhammer, 1907), 191.

16. Note Edgar Wind's discussion of Augustinian triads, *Pagan Mysteries in the Renaissance* (New Haven, CT: Yale University Press, 1958), 42–51.

17. Augustine, *On the Trinity*, in *A Select Library of the Nicene and Post-Nicene Fathers of the Christian Church*, ed. Philip Schaff (orig. 1887; Grand Rapids, MI: Eerdmans, 1980), 124 (8.10).

18. Ibid., 127–29 (9.4–5); 148–50 (11.5).

19. Ibid., 191–92 (14.12).

20. See Daniel Nodes, "Homeric Allegory in Egidio of Viterbo's Reflections on the Human Soul," *Recherches de Théologie et Philosophie Médiévales* 65, no. 2 (1998): 321. See also Brian Stock, *Augustine the Reader: Meditation, Self-Knowledge, and the Ethics of Interpretation* (Cambridge, MA: Harvard University Press, 1996).

21. Augustine, *On the Trinity*, 144–48 (11.1–4).

22. Jens Halfwassen, *Hegel und der spätantike Neoplatonismus* (Hamburg: Meiner, 1999).

23. See Peter Dronke, "Thierry of Chartres," in *A History of Twelfth-Century Western Philosophy* (Cambridge: Cambridge University Press, 1998), 383. Alain de Lille, *De articulis Catholicae fidei*, PL 210 (Paris: Migne, 1855), 0603B–0604A.

24. Egidio da Viterbo, a cardinal and general of the Augustinian order (from 1503), elaborated on Augustine's ideas in his commentaries on Peter Lombard's *Sententiae.* See Nodes, "Homeric Allegory," and J. W. O'Malley, "Egidio da Viterbo and Renaissance Rome," in *Egidio da Viterbo, O.S.A. e il suo tempo: Atti del V Convegno dell'Istituto Storico Agostiniano, Roma* (Rome: Analecta Augustiniana, 1983), 67–84.

25. Nicholas of Cusa, *De beryllo*, 806–7 (¶33).

26. For an overview of Nicholas of Cusa on mathematical theology, see J.-M. Counet, "Mathematics and the Divine in Nicholas of Cusa," in T. Koetsier and L. Bergmans, eds., *Mathematics and the Divine: A Historical Study* (Amsterdam: Elsevier, 2005), 273–90.

27. Nikolaus Stuloff argues for Byzantine sources for Cusanus's mathematical theology in Stuloff, "Mathematische Tradition in Byzanz und ihr Fortleben bei Nikolaus von Kues," *Mitteilungen und Forschungsbeiträge der Cusanus-gesellschaft* 4 (1964): 420–36.

28. On this topic see Michel Jeanneret, *Perpetuum mobile* (Paris: Macula, 1997).

29. Nicholas of Cusa, *De docta ignorantia* (*On Learned Ignorance*), trans. Jasper Hopkins (Minneapolis: Arthur J. Banning, 1981), "Mathematics assists us very greatly in apprehending various things divine" 1.18 (¶30).

30. Ibid., 1.33 (¶60). Any larger number or polygon with more sides "would be inconsistent with the simplicity and the perfection of the Maximum."

31. Nicholas of Cusa, *De theologicis complementis*, 747 (¶2).

32. Ibid.,747–48 (¶2).

33. Nicholas of Cusa, *De docta ignorantia*, 1.16 (¶26).

34. Nicholas of Cusa, *Cribratio Alkorani*, in *Nicholas of Cusa's "De pace fidei" and "Cribratio Alkorani,"* trans. Jasper Hopkins (Minneapolis: Arthur J. Banning, 1994), 2.1024 (¶9). This triad was later illustrated in Bovelles, *Liber de sapiente*, 133r.

35. Anonymous, *Vitis mystica*, PL 184 (Paris: Migne, 1854), 0707B.

36. Anne-Marie Lecoq, *François Ier imaginaire: Symbolique et politique à l'aube de la Renaissance française* (Paris: Macula, 1987), 398; the connection with lilies was made by Giovanni Ludovico Vivaldi de Monte Regali, and others, cited in ibid., 398–99; Lecoq discusses the triangular arms of France on medals, 299.

37. E.g., on the tomb slab of Giovanni d'Alamanno in the crypt of the Florentine Duomo.

38. Bovelles, *Liber cordis* (1519), in *In hoc opere Caroli Bovilli*... ([Paris]: Josse Bade, 1523), fol. XIr, "Trigona enim pyramis, cum sit post sphaeram solidarum & corporearum omnium figurarum, simplicissima & prima:haec nimirum in suae proprietate figurae, perlucida & clara diuinae Trinitatis vestigia seruat..."

39. Othloh of St. Emmeran, an eleventh-century monk who wrote on the "mysteries" of numbers, suggests a triangular figure with a cross superimposed on it as a tool for teaching parishioners about the Trinity: Monachus S. Emmerami Othlonus, *Dialogus de tribus quaestionibus*, PL 146 (Paris: Migne, 1853), 0111C. See also Thiofrid of Efternach (Thiofridus Efternacensis), *Flores epitaphii sanctorum*, PL 157 (Paris: Migne, 1854), 0358A, likening the Trinity to the triangle's status as first plane figure.

40. Garnerius of Rochefort (Lingonensis), *Sermones*, PL 205 (Paris: Migne, 1855), 0718D: "patet quod in trianguli ratione familiarius trinitatis invenies sacramentum. Sic igitur per triangulum in fide confirmamur..." See also Gerhoh of Reichersberg (Gerhohus Reicherspergensis), *Syntagma de Henricis IV et V et Gregorio VII*, PL 194 (Paris: Migne, 1855), 1450C: "Et quid hujus mysterio numeri pulchrius? In quo et trianguli species fidem sanctae trinitatis informat." Also Rupert of Deutz, *De trinitate*: "Ergo denarius secunda unitas triangula unam trinitatis unicae fidem significat" (0709A).

41. "Quodam osculo linearum... sic tres persone... se quodammodo osculantur." Right angles have neither the "generation" of the obtuse nor the "corruption" of the acute. Alain de Lille, "Sermo de Sphaera intelligibili," in Marie-Thérèse d'Alverny, ed., *Textes inédits avec une introduction sur sa vie et ses oeuvres* (Paris: J. Vrin, 1965), 305.

42. He cites one Claudianus (probably incorrectly): "De hoc triangulo ait Claudianus [Claudius Claudianus, or Claudien Mamert] in libro de Anima: Deus triangulum suum impressit mundo. Vere mundo Deus impressit triangulum, quia in eo trinitatis resultat vestigium" (ibid., 305 and 305n56).

43. Ibid., 306. "Hunc triangulum menti imprimamus. Habeamus potentiam resistendi vitiis, sapientiam in moribus ordinandis, benignitatem in sustinendis molestiis."

44. "Impossibile est beatam trinitatem et essentiae unitatem aptius a nobis repraesentari in exemplo creaturae sensibilis quam per res geometricas. Nam in solo triangolo inter omnes res factas invenitur unitas essentiae cum distinctione trium occupantium eandem essentiam." Roger Bacon, "Mathematicae in Divinis Utilitas," in Bacon, *Opus Maius*, ed. John Henry Bridges (Oxford: Clarendon, 1897), 218–19 (1.4).

45. The Uta Codex is Munich, BSB, Clm. 13601. In the Rothschild Canticles (Beinecke

Ms. 404), symbolic triangles appear on f. 42r and 79r. Jeffrey Hamburger discusses these images, the abstract image of God as triangle in the Breviary of Charles V (BN Ms. lat. 1288, f. 617r), and the triangle as a trinitarian symbol in *The Rothschild Canticles: Art and Mysticism in Flanders and the Rhineland* (New Haven, CT: Yale University Press, 1990), 132, plus figures 28, 49, and 209.

46. See John Tolan, *Petrus Alfonsi and His Medieval Readers* (Gainesville: University Press of Florida, 1993), 37–39, and Bernard McGinn, "Cabalists and Christians: Reflections on Cabala in Medieval and Renaissance Thought," in Richard H. Popkin and Gordon M. Weiner, eds., *Jewish Christians and Christian Jews* (Dordrecht: Kluwer Academic, 1994), 12–13.

47. See Alexander Patschovsky, "Die Trinitätsdiagramme Joachims von Fiore (†1202); Ihre Herkunft und semantische Strucktur im Rahmen der Trinitätsikonographie, von deren Anfängen bis ca. 1200," in *Die Bildwelt der Diagramme Joachims von Fiore: Zur Medialität religiös-politischer Programme im Mittelalter* (Ostfildern, Germany: Thorbecke, 2003), 55–114, esp. "Anlage," 106–14.

48. See ibid., esp. plates 33a–34, and McGinn, "Theologians."

49. Zöckler notes the iconography in a thirteenth-century French manuscript ("Religiöse Sinnbilder," 296). See also Michael Evans, "An Illustrated Fragment of Peraldus's *Summa* of Vice: Harleian MS 3244," *JWCI*, vol. 45 (1982), 14–68, and Kumler, *Translating Truth.*

50. Evans, "Illustrated Fragment," 24 and n67.

51. See Eamon Duffy, *Marking the Hours: English People and Their Prayers, 1240–1570* (New Haven, CT: Yale University Press, 2006), 139.

52. Bovelles may not have known Alberti's text, but he may well have probably had access to Jean Pèlerin's *De artificiali perspectiva*.

53. Pierre Magnard's introduction to Bovelles, *Livre du sage*, 20 and 43n32, citing Bovelles's *Divinae caliginis liber*.

54. See chapter 1, n39.

55. Thanks to Michael I. Allen for help with this translation. Bovelles, *Ma. Corp.*, in *Que hoc volumine*, f. 189r: "Hec mathematicam assurectionem proponit; quam tamen his relinquimus perscrutanda[m] qui ex sensibilibus ac mathematicis signis: arcana / profundaque mente: sublimes intelligibilium rerum veritates / capescere sufficiunt." The *assurectio* (a type of mathematical explanation) is also discussed in Bovelles's *Divinae caliginis liber.* See Joseph M. Victor, *Charles de Bovelles, 1479–1553: An Intellectual Biography* (Geneva: Droz, 1978).

56. Cusa, *De beryllo*, 796 (¶10). Bovelles, *Liber de sapiente*, 119v (*recte* 117v).

57. J. N. Hillgarth, *Ramon Lull and Lullism in Fourteenth-Century France* (Oxford: Clarendon, 1971), 284.

58. Augustin Renaudet, *Préréforme et humanisme à Paris pendant les premières guerres d'Italie, 1494–1517* (Paris: Librairie d'Argences, 1953), 410, 506.

59. "Trinitas lumen luminum / illumina cor hominum, / tuam similitudinem / reformans ad imaginem." See Kathryn Ann Smith, *Art, Identity and Devotion in Fourteenth-century England: Three Women and Their Books of Hours* (London: British Library, 2003), 105, and Franz Joseph Mone, ed., *Lateinische Hymnen Mittelalters* (Freiburg im Breisgau, Germany: Herder'sche Verlagshandlung, 1854), 2:46.

60. Franz Winzinger ("Albrecht Dürers Münchener Selbstbildnis," *Zeitschrift für*

Kunstwissenschaft 8 [1954]: 43–64) also argued that Giotto's *Blessing Christ* in the Arena Chapel and Jan van Eyck's Holy Face panel followed this system. According to this argument, Dürer drew on ideas derived ultimately from Nicholas of Cusa—to extend the ideal proportions embodied in Christ to his own image as *imago Dei.*

CHAPTER FOUR

1. Focusing on the tavola quadrata, John Canaday claims that triangular composition arose from increasing emphasis on the altarpiece. "The altar itself is the climax of the symmetrical scheme of a church" (*What Is Art?* 175). But an altarpiece capping a dramatic view down the nave of a church was atypical; most were in small chapels or hidden behind choir screens that broke up the long view. Triangular composition might echo the line of sight down a nave but would rarely have provided its climax.

2. White, *Birth and Rebirth of Pictorial Space*, 170.

3. See Megan Holmes on the tavola quadrata: *Fra Filippo Lippi the Carmelite Painter* (New Haven, CT: Yale University Press, 1999), 121–28 (as influenced by Alberti, 121) and 210.

4. See Rona Goffen, " 'Nostra conversatio in caelis est': Observations on the *sacra conversazione* in the Trecento," *AB* 61, no. 2 (June, 1979): 198–222, for some triangularly composed works by Agnolo Gaddi and Orcagna.

5. Diana Norman, " 'In the beginning was the Word': An Altarpiece by Ambrogio Lorenzetti for the Augustinian Hermits of Massa Marittima," *Zeitschrift für Kunstgeschichte* 58, no. 4 (1995): 478–503.

6. Uta Neidhardt and Christoph Schölzel, "Jan van Eyck's *Dresden Triptych*," in Susan Foister and Delphine Cool, eds., *Investigating Jan van Eyck* (Turnhout, Belgium: Brepols, 2000), 30, citing H. Gronau, "Zur Geschichte des van Eyckschen Flügelältarchens in der Dresdner Galerie," *Kunstchronik*, n.f. 27 (1916): col. 445–46.

7. On Heymericus, see Antony Black, *Council and Commune: The Conciliar Movement and the Council of Basle* (Shepherdstown, WV: Patmos, 1979), 58–84; Michael Evans, "Geometry of the Mind," *Architectural Association Quarterly* 12, no. 4 (1980): 32; Maarten Hoenen, "Academics and Intellectual Life in the Low Countries: The University Career of Heymeric de Campo (†1460)," *Recherches de Théologie Ancienne et Médiévale* 61 (1994): 173–209; Klaus Reinhardt, "Werke des Heymericus de Campo (+1460) im Codex Cusanus," *Traditio* 50 (1995): 295–310; Christel Meier, "Figura ad oculum demonstrativa: Visuelle Symbolik und verbale Diskursivitat nach Heymericus de Campo," in Nine Miedema and Rudolf Suntrup, eds., *Literatur, Geschichte, Literaturgeschichte* (Frankfurt: Peter Lang, 2003); Heymericus de Campo, *Opera selecta*, ed. and trans. Ruedi Imbach and Pascal Ladner (Freiburg, Germany: Universitätsverlag, 2001).

8. See Antony Black, *Council and Commune: The Conciliar Movement and the Council of Basle* (Shepherdstown, WV: Patmos, 1979), 35–37.

9. Antony Black, "The Realist Ecclesiology of Heimerich van de Velde," in *Church, State, and Community* (Aldershot, UK: Ashgate, 2003), 275–76.

10. See Florian Hamann's exhaustive study of the sigillum: *Das Siegel der Ewigkeit. Universalwissenschaft und Konziliarismus bei Heymericus de Campo* (Münster: Aschendorff, 2006), including his edition of Heymericus's *Disputatio de potestate ecclesiastica*, a development of the sigillum's applications to ecclesiology, 264–335.

11. Heymericus de Campo, "De sigillo eternatis," in *Opera selecta*, 99–100 (chap. 2).

12. Ibid., 101 (chap. 4).

13. Meier, "Figura ad oculum," 760–61.

14. See Hamann, *Das Siegel*, p. 338, fig. III, from Codex Cusanus 106, f. 187v.

15. Ibid., 197–99.

16. Joseph Gill, *The Council of Florence* (Cambridge: Cambridge University Press, 1959), 67, citing a letter by Traversari to Sigismund, no. 176 in Ambrogio Traversari, *Latinae epistolae . . .* (Bologna: Forni, 1968), vol. 2, col. 238.

17. See Antony Black, "Heimericus de Campo: The Council and History," *Annuarium Historiae Conciliorum* 2, no. 1 (1970): 78–86.

18. At the Council, Torquemada and Heymericus worked closely together to establish a position on the controversy surrounding the *Revelations* of St. Birgitta. See A. Fredriksson Adman, "Heymericus de Campo, *Dyalogus super Reuelacionibus beate Birgitte*: A Critical Edition with an Introduction," PhD diss., University of Uppsala, 2003, 24–27; Hamann, *Das Siegel*, 34–35. Like Torquemada, Heymericus also opposed the Hussites: Hamann, *Das Siegel*, 26; Adman, "Heymericus de Campo," 76.

19. The author now known as Pseudo-Dionysius claimed to be Dionysius of Areopagus, an obscure figure mentioned in the Gospel of St. Luke; French theologians often associated him with St. Denis, which lent a nationalist tinge to their interest in his ideas. Georges Didi-Huberman has persuasively used Dionysius's writings to explain the devotional function of the fictive marbles painted by Fra Angelico. See Didi-Huberman, *Fra Angelico: Dissemblance and Figuration*, trans. Jane Marie Todd (Chicago: University of Chicago Press, 1995). I focus rather on the more public figurative devotional image and the political uses of Dionysian ideas.

20. See Antony Black, *Monarchy and Community: Political Ideas in the Later Conciliar Controversy, 1430–1450* (Cambridge: Cambridge University Press, 1970), 62–65, for a clear exposition of the centrality of Neoplatonic emanationism to Torquemada's political arguments.

21. At the Knights of Columbus Vatican Film Library in St. Louis, I consulted the microfilms of twenty-six late medieval Vatican manuscripts of Aristotle's selected works in Latin that include the *De memoria et reminiscentia*. Of the sixteen that had the diagram, four were mountain-style, eleven were shield-style, and one had both.

22. Hamann, *Das Siegel*, 209–12; Black, "Heimericus de Campo," 78–79.

23. Latin text cited in Ingrid D. Rowland, "Render unto Caesar the Things Which Are Caesar's: Humanism and the Arts in the Patronage of Agostino," *Renaissance Quarterly* 39, no. 4 (Winter 1986): 710n126. See also Marsilio Ficino, *Platonic Theology*, trans. Michael J. B. Allen, ed. James Hankins (Cambridge, MA: Harvard University Press, 2004), 4:137 (13.2): the highest souls next to God are "twelve idola of the twelve souls moving the twelve spheres."

24. Jean Thenaud, *Cabale metrifiee*, BNF ms. fr. 882, f21r.

25. Jean Thenaud, *Troys resolutions et sentences,* Oesterreichische Nationalbibliothek Cod. 2645, 3v: "deux triangles moult semblable et conformes a ceulx de gloire et triumphe qui sont es cieulx surimperiaulx esquelz estoyent six escuz de liz. Desquelz resultoit le tresparfaict plusque pur & argente quadrangle de pardurable amour et inseparable unyon royalle."

26. Lecoq, *François Ier imaginaire*, 400–409. The upper triangle—with "Father" and

"Son" at the same height—might reflect the fact that Louise was still exercising authority over her son.

27. Nicholas of Cusa, *De docta ignorantia* 1.21 (¶64–66). Florian Hamann argues in "Koran und Konziliarismus: Anmerkungen zum Verhältnis von Heymericus de Campo und Nikolaus von Kues," *Vivarium* 43, no. 2 (2005): 276–91, that Heymericus was not (as was previously believed) Nicholas's teacher but that the two kept up an ongoing intellectual exchange over many years. On their relations and joint interest in Llull, see Eusebio Colomer, *Nikolaus von Kues und Raimund Llull aus Handschriften der Kueser Bibliothek*, Quellen und Studien zur Geschichte der Philosophie 2 (Berlin: Walter de Gruyter, 1961); Eusebio Colomer, "Nikolaus von Kues und Heimeric van den Velde," *Mitteilungen und Forschungsbeiträge der Cusanus-gesellschaft* 4 (1964): 198–213; and Ruedi Imbach, "Das Centheologicon des Heymericus de Campo und die darin enthaltenen Cusanus-reminiszenzen: Hinweise und Materialien," *Traditio* 39 (1983): 466–77. See also Hamann, *Das Siegel*, 50–54.

28. He was known later as the Hercules of the Eugenians (after Pope Eugenius IV). Peter L. McDermott ("Nicholas of Cusa: Continuity and Conciliation at the Council of Basel," *Church History* 67, no. 2 [June 1998]: 254–73) argues that Cusa's support of the pope was not a real departure from his prior outlook.

29. Nicholas of Cusa, *De docta ignorantia* 1.12 (¶34).

30. Nicholas of Cusa, *De theologicis complementis*, 747 (chap. 2).

31. Nicholas of Cusa's desire for union accorded with his general ideas about the reconciliation of faiths and had a practical, scholarly aspect: contact with the Greeks fostered intellectual exchange and study. His Platonism involved deep engagement with the Greek author himself, by contrast with Heymericus's reliance on Dionysius (and skepticism about Plato).

32. Bessarion, *In calumniatorem Platonis* 2.8.21. He cites the Pythagoreans, who, he argues, believed nothing was more apt than mathematics for "establishing the memory of divine things." Charles de Bovelles echoed this idea in the early sixteenth century; see chapter 1, n39.

33. Plato, *Epinomis*, in *Collected Dialogues of Plato*, 977c–978a. Bessarion, *In calumniatorem Platonis* 4.12.1–2. Also cited by Guillaume Postel, *Des admirables secrets des nombres platoniciens*, ed. Jean-Pierre Brach (Paris: Vrin, 2001), f. 181r, 66.

34. Bessarion, *In calumniatorem Platonis* 4.12.1–2.

35. Ibid.

36. Nicholas of Cusa, *De theologicis complementis*, 755 (chap. 5).

37. Nicholas of Cusa, *De beryllo*, 818 (¶56).

38. Nicholas of Cusa, *De theologicis complementis*, 748 (chap. 2). This speculation is "the most delightful and most inexhaustible nourishment for the mind . . . the mind's movement from *that it is* toward *what it is*."

39. Nicholas of Cusa, *De coniecturis* 1.2.7, 1.10.52, in *Metaphysical Speculations*, 2:166, 187.

40. Ibid., 241 (2.5.154).

41. Nicholas of Cusa, *De possest*, 936 (¶44).

42. Ibid., ¶43.

43. Cassirer, *Individual and the Cosmos*, 152.

44. Alberti probably knew Cusanus, perhaps through their mutual friend Paolo Toscanelli, but nothing suggests they were close. This possibility is discussed in Karsten

Harries, *Infinity and Perspective* (Cambridge, MA: MIT Press, 2001), 68–69. Gadol made this connection in *Leon Battista Alberti*, 196n68.

45. Alberti, *De pictura* 2.47: "arte superficiem illam fonti amplecti."

46. See Catherine Lawless, "Myth, Ritual and Orthodoxy: Cosimo de' Medici and St Peter Martyr," *Cultural and Social History* 2, no. 3 (September 2005): 292–95; Allie Terry, "Politics on the Cloister Walls: Fra Angelico and His Humanist Observers at San Marco" (PhD diss., University of Chicago, 2005), 108–10.

47. See Carol Richardson, *Locating Renaissance Art* (London: Yale University Press, 2007), 32.

48. George Holmes, "How the Medici Became the Pope's Bankers," in Nicolai Rubinstein, ed., *Florentine Studies* (London: Faber and Faber, 1968), 357–80. For an example of possible Medici patronage in direct relationship to the Council of Florence, see Francis Ames-Lewis, "Domenico Veneziano and the Medici," *Jahrbuch der Berliner Museen* 21 (1979): 67–90.

49. Terry, "Politics," 135–36, and Terry, "Meraviglia on Stage: Dionysian Visual Rhetoric and Cross-Cultural Communication at the Council of Florence," *Journal of Religion and Theatre* 6, no. 2 (Fall 2007): 97, available at http://www.rtjournal.org/vol__6/no__2/terry.html (accessed September 28, 2010). Terry also speculates that Fra Angelico knew Traversari, Cosimo, and others before he became an Observant Dominican ("Politics," 81–84). Terry's examination of the way the cloister's library and frescoes functioned for lay users makes a key intervention to our understanding of the fifteenth-century convent and parallels my suggestions about the political significance of San Marco's altarpiece.

50. Other saints do float in "our" space, that is, on the columns that make up the frame, not depicted in the plate.

51. Anthony Molho, "The Brancacci Chapel: Studies in its Iconography and History," *JWCI* 40 (1977): 50–98, 322.

52. Noël Valois connects Fra Angelico and Torquemada, but on slim evidentiary grounds, in "Fra Angelico et le Cardinal Jean de Torquemada," Société Nationale des Antiquaires de France, ed., *Centénaire 1804–1904: Recueil de Mémoires* (Paris: C. Klincksieck, 1904), 461–70.

53. Daniel Arasse, "Annonciation/énonciation," *VS, Quaderni di studi semiotici* 37 (January–April 1984): 9.

54. My echo of Walter Benjamin's phrase on the politicization of aesthetics is intentional, though the specific political circumstances are quite different. Benjamin, "The Work of Art in the Age of Mechanical Reproduction," in *Illuminations*, trans. Harry Zohn (New York: Schocken Books, 1968), 244.

55. Jan van Eyck's *Van der Paele Madonna* has multiple vanishing points, but the primary ones can be located in the nether folds of Mary's robes, suggestively indicating her body as the source of divine incarnation. See also Puttfarken, *Discovery of Pictorial Composition*, 80.

56. The Virgin is the *templum* of the Trinity (in the hymn of the seven joys of Mary); also the room of the Trinity, "Trinitatis triclinium," in Adam of St. Victor's *Salve mater Salvatoris* and inscribed on the threshold of Fra Angelico's San Marco *Annunciation.* William Hood, *Fra Angelico at San Marco* (New Haven, CT: Yale University Press, 1993), 276.

57. Arasse, "Annonciation/énonciation," 10n13.

58. James Elkins, *The Poetics of Perspective* (Ithaca, NY: Cornell University Press, 1994), 146. Michael Kubovy (*The Psychology of Perspective and Renaissance Art* [Cambridge: Cambridge University Press, 1986], 3) points out that the inverted triangle echoes the triangular forms in the niche above.

59. One might compare the position of Danaë's legs in Jan Gossaert's painting, where a draped triangle of cloth echoes the unseen pubic triangle.

60. The first level's Latin text includes the painter's signature: "Opus Dominici de Venetiis | Mater Dei Miserere Mei | Datum est"; above, the center step contains the faint words "Ave Maria," but the oblique sides of the pedestal present pseudo-Hebrew (?) inscriptions; the third level is completely veiled with a Turkish (?) carpet. Joseph Archer Crowe and Giovanni Battista Cavalcaselle publish only the Latin in *A New History of Painting in Italy, from the II to the XVI Century* (London: Dent, 1909), 317.

61. Wohl, *Paintings of Domenico Veneziano*, 46.

CHAPTER FIVE

1. See Jean Louis Schefer, *Paolo Uccello, "Le Deluge"* (Paris: P.O.L., 1999). Eugenio Marino, *Il "Diluvio" di Paolo Uccello in S. Maria Novella ed il Concilio di Firenze (1439–1443): Saggio d'iconoteologia storica* (Pistoia, Italy: Centro Riviste della Provincia Romana, 1992). For Marino, the two arks represent the "bad" council of Basel (left) and the "good" council of Florence (right). The *triangolo-naufragio* is a central component of his descriptive vocabulary.

2. Harald Keller, *The Renaissance in Italy: Painting, Sculpture, Architecture* (New York: H. N. Abrams, 1969), 83.

3. Edgar Wind writes in "The Revival of Origen," in *The Eloquence of Symbols: Studies in Humanist Art* (Oxford: Clarendon, 1983), 42–55, of the revival of interest in the late antique heterodox author Origen. Origen believed the ark was a pyramid, the shape that appears in Ghiberti's depiction of the ark on the bronze Gates of Paradise on the Florentine Baptistery, as well as in Uccello's fresco. The more orthodox view was that it was a rectilinear three-story structure topped with a peaked roof. See also Brian Curran, *The Egyptian Renaissance: The Afterlife of Ancient Egypt in Early Modern Italy* (Chicago: University of Chicago Press, 2007), 66–67. Origen's *Homilies on Genesis* were in the library of S. Maria Novella in 1489, according to that year's inventory. Stefano Orlandi, *La biblioteca di S. Maria Novella in Firenze dal sec. XIV al sec. XIX* (Florence: Ed. Il Rosario, 1952), 29. Ghiberti's measured pyramid seems almost a version of the perspective grid, while Uccello is more interested in the poignant disorder unfolding between the two ordered bodies.

4. Robert Klein, "Pomponius Gauricus on Perspective," in *Form and Meaning*, trans. Madeline Jay and Leon Wieseltier (New York: Viking, 1979), 102–28. Jean Pèlerin (Viator) described it as *cornuta* in his *De artificiali perspectiva* (1505), A2v.

5. Alessandro Parronchi, "Punctum dolens," in *Studi su la dolce prospettiva* (Milan: A. Martello, 1964), 296–312. Parronchi also questions Klein's assertion that Alberti did not understand the distance point.

6. See Marsilio Ficino, *Commentary on Plato's Symposium on Love*, trans. Sears Jayne (Dallas: Spring, 2000), 141–45 (speech 6, chap. 18).

7. See, for instance, Marsilio Ficino, *De triplici vita* (*Three Books on Life*), ed. Carol Kaske and John Clark (Binghamton, NY: Medieval and Renaissance Texts and Studies, 1989).

8. Michael J. B. Allen, "Marsilio Ficino: Daemonic Mathematics and the Hypotenuse of the Spirit," in Anthony Grafton and Nancy Siraisi, eds., *Natural Particulars: Nature and the Disciplines in Renaissance Europe* (Cambridge, MA: MIT Press, 1999), 128.

9. Marsilio Ficino, *Platonic Theology*, trans. Michael J. B. Allen, ed. James Hankins (Cambridge, MA: Harvard University Press, 2004), 6:21 (17.2).

10. Karl R. Popper, "Addendum," in *The Open Society and Its Enemies* (Princeton, NJ: Princeton University Press, 1963), 1:319.

11. Allen, "Marsilio Ficino," 123.

12. Ibid., 125. See also Proclus, *Commentary on the First Book of Euclid's "Elements,"* regarding the "generative fertility" of the objects of Nous, producing mathematicals as "offspring" (5); meanwhile math provides the "specific numbers by which the periods of fruitful birth and of its opposite, unfruitfulness, can be discerned" (7; see also 19–21). On the contested topic of whether the mathematical figures preexist in the soul before birth and are *remembered* in the course of acquiring experience (a debate that pitted Plato against Aristotle, who believed mathematical figures were abstractions from experience), Proclus took Plato's side: "We must . . . posit the soul as the generatrix of mathematical forms and ideas. And if we say that the soul produces them by having their patterns in her own essence and that these offspring are the projections of forms previously existing in her, we shall be in agreement with Plato" (11).

13. Ficino wrote: "Many brides, with as many husbands, are made to copulate in public ceremonies. Strong passions are mixed with softer ones, and the rich mate with the poor, so that the whole city is made even-numbered after being odd-numbered, as wine is mixed with water to make a temperate drink" (*Opera omnia* [Turin: Bottega d'Erasmo, 1962], vol. 2, pt. 1, p. 1424). See also Roussel, commentary in Boethius, *Arithmetica*, 123v, describing the universal fecundity associated with the number.

14. George L. Hersey, *Pythagorean Palaces: Magic and Architecture in the Italian Renaissance* (Ithaca, NY: Cornell University Press), 38.

15. Ibid., 38, 104.

16. Proclus, *In Platonis Rem Publicam Commentarii* (Leipzig: B. G. Teubner, 1901), 2:47–48; *Commentaire sur la République*, trans. A. J. Festugière (Paris: Vrin, 1970), 2:155–56 (diss. 13). Citing Ficino, Allen describes the "triangular 'powers' of the human spirit" in "Marsilio Ficino," 121.

17. A text titled "De daemonibus," attributed to Michael Psellos and interpreted by Ficino, alters this typology of triangles: *Index eorvm, qvae hoc in libro habentvr. Iavmlichus de mysteriis . . .* (Venice: Aldus Manutius and Andrea Soceri, 1516), 50v. This idea was taken up by G. P. Valeriano Bolzani in his *Hieroglyphica* (orig. Lyon: Paul Frelon, 1602; New York: Garland, 1976), 418–19. See Leonardo Tarán, "Proclus and the Old Academy," in *Collected Papers* (Boston: Brill, 2001), 598–99. Also Proclus, as in previous note.

18. Ficino, *Platonic Theology*, 6:23 (17.2).

19. From Ficino's commentary on Plato's *Timaeus*, in *Platonis Opera omnia*, trans. Ficino (Venice, 1491), 246v, cited and translated by Allen, "Marsilio Ficino," 128, also appearing in Ficino, *Opera omnia*, vol. 2, pt. 1, p. 1452.

20. Allen, "Marsilio Ficino," 126.

21. Proclus, *In Platonis Rem Publicam*, 47–48; *Commentaire*, 155–56.

22. For Proclus, pseudo-Psellos, and Xenocrates, the equilateral triangle is the most equal; the scalene, the most unequal. The equilateral triangle represents God, the scalene

represents the human, and the in-between isosceles represents the daimonic. In Valeriano Bolzani's version, undoubtedly due to a misunderstanding of the Greek notion of the daimon, the scalene triangle becomes "demonic" and bad. Bolzani, *Hieroglyphica*, 419.

23. Cassirer, *Individual and the Cosmos*, 88–92. Most recently, Anita Albus (*The Art of Arts: Rediscovering Painting* [New York: Alfred A. Knopf, 2000]) viewed Bovelles as a crucial figure in Renaissance humanism. See my review (*AB* 84, no. 4 [December 2002]: 685–86).

24. Sven Dupré, "Printing Practical Mathematics: Oronce Fine's *De speculo ustorio* between Paper and Craft," in Alexander Marr, ed., *The Worlds of Oronce Fine: Mathematics, Instruments and Print in Renaissance France* (Donington, UK: Shaun Tyas, 2009), 66 (note that a misprint has the 1542 book as Fine's); in addition see Victor, *Charles de Bovelles*, 44.

25. Bovelles, *Liure singulier*, 7r, 8r: "LAngle droict est le plus noble, & le principal des angles . . ."; "Le cercle est la plus belle & plus noble figure . . ."

26. Ibid., 55v. "Et nest aucunement possible, que l'engin humain puist bien prouffiter en la philosophie & sciences des choses naturelles, sans laide des arts mathematiques: esquelles sont contenues plusieurs mystiques, sur lesquelles se sont fondez & reiglez les antiens philosophes, pour inuenter & descripre les occultes, proprietez de toutes choses naturelles. Car comme on dict en prouerbe philosophique: Species rerum sunt, vt species magnitudinum & numerorum. Cestadire que les especes des choses naturelles, sont comme les especes des quantitez & des nombres."

27. Bovelles, *Liber de sapiente*, in *Que hoc volumine*, 123r (*recte* 125r).

28. Ibid., 131v.

29. Ibid., 131r.

30. Bovelles, *Liber de sensu*, in *Que hoc volumine*, 32v. "Aures due / plus distant quem vtriusque oculi, hi quoque / plusquam ambe nares / he plusquam lingua:que coniuncta est atque indiuisa. Hi igitur sensus quattuor auditus / visus / olfacus / & gustus / qui in animalis capite sunt siti: haud inuenustam claudant (sed eversa) pyramidem." Bovelles's diagram of the senses, with its disembodied parts, suggests a kind of unsatisfying incoherence between the mathematical model and the physical reality, one that blossoms into a grotesque image in a woodcut in the *Geometrie pratique*.

31. Bovelles, *Liber de sensu*, 44r-v. Thomas Frangenberg studies relations among the senses in Bovelles in his "*Auditus visu praestantior*: Comparisons of hearing and vision in Charles de Bovelles's *Liber de sensibus*," in Charles Burnett, Michael Fend, and Penelope Gouk, eds., *The Second Sense: Studies in Hearing and Musical Judgment from Antiquity to the Seventeenth Century* (London: Warburg Institute, 1991), 71–94, esp. 83–84.

32. Bovelles, *Liure singulier*, 5r: "Ie dis doncques, que ces cinq sens honorables & haultains, situéz & organiséz en la teste, ne sont exemptz de figure Geometrique. Car lordre selon equel ilz sont naturellement posez & situez en la teste, garde & observe la belle figure triangulaire: laquelle est la premier, & la plus mystique de toutes les figures angulaires."

33. Bovelles, *Liber propriae rationis*, in *In hoc opere . . .*, LVIIIr. "Quemadmodum in sublimi pyramidis cono, concurrunt ea puncta, quae in ima basi numerantur: ita & in consustantiali mentis speculo, siue in potentia animi, tanq[uam] in apice & cono mundi: omnes mundan[a]e substanti[a]e, ad vnitatem & interrigimam coincidentiam reuocantur."

34. Bovelles, *Liber de sapiente*, 132v. This notion also appears in Bovelles's *Physicorum elementorum*, fols. LVr–LVIIr, where man is a triple eye and triple mirror of the world.

35. Bovelles, *Liber de sapiente*, 131v.

36. Pacioli, *De divina proportione*, 17v (*recte* 25v). "El suo nobilissimo membro exteriore cioe testa se ben si guarda se trouera formata in su la forma dela prima figura in le recte linee cioe trangula equilatera dicta ysopleuros posta per fondamento e principio de tutti li altri sequenti libri dal nostro Euclide nel primo luogo del suo primo libro." "[C]o[m] me vedi desegnato in dicta figura la cui quantita a noi no[n] e nota precisa ma solo q[ue] lla li egregii pictori lano dala natura reseruata ala gratia e albitrio delochio. E questa sia vna spe[cie?] dele p[ro]portioni irrationali qual p[er] numero non e possibile annominare. . . . Ma doue interuene la irrationalita dele proportioni cioe che p[er] alcu[n] mo[do] non se possono nominare per numero restano al degno arbitrio del prospectiuo qual con sua gratia le ha aterminare. Peroche larte imita la natura quanto li sia possibile. E se apo[n] to larteficio facesse q[ue]llo che la natura ha facto non chiamaria arte ma vnaltra natura totaliter ala prima simile cheverebe a essere lamedesima. Questo dico acio non vi dobiate marauegliare se tutte cose aponto non r[e]ndano ale mani delopefice peroche non e possibile. E di qua nasci che li sauii dicano le sci[enti]e e discipline mathematici esser abstracte e mai actualiter no[n] e possibile ponerle in esse visibili." Following a discussion similar to Alberti's on the "pinguior Minerva," he continues, "Lo faciamo [i.e., use imprecise vocabulary] perche non habiamo vocabuli piu proprii a exprimer lor co[n]cepti & cetera. E questo basti quanto ala proportionale diuisione del profilo dela testa humana debitamente formata lasciando el sup[er]fluo ala gratia delopefice co[m]me la tomba del ceglio e ponta del naso." Firenzuola's description of female beauty uses a triangle very much like this one to establish the proportions of the face in profile. Agnolo Firenzuola, *On the Beauty of Women*, trans. Konrad Eisenbichler and Jacqueline Murray (Philadelphia: University of Pennsylvania Press, 1992), 24–25. See Elizabeth Cropper, "On Beautiful Women, Parmigianino, Petrarchismo, and the Vernacular Style," *AB* 58, no. 3 (September 1976): 374–95.

37. Bovelles, *Liure singulier*, 2r.

38. "La Geometrie co[n]sidere les mesures, les quantitez & dimensions corporelles, lesquelles sont posees & situees au corps, & en toute chose solide & materiele" (ibid., 3v).

39. "Le corps se prent en Geometrie, non pour la substance du corps humain subiect & seruant à lame, mais pour toute mesure corporelle aiant trois dimensions, cestascauoir longueur largeur & profundite" (ibid., 5v).

40. Ibid., 5r: "La plaine, autrement dicte superficie, resemble par iuste proportion au nombre de trois: car pour le moins sont necessaires trois poinctz, pour clorre & fermer vne plaine. Au moindre champ de terre, quel quil soit, fault trois lisieres pour le fermer: comme il appert au triangle A B C."

41. The Nile also famously terminates in a delta, a triangular letter, and the triangle is also associated with the shape of the Great Pyramids. See J. Gwyn Griffiths, "Aristotle and Herodotus on Egyptian Geometry," *Classical Review*, n.s. 2, no. 1 (March 1952): 10–11. See Aristotle *Metaphysics* 981b. These authors were echoed in the writings of Plutarch and Proclus. Proclus cites the division of fields required by the flooding of the Nile, *Commentary on the First Book of Euclid's "Elements,"* 52. The account in Herodotus bears little resemblance to Bovelles's account. In it, the pharaoh Sesostris divided the land of Egypt so as to equalize access to good water. Each paid a certain tax. Measurement was devised not

because boundaries became troubled but because the river's flooding eroded particular parcels, making them smaller (hence requiring revaluation). Herodotus wrote of Sesostris that "the king used to send men to examine and to find out by measurement how much less the piece of land had become, in order that for the future the man might pay less, in proportion to the rent appointed: and I think that thus the art of geometry was found out and afterwards came into Hellas also." Herodotus, *The History*, trans. G. C. Macaulay, ed. Donald Lateiner (New York: Barnes and Noble, 2004), 2.110 (p. 113). Polydore Vergil, *On Discovery* (*De inventoribus rerum*), ed. Brian P. Copenhaver (orig. 1470; Cambridge, MA: Harvard University Press, 2002), 1.18, tells the same story, adding that Josephus argues the Hebrews, blessed with long life by God, had the extra time to invent geometry (146–49); this is echoed in the prelude to Johannes Cesarius Juliacensis's (John of Jülich's) epitome of Bovelles's *Introduction to Geometry*, in Jacques Lefèvre d'Étaples, ed., *Introductio . . . in arithmecam divi Severini Boetii* (n.p., 1507), f. E2v.

42. Bovelles, *Liure singulier*, 3r-v. "En lieu de pluye, pour la fertilité des champs, par la provide[n]ce de Dieu, le Nil chascun an en temps desté se deriue, & arrouse les champs, & quelque espace de temps demeure sur les terres. Puis qua[n]t il se retire, les lisieres & bornes des champs sont troublees & co[n]fondues, dont sortissoient anciennement grandes noyses & questions entre les AEgyptiens. Parquoy pour oster les controuersies populaires, fut ordonné par les Roys Daegypte, que par les prebstres lesquelz estoient oysifs & sans payer tribut aux Roys, fust trouué quelque art de si bien mesurer & bourner les cha[m]ps, que par le annuel desriuement du Nil, les champs ne fussent plus confondus ne troublez."

43. Bernard Palissy, *Œuvres*, ed. Fillon (Niort, France: L. Clouzot, 1888), 2:44 Priapus "a invente le partage des terres, afin que chacun cultivat sa part."

44. This painting was made as the cover of a portrait of Bishop Bernardo de' Rossi, now at the Museo Capodimonte in Naples, and likely served as a kind of enigmatic moral portrait of the bishop. Lorenzo Lotto also used the branching tree with fertile and infertile sides in a 1524 intarsia panel in Bergamo representing Cain's murder of Abel. See Diana Gaalis, "Concealed Wisdom: Renaissance Hieroglyphic and Lorenzo Lotto's Bergamo *Intarsie*," *AB* 72, no. 3 (September 1980): 363–75. Galis also discusses Lotto's possible acquaintanceship with Pierio Valeriano, author of the 1556 *Hieroglyphica*. See also Erwin Panofsky's *Hercules am Scheideweg*, Studien der Bibliothek Warburg 18 (Leipzig: Teubner, 1930) for the Pythagorean *Y* and its juxtaposition of moral alternatives, and Geoffroy Tory, "Y," in *Champ fleury* (orig. Paris, 1529; Yorkshire: S.R. Publishers, 1970), M1v–M3v, where he illustrates the split between virtue and vice). Tory offers fascinating, sometimes bizarre reflections on combined numerical and alphabetic symbolism.

45. Bessarion, *In calumniatorem Platonis* 3.15.4.

46. Pacioli, *De divina proportione*, 214r: "Cioe si commo indiuinis vna medesima substa[n]tia sia fra tre persone padre figlio e spirito sancto. Cosi vna medesima proportione de questa sorte sempre conuen se troui fra tre termini." See also Bovelles, *Liure singulier*, 5v: "A la semblance samblance & imitation de la treshaulte & tressaincte trinité diuine, n'y ha en toute science de mathematique que trois mesures, & corporelles dimensions: longueur, largeur, & profundité."

47. Thomas Aquinas, *Summa theologica* q.36, a.3, reply to question 1.

48. Augustine, *On the Trinity*, 156–57 (12.5).

49. Nicholas of Cusa, *De docta ignorantia* 2.9 (¶150).

50. Plutarch, *Moralia*, vol. 5, trans. Frank Cole Babbitt (Cambridge, MA: Harvard University Press, 1928), 56 (373F–374A).

51. Barbara Newman, *God and the Goddesses: Vision, Poetry, and Belief in the Middle Ages* (Philadelphia: University of Pennsylvania Press, 2003), 248.

52. Bovelles, *Liber de sapiente*, 133r.

53. Ibid., 118v–121r.

54. Ibid., 137r-v.

55. Proclus, *Commentary on the First Book of Euclid's "Elements,"* 131.

56. Murdoch, *Album of Science*, 47, figs. 37–38, and 379.

57. Boethius describes the original number, one, as "mother of all numbers" and the first triangular number (1, 3, 6, 10, 15, etc.). *Boethian Number Theory*, 141.

58. Bovelles, *Ma. Corp.* The text has been studied by P. M. Sanders, "Charles de Bovelles's Treatise on the Regular Polyhedra (Paris, 1511)," *Annals of Science* 41 (1984): 513–66. Sanders also addresses Luca Pacioli's use of the extreme and mean divisions of a line as symbolic of the Trinity—presented as evidence of Plato's knowledge of the Trinity (518).

59. See Bovelles, *Ma. Corp.*, 186v–188r. Similar language appears in Johannes Reuchlin's *On the Art of the Kabbalah* (*De arte cabbalistica*), trans. Martin and Sarah Goodman (Lincoln: University of Nebraska Press, 1983), 207, 209 (e.g., "numerum foecundum," "nasci"). For him, the tetragon is the figure of Form and the cube that of Matter. Bovelles's language extends Proclus's, which, however, he may not have known directly; Proclus's *Commentary on Euclid* was first printed in the West, in Greek and Latin, in 1533.

60. Bovelles, *Ma. Corp.*, 189r: "Trigonus igitur: suo se numero inuoluit . . ."

61. Ibid., 189r: "Ex his manifestum est trigoni ysopleuri potentiam esse trinam. . . . Trinus est: tria potest. Vt trinitatis insignitur nomine: ita et trium fecundus est triaque gignit."

62. Sanders, "Charles de Bovelles's Treatise," 559.

63. Ibid., 190v.

64. Ibid., 189v. The same idea appears in Bovelles's *Liure singulier*, 39v.

65. Marielene Putscher, "Das *Buch der Heiligen Dreifaltigkeit* und seine Bilder in Handschriften des 15. Jahrhunderts," in Christoph Meinl, ed., *Die Alchemie in der europäischen Kultur- und Wissenschaftsgeschichte*, Wolfenbütteler Forschungen 32 (Wiesbaden: Otto Harrassowitz, 1986), 151–78, 155.

66. Newman, *God and the Goddesses*, 240–44, and Putscher, 154–55.

67. An unusual example with even more complex geometry is Israhel van Meckenem's *Coronation of the Virgin*. This strange engraving offers an especially pronounced triangular composition with three nearly identical members of the Trinity and the Virgin in a sense imbricated within the Trinity (on the same gridded cloth). The composition as a whole tends to reverse expectations of hierarchy: if the angels are above the Trinity and the Trinity is above the Virgin, who is really being given the most honor?

68. Peter Kern, *Trinität, Maria, Inkarnation* (Berlin: E. Schmidt, 1971), 106, AH 32, 30: 3, 2.

69. Ibid., 132, 133, 136, *In Annuntiatione Beatae Mariae Viriginis*.

70. See Barbara Catherine Raw, *Trinity and Incarnation in Anglo-Saxon Art and Thought* (Cambridge: Cambridge University Press, 1997), fig. XVb (the AElfwine Prayerbook, BL Cotton Titus D xxvii, 75v), 150–60, and Rik van Nieuwenhove, *Jan van Ruusbroec, Mystical*

Theologian of the Trinity (Notre Dame, IN: University of Notre Dame Press, 2003), 50. Millard Meiss ("An Early Altarpiece from the Cathedral of Florence," *Metropolitan Museum of Art Bulletin* n.s. 12, no. 10 [June 1954]: 316) notes that a painting depicting a Virgin of intercession with Christ and God the Father was on the altar of the Trinity in the Florentine Duomo; he cites G. Poggi, *Il Duomo di Firenze* (orig. Berlin: Bruno Cassirer, 1909; Florence: Edizioni Medicea, 1988), CIX–CX, documents 371–72, 1006–16, 1019–36, 1056–61.

71. See Nigel Morgan, "The Coronation of the Virgin by the Trinity and Other Texts and Images of the Glorification of Mary in Fifteenth-Century England," in Nicholas Rogers, ed., *England in the Fifteenth Century* (Stamford, Lincolnshire, UK: P. Watkins, 1994), 223–41. The Latin hymn "Septem Gaudia beatae Mariae Virginis" (The Seven Joys of the Virgin), with text attributed to Philip de Grève, opens with the words "Virgo templum trinitatis." Mone, *Lateinische Hymnen*, 2:165.

72. Newman, *God and the Goddesses*, 283.

73. Several commentators have pointed out the "triangularity" of this depiction of Mary. Rona Goffen (*Giovanni Bellini* [New Haven, CT: Yale University Press, 1989], 60–62) argues that the geometric treatment severs the figure from the landscape; see more generally her "Icon and Vision: Giovanni Bellini's Half-Length Madonnas," *AB* 57, no. 4 (December 1975): 487–518. See also Peter Humfrey, introduction to Humfrey, ed., *The Cambridge Companion to Giovanni Bellini* (Cambridge: Cambridge University Press, 2008), 3.

74. Hans Belting discusses this painting briefly as an instance of synthesis between "medieval" icon and new, Renaissance, "painted poetry" (*Likeness and Presence: A History of the Image before the Era of Art* [Chicago: University of Chicago Press, 1994], 473–74). On Poussin's painting, see Erwin Panofsky, "Et in Arcadia Ego," in *Meaning in the Visual Arts* (Chicago: University of Chicago Press, 1993).

75. For Honorius Augustodunensis, the cow and donkey are melancholic animals. Honorius Augustodunensis, *De philosophia mundi*, PL 172 (Paris: Migne, 1854), 0055D: "et si in aliqua plus abundaverit ignis, cholerica nata sunt animalia ut leo; si terra, melancholica, ut bos et asinus."

76. Millard Meiss, *Painting in Florence and Siena after the Black Death* (Princeton, NJ: Princeton University Press, 1951), 132. See also Beth Williamson, *The Madonna of Humility: Development, Dissemination and Reception, c. 1340–1400* (Woodbridge, UK: Boydell, 2009).

77. It does not appear in the early Apocrypha. Pamela C. Berger, *The Goddess Obscured* (Boston: Beacon, 1985), 91.

78. Ibid., 93 and 161n24, citing L. Douët-d'Arcq, ed., *Chronique d'Enguerrand de Monstrelet* 5 (Paris: Madame Veuve Jules Renouard, 1861), 3–4. It appeared in church mural cycles (Berger, *Goddess Obscured*, 112–18) and receives a full-page illustration in the Hours of the Rohan Master.

79. Charles de Bovelles cites this in his *Proverbes et dicts sententieux avec l'interpretation d'iceux* (Paris: Guillaume Le Noir, 1557), 6r-v.

80. Bolzani, *Hieroglyphica*, 173: "Habet & hoc Ciconia commune cum superis, quod, vti dicebamus, bellum illi assidue cum Serpentibus est: quod quidem animantium genus ita prorsus terrestre est, vt humi repat, humi semper vel adhaerescat, vel in abditissmas se terrae latebras occultet. Ita animus qui mortalium delicias auersetur, terrenóque affectus humi propemodum obsolutos longe summoueat, & penitus amoliatur, per hieroglyphi-

cum huius alitis, quae Serpentem dilaceret, apte significabitur." Bolzani's text insists on the materiality of the earth: it is not "terra" but "humus" that he repeats three times—the same root as *humility*. Nicholas of Cusa refers to the etymology of *man* (*homo*) as from *humus*, the earth (*De coniecturis* 213 (2.6.101). Plutarch, *Moralia*, trans. Paul Clement (Cambridge, MA: Harvard University Press, 1969), 8:355 (670A), also states that the ibis (another wading bird) has the weight of a newborn child's heart at birth and forms an equilateral triangle with its bill and feet.

81. Pèlerin, *De artificiali perspectiva*, as in chapter 2, n44.

82. On the seven sorrows of Mary, see Sandro Sticca, *The Planctus Mariae in the Dramatic Tradition of the Middle Ages* (Athens: University of Georgia Press, 1988).

83. Saturnian men are also builders (of "magna aedificia"). Ramon Llull, "De astronomia" 1.1, in *Raimundi Lulli Opera Latina* 17, ed. M. Pereira and T. Pindl-Büchel (Turnhout, Belgium: Brepols, 1989), 108–9. The downcast melancholy temperament is associated with memory and mathematics. Saturn, also the paternal god of time who cannibalizes his children—time devouring, in the end, the products of time—carries a sense of solipsism and circularity, like the ouroboros, which is also often used as a symbol of time.

84. Raymond Klibansky, Erwin Panofsky, and Fritz Saxl, *Saturn and Melancholy: Studies in the History of Natural Philosophy, Religion, and Art* (London: Nelson, 1964), esp. 306–65. Carla Mazzio has followed up on this in "The Three Dimensional Self: Mathematics, Melancholia, Drama," in David Glimp and Michelle R. Warren, eds., *The Arts of Calculation: Quantifying Thought in Early Modern Europe* (New York: Palgrave, 2004), 39–65, also pointing out that Burton's *Anatomy of Melancholy* mathematics is both a cause of melancholy and a kind of therapy for it. See also T. Lynch, "The Geometric Body in Dürer's Engraving *Melencolia I*," *JWCI* 45 (1982): 226–32.

85. See Elkins, *Poetics of Perspective*, 167–70.

86. Daniel Arasse has argued that the figure of the melancholic Michelangelo-Heraclitus in Raphael's *School of Athens* is intended as the negative side of Platonic idealism (falling under the figure of Plato and corresponding to the figure of Diogenes opposite him as the negative of Aristotle's materialism). Arasse argues that for Raphael a negative element to Platonism became conceivable only in the course of making the painting. See Arasse, "La mémoire de Raphaël," in Wessel Reinink and Jeroen Stumpel, eds., *Memory and Oblivion: Proceedings of the XXIXth International Congress of the History of Art* (Dordrecht: Kluwer, 1999), 787–93.

87. Thomas More, *Utopia*, trans. Clarence Miller (New Haven, CT: Yale University Press, 2001), 22–23.

88. Emilio Sereni, *History of the Italian Agricultural Landscape*, trans. R. Burr Litchfield (Princeton, NJ: Princeton University Press, 1997). Bellini's *Madonna of the Meadow* appears on p. 151.

89. Denis Cosgrove addresses the issue of geometry in relation to the *terraferma* in his "The Geometry of Landscape: Practical and Speculative Arts in Sixteenth-Century Venetian Land Territories," in Denis Cosgrove and Stephen Daniels, eds., *The Iconography of Landscape: Essays on the Symbolic Representation, Design and Use of Past Environments* (Cambridge: Cambridge University Press, 1989), 254–75.

90. Joseph Koerner, "Representative Man," in *The Moment of Self-Portraiture in German Renaissance Art* (Chicago: University of Chicago Press, 1993), 187–202.

91. Hersey (*Pythagorean Palaces*, 104) says of Cesariano's "sadistic" Vitruvian Man:

"Geometrically he may be having an orgasm but anatomically he is on the rack." Artists also used perspective to emphasize, or echo, bodily harm; instances of this include Uccello (*The Battle of San Romano*) and Hans Baldung Grien (*Bewitched Groom*, in which a sharply angled perspective connotes sexual potency stolen by the witch).

92. Reisch's geometry man might be compared to the "zodiac man" in the same book; it is also a way of understanding and diagramming the body via a numerical system. Astrology had long been providing a mathematical order for the body in the form of the zodiac man that continued to appear in medical treatises, books of hours, and almanacs. For the best-known example, see Harry Bober, "The Zodiacal Miniatures of the *Très Riches Heures* of the Duke of Berry: Its Sources and Meaning," *JWCI* 11 (1948): 1–34. Personality, too, from a late medieval standpoint, was strongly influenced by both the balance (or imbalance) of the bodily humors and by the conjunctions of the stars. For authors like Ficino these two systems were musically synchronized; particular celestial bodies rule over the humoral types (Saturn, in the case of melancholics—the disposition to which scholars and artists were believed to be prone).

93. Stephen Greenblatt, "Murdering Peasants: Status, Genre, and the Representation of Rebellion," *Representations* 1 (February 1983): 1–29.

94. Some of his associates did, at least temporarily, hold radical views at least partly aligned with the peasants'. See ibid., 26n10, and Herbert Zschelletzschky, *Die "drei gottlosen Maler" von Nürnberg* (Leipzig: Seemann, 1975).

95. See Michael Camille, *The Gothic Idol: Ideology and Image-Making in Medieval Art* (Cambridge: Cambridge University Press, 1989), and Michael Cole, "Perpetual Exorcism in Sistine Rome," in Michael Cole and Rebecca Zorach, eds., *The Idol in the Age of Art: Objects, Devotions, and the Early Modern World*, (Farnham, UK: Ashgate, 2009), 57–76.

96. Vitruvius, *Ten Books on Architecture*, trans. Ingrid D. Rowland (Cambridge: Cambridge University Press, 1999), 55 (4.1): "The third type, which is called Corinthian, imitates the slenderness of a young girl."

97. Franz Winzinger, "Albrecht Dürers Münchener Selbstbildnis," *Zeitschrift für Kunstwissenschaft* 8 (1954): 44.

98. Vitruvius, *Ten Books*, 91 (7.5).

99. Bovelles, *De intellectu*, in *Que hoc volumine*, 12r: "Ait [Ramon Llull] enim in ipsa imaginatione sensibiles huismodi species defecari & euadere puriores. indeque transire cominus ad intellectum nudatas omni materia & prisca accidentium eluuie spoliatas. . . . Nichil enim intellectuale mundus gignit. In eo nil intellectualis spiritalisue nature natum est subsistere."

100. Roussel, commentary in Boethius, *Arithmetica*, 86r. "Porro trigoni universorum basim sua crassitie obtinet materies [sic]: turbinem autem & conum purus actus isque deus." Roussel explains that a triangle remains if the base is removed from the "cone," but not the cone from the base, 86v.

101. See n33 in this chapter.

CHAPTER SIX

1. Thanks to Stephanie Leitch for originally pointing out this image to me. Vespucci begins from the point of view of "we who inhabit Lisbon" and explains that "a perpendicular line, while we stand upright, if suspended from a point of the heavens exactly vertical,

hangs over our heads; but it hangs over them sideways. . . . An orthogonal triangle is thus formed, of which we have the right line." Amerigo Vespucci, *The Letters of Amerigo Vespucci, and Other Documents Illustrative of His Career*, ed. and trans. Clements Markham (London: Hakluyt Society, 1894), 51.

2. Beginning with Samuel Edgerton, *The Renaissance Rediscovery of Linear Perspective* (New York: Basic Books, 1975), esp. 91–123.

3. Philip Sohm, "Dürer's *Melencolia I*: The Limits of Knowledge," *Studies in the History of Art* 9 (1980): 13–32.

4. Archimedes was held to have been obsessed with mathematics with an intensity comparable to that of erotic desire: "under the lasting charm of some familiar and domestic Siren, he forget even his food and neglected the care of his person. . . . [W]hen he was dragged by main force . . . to the place for bathing or anointing his body, he would trace geometrical figures in the ashes, and draw lines in the oil with which his body was anointed, being possessed by a great delight, and in very truth a captive of the Muses." Plutarch, *Life of Marcellus*, in *Lives*, trans. Bernadotte Perrin (Cambridge, MA: Harvard University Press, 1961), 5:481 (17.6).

5. Bernard Palissy, *Recette veritable* [*par laquelle tous les hommes de la France pourront apprendre a multiplier et augmenter leurs tresors*], ed. Frank Lestringant and Christian Barataud (orig. 1563; Paris: Macula, 1996), 187. "Comment, dirent-ils, à l'homme? Et faut-il que nous obéissions et servions à l'homme, qui est si méchant et plein de folie? . . . Permettez-nous mesurer la tête de l'homme . . . et vous connaîtrez que l'homme n'a aucune ligne directe . . . en toutes ses parties."

6. Ibid. "Je n'y sus jamais faire une mesure assurée, parce que les folies qui étaient en ladite tête lui faisaient changer ses mesures. Adonc je fus confus, parce que je trouvais ladite tête tantôt d'une sorte, tantôt d'une autre, et combien qu'aucunes fois il y eût quelque apparence de lignes directes, ainsi que j'apprêtais mes outils pour les figurer, soudain et un un moment je trouvais que les lignes directes s'étaient rendus obliques, dont je fus fort étonné, voyant qu'il n'y avait aucune ligne directe en la tête de l'homme, à cause que sa folie faisaient fléchir toutes les lignes directes et les rendaient obliques."

7. Harries, *Infinity and Perspective*, 42.

8. Arasse, "Annonciation/énonciation," 9.

9. See Michael Fried, *Absorption and Theatricality: Painting and Beholder in the Age of Diderot* (Chicago: University of Chicago Press, 1980).

10. Arasse, "Annonciation/énonciation," 16.

11. Sometimes God the Father is also found in a position reminiscent of the vanishing point, as in some paintings by Pinturicchio (e.g., at Santa Maria Maggiore in Spello, where Pinturicchio also creates a competing perspective—in the round window on the left whose crossbars form a cross—keyed to the point of view of his inserted self-portrait).

12. Damisch, *Origin of Perspective*, 446. Arasse also cites Damisch's seminar on this topic in his "Annonciation/énonciation," 14n18.

13. Damisch, *Origin of Perspective*, 446.

14. Iamblichus (attributed), *Theology*, 45.

15. It has been argued that Alberti is speaking of a separate drawing entirely. See, e.g., Golsenne and Prévost, introduction to Alberti, *Peinture*, 14. The passage is unclear on whether the distance point is to be placed at the precise distance from the vanishing point that corresponds to the viewer's distance from the picture plane; this has provoked a great

deal of debate. Alessandro Parronchi calls this the *punctum dolens*, the sore point, of discussions of perspective. I agree with Parronchi that Alberti more likely misspoke rather than truly misunderstanding this system, but his discussion of it is strange—as often when he confronts the mathematical system with human vision. The diminutive "areola" (in Italian *picciolo spatio*) seems to insist on the subsidiary nature of the space (that is, adjacent to the primary pictorial surface). Edgerton also believes Alberti refers to a distance point positioned to one side in this "areola"; see 44–45 in *Renaissance Rediscovery.* Filarete for his part states clearly that the distance point represents the ideal viewing distance, and that the lines drawn diagonally through the floor tiles to the distance point correspond to the viewer's "visual rays." (Filarete, *Trattato*, 651–52 [bk. 23]).

16. Lyle Massey, "Configuring Spatial Ambiguity: Picturing the Distance Point from Alberti to Anamorphosis," in Lyle Massey, ed., *The Treatise on Perspective: Published and Unpublished* (Washington, DC: National Gallery of Art, 2003), 166.

17. Ibid.

18. Daniel Arasse, *L'annonciation italienne: Une histoire de perspective* (Paris: Hazan, 1999), 13.

19. Ibid., 177.

20. Edgar Wind, *Pagan Mysteries in the Renaissance* (New Haven, CT: Yale University Press, 1958), 106, 108.

21. Rebecca Zorach, "Love, Truth, Orthodoxy, Reticence, or What Edgar Wind Didn't See in Botticelli's *Primavera*," *Critical Inquiry* 34, no. 1 (2007): 190–224. Much of the following is excerpted and adapted from the historical portions of the essay, leaving aside the historiography with some regret. Though there is a patch of puzzling material loss in the lower right of the panel, my argument is not that the painting has been cut down but rather that the figure's absence from the composition is a meaningful one.

22. Wind, *Pagan Mysteries*, 112. Ficino, *Opera omnia*, vol. 2, pt. 2, p. 1559.

23. Gombrich, suggesting the painting is a kind of educational tool for the younger Lorenzo di Pierfrancesco de' Medici, cites a letter by Ficino to the young man that identifies the three graces as Mercury-Jupiter, Venus, and Sol (Apollo). For Gombrich, this poses a nagging question: if the graces are already equivalent to these three gods, why are Venus and Mercury depicted too?

24. Ernst Gombrich, "Botticelli's Mythologies: A Study in the Neoplatonic Symbolism of His Circle," *JWCI* 8 (1945): 57. Ficino, vol. 1, pt. 2, pp. 845–56. Another letter mentions the same three (vol. 1, pt. 2, pp. 861–62). The triad echoes Ficino's discussion of "Splendor, Youth, and Abundant Happiness" corresponding to the three higher senses (reason, sight, and hearing) in his *Commentary on Plato's "Symposium"* 5.2.

25. Ficino, *De triplici vita*, 125.

26. Ficino, *Commentary on Plato's "Symposium"* 1.4.

27. Gombrich also points out that Alberti recommended nine or ten figures as the maximum for a painting (and here, with Cupid as "not much of a person," we have eight and a half; Gombrich, "Botticelli's Mythologies," 48). Archival documents also equivocate about the number of figures. John Shearman published archival documents relating to the Medici collections in which a discrepancy appears between two copies (from 1499 and 1503) of a 1498 document: in the first there are nine figures, in the second, eight. See John Shearman, "The Collections of the Younger Branch of the Medici," *Burlington Magazine* 117, no. 862 (January 1975): 17.

28. For insights into Mantegna's painting I am indebted to Stephen Campbell's *The Cabinet of Eros: Renaissance Mythological Painting and the Studio of Isabella d'Este* (New Haven, CT: Yale University Press, 2004), chap. 4. See also Edgar Wind, *Bellini's Feast of the Gods: A Study in Venetian Humanism* (Cambridge: Harvard University Press, 1948).

29. Charles M. Rosenberg, *The Este Monuments and Urban Development in Renaissance Ferrara* (Cambridge: Cambridge University Press, 1997), 102. See also Rosenberg, "The Iconography of the Sala degli Stucchi in the Palazzo Schifanoia in Ferrara," *AB* 61 (1979): 377–83. Rosenberg's argument was brought to my attention by Iva Olah, "Tarocchi and the Depiction of Christian Princely Virtues in the Schifanoia Frescoes," MA thesis, Carleton University, 2005.

30. Gombrich, "Botticelli's Mythologies," 61.

31. See, e.g., Lauro Martines, *Power and Imagination: City-States in Renaissance Italy* (New York: Alfred A. Knopf, 1979), 158. In one sign of covert exercise of power, Lorenzo monitored voting in the governing council, while not explicitly intervening in it.

32. Gombrich, "Botticelli's Mythologies," 15.

33. Alberti's number of figures also fits the approximate number of men (eight to ten) in the small councils through which political business was conducted in Florence. For a concise discussion of the "small councils," see Martines, *Power and Imagination*, 148–61.

34. See Alison Brown, "Pierfrancesco de' Medici, 1430–1476: A Radical Alternative to Elder Medicean Supremacy?" *JWCI* 42 (1979): 85 and n18, 97 and n83.

35. See Ficino, letter to Lorenzo di Pierfrancesco de' Medici, October 28, 1482, in *Opera omnia*, vol. 1, pt. 2, pp. 845–46.

36. See Michael Rocke's *Forbidden Friendships: Homosexuality and Male Culture in Renaissance Florence* (Oxford: Oxford University Press, 1996). Lorenzo protected men of his entourage and seems to have encouraged, at least tacitly, expressions of homoeroticism in cultural productions (most notably Poliziano's *Orfeo* and other works). See ibid., 198.

37. Jacques Lacan, "Of the Gaze as Objet Petit a," in *The Four Fundamental Concepts of Psychoanalysis*, ed. Jacques-Alain Miller, trans. Alan Sheridan (New York: W. W. Norton, 1998), 92.

38. François Gantheret, "Du coin de l'oeil," *Nouvelle Revue de Psychanalyse* 44 (Autumn 1991): 117. "C'est latéralement, de coin, que nous vient la perception sauvage qui nous met en présence. . . . Mais Signorelli nous regardant regarder, nous sommes, du coin de l'oeil d'où nous le voyons nous voir voyant, dans la représentation iconique, byzantine, celle où, dans le même espace, se côtoient les dieux et celui qui n'est pas le spectateur, mais en présence."

39. Benjamin, "On Some Motifs in Baudelaire," in *Illuminations*, 155–200.

40. Lucien Goldmann, "Le sujet de la creation culturelle," in *Marxisme et sciences humaines* (Paris: Gallimard, 1970), 94–120.

41. Francis W. Kent, *Household and Lineage in Renaissance Florence: The Family Life of the Capponi, Ginori, and Rucellai* (Princeton, NJ: Princeton University Press, 1977). While scholars like Richard Goldthwaite have argued that the Florentine Renaissance saw a progressive diminution of family size to a twentieth-century-style nuclear family, Kent suggests that households were composites involving an extended kin group.

42. Thanks to Dawna Schuld for this insight.

43. Perisauli, *De triumpho stultitiae*, 77–79.

CHAPTER SEVEN

1. See, e.g., Willard McCarty, "The Shape of the Mirror: Metaphorical Catoptrics in Classical Literature," *Arethusa* 22, no. 2 (Fall 1989): 167n11.

2. Lomazzo, *Trattato*, 99. "Ritrouarono i Greci ad immitatione de i piú antichi la vera proportione venerabile, nellaquale non si poteua scorgere se non estrema bellezza, & leggiadria è la diedero nello specchio tria[n]gulare a Venere Dea della celeste bellezza dalla quale tutte le altre deriuano. Mà noi lasciando lo specchio la dimostraremo ne la figura tria[n]golare, & nel triangolo issopleuro, che è vn triangolo che hà due linee uguali; & la terza disuguale, & di forma piramidale."

3. This pyramidal form has a long history following Lomazzo. The *figura serpentinata* (strongly associated with Michelangelo) is opposed by André Félibien (*Conférences de l'Académie royale de peinture et de sculpture* [Paris: Frederic Leonard, 1669], 13), who finds it "extravagant," associating it with Florentine school of art (he prefers the Roman).

4. John Groves, *A Greek and English Dictionary* (Boston: Hilliard, Gray, 1834), 135. Aristophanes *Lysistrata* 1.64. The *Lysistrata* was first printed in Italy (Florence) in 1515 by Filippo Giunta. I thank Chris Faraone for suggesting that I investigate this meaning of the Greek letter delta. Thomas Inman (*Ancient Faiths Embodied in Ancient Names* [London: S.P., 1868–69], 827) suggests that along with the "Venus delta" (also the title of a novel by Anaïs Nin), the triangle relates to the "triple" male genital organs.

5. Vincenzo Cartaro, *Immagini* (Venice: Ziletti, 1571), 341.

6. Tacitus, *Histories* 2.2–3. Maximus of Tyre also wrote (8.8) that the statue resembles "a white pyramid, the substance of which is unknown." Cited in Philip H. Young, "The Cypriot Aphrodite Cult: Paphos, Rantidi, and Saint Barnabas," *Journal of Near Eastern Studies* 64, no. 1 (2005): 25n10.

7. Quintus Curtius (*Historiae Alexandri Magni*, ed. Carlo M. Lucarini [Berlin: Walter de Gruyter, 2009], 4.7) described the appearance of the Jupiter Ammon object as similar to an umbilicus ("Vmbilico maxime similis est habitus"). See also Edouard Naville, "The Origin of Egyptian Civilization," *Journal of the Royal Anthropological Institute of Great Britain and Ireland* 37 (January-June 1907): 213.

8. Cartaro, *Immagini*, 341.

9. Ibid., 348.

10. Ibid.

11. For this type see Angelo Geissen and Wolfram Weiser, *Katalog Alexandrinischer Kaisermünzen der sammlung des Instituts für Altertumskunde der Universität zu Köln*, Papyrologica Coloniensa 5 (Opladen, Germany: Westdeutscher Verlag, 1983), pt. 4 (Claudius Gothicus—Bleimünzen), 74–75, nn3192, 3193, 3194; also with a standing Nike, n3195.

12. On Tanit see F. O. Hvidberg-Hansen, *La déesse TNT: Une étude sur la réligion canaanéo-punique* (Copenhagen: Gad, 1982).

13. Lomazzo, *Trattato*, 471: "Lo specchio di Venere triangolare la quale è Dea che dà il desiderio & la facilità delle cose, significa la prudenza: percioche solo il prudende essempio in tutte le attioni di se stesso si come nello specchio si vede l'imagine propria."

14. Ibid., 23: "douesse sempre fare la figura piramidale, serpentinata, & moltiplicata per vno doi e trè." See also David Summers, *Michelangelo and the Language of Art* (Princeton, NJ: Princeton University Press, 1981), 411–14.

15. Lomazzo, *Trattato*, 471.

16. Tacitus, *Histories* 2.2–3. While the "umbilical" relationship between the Paphian Venus and the *figura serpentinata* is submerged in Lomazzo's account, it reemerges in Hogarth's "line of beauty": a spiral that produces a conical shell shape.

17. Thanks to Charles Dempsey for suggesting the connection of the triangular number 6 with Lomazzo's notion of the *figura serpentinata*.

18. Tamar Garb, *Bodies of Modernity: Figure and Flesh in Fin-de-siècle France* (London: Thames and Hudson, 1998).

19. Koerner, *Moment of Self-Portraiture*, 446.

20. A. Richard Turner registers a general view when he writes (of another example) that "it seems surprising that an important part of a major commission by a leading artist should derive from such a humble source as prints." A. Richard Turner, *The Vision of Landscape in Renaissance Italy* (Princeton, NJ: Princeton University Press, 1966), 211.

21. Rona Goffen, "Sex, Space, and Social History in Titian's *Venus of Urbino*," and Daniel Arasse, "The *Venus of Urbino*, or the Archetype of a Glance," both in Rona Goffen, ed., *Titian's "Venus of Urbino"* (Cambridge: Cambridge University Press, 1997), 63–90 and 91–107.

22. Goffen, "Sex, Space," 83; Arasse, "*Venus of Urbino*," 96.

23. In one of Charles Estienne's anatomical woodcuts, parallel floor lines converge on the obstetric model's crotch. On the *modi* and Estienne, see Bette Talvacchia, *Taking Positions: On the Erotic in Renaissance Culture* (Princeton, NJ: Princeton University Press, 2001).

24. Baldassarre Castiglione, *The Book of the Courtier*, trans. George Anthony Bull (Baltimore: Penguin Books, 1967), 108.

25. Filenio Gallo, "A Lilia," no. 41 in *Rime di Filenio Gallo*, ed. Maria Antonietta Grignani (Florence: Leo S. Olschki, 1973), 117.

26. Gallo, "A Safira," no. 62 in ibid., 248, and no. 205 (314).

27. This is a misleading term; see my discussion in "Milk," in Zorach, *Blood, Milk, Ink, Gold*.

28. Victor Burgin, "Perverse Space." in *In/Different Spaces: Place and Memory in Visual Culture* (Berkeley: University of California Press, 1996), 57–76.

29. Gianfrancesco Pico della Mirandola, *On the Imagination*, ed. and trans. H. Caplan (New Haven, CT: Yale University Press, 1930), 37.

30. Ibid., 45. False imaginations come from bodily humors, from "disposition of the body, which we obtain from our parents, from our native land, and from our manner of living" (53).

31. Aristotle, *De anima*; Symphorien Champier, *Periarchon* (Lyon: Trechsel, 1533), 62. "Sunt in coitu tria necessaria, appetites uidelicet imaginatiuus, spiritus, & humor."

32. On jealousy in literary contexts, see Erika Milburn, " 'D'Invidia e d'Amor figlia sì ria': Jealousy and the Italian Renaissance Lyric," *Modern Language Review* 97, no. 3 (July 2002): 577–91.

33. Juan Luis Vives, *The Passions of the Soul*, ed. and trans. Carlos Noreña (Lewiston, NY: Edwin Mellen, 1990), 81.

34. Benedetto Varchi, "De invidia," in *Opere* (Trieste: Sezione letterario-artistica del Lloyd austriaco, 1858–59), 2:586: "Tu sola senza modo e misura alcuna, anzi contra ogni uso e ragione umana impoverisci dell'altrui ricchezze, e vai ricca dell'altrui mendicità."

35. René Girard, *A Theater of Envy: William Shakespeare* (Oxford: Oxford University Press, 1991); see esp. 3–7 (where he defines envy more rigidly than I do).

36. Varchi defines *zelotypia* as "invidia di forma, o vero bellezza" (*Opere*, 2:571).

37. Ibid.: "un pauroso sospetto dell amante che la cosa amata, la quale egli non vorrebbe avere commune con alcuno, non faccia copia di sè a niuno altro." For usage of "fare copia di se," see Guido da Pisa, *I fatti d'Enea di frate Guido da Pisa*, ed. Francesco Foffano (Florence: Sansoni, 1900), 66n18, defining *dare copia*: "dare opportunità, dar modo. L'espressione e da avvicinare a l'altro, comunissima nel trecento, *fare copia di sé*." See also Paolo Costa and Francesco Cardinali, *Dizionario della lingua italiana* (Bologna: Fratelli Masi, 1821), 497: "Far il piacere di alcuno, o il tuo piacere, in senso disonesto, vale Concedersi all' altrui disoneste voglie, Fare copia di se." Ludwig Gottfried Blanc and Giunio Carbone, *Vocabolario dantesco* (Florence: Barbèra, 1890), definition of *puttana*, "Una donna che fa copia di sè a prezzo," 279.

38. Paleotti, *Discorso*, 156v: "Ecco avviene spesso che uno ricerca l'altro a fare copia di se stesso & lasciarsi ritrarre."

39. Varchi, *Opere*, 2:571: "la gelosia è una paura o sospetto che alcuno, il quale noi non voremmo, ne goda alcuna bellezza; e questo per due cagioni, o per goderla noi soli, o perchè la goda solo quegli cui volemo noi."

40. Ibid., 592: "grandissimi e simulatori e dissimulatori."

41. On artists as envious, see Mark Jacobsen, "The Meaning of Mantegna's Battle of Sea Monsters," *AB* 64, no. 4 (December 1982): 623–29.

42. Matteo Maria Boiardo, *Tarocchi*, ed. Simona Foà (Rome: Salerno, 1993), 68.

43. Daniel Arasse, " 'Petit pinceau deviendra grand': Parmigianino et la scène de Vulcain," in *Le sujet dans le tableau* (Paris: Flammarion, 1997), 93–105.

44. It was also copied, with an elaborate frame, by the etcher known as IΦV.

45. Vasari, *Le vite dei più eccellenti pittori, scultori e architetti* (Turin: Einaudi, 1986), 2.797.

46. Henri Zerner, *The School of Fontainebleau: Etchings and Engravings*, ed. Stanley Baron (London: Thames and Hudson, 1969), 17–21, and Zerner, "L'eau-forte à Fontainebleau: Le rôle de Fantuzzi," *Art de France* 4 (1964): 70–85.

47. See Arthur E. Popham, *Catalogue of the Drawings of Parmigianino* (New Haven, CT: Yale University Press, 1971), cat. 320; 107; 203. In the case of the Archimedes drawing (British Museum 1952,0121.66), the window has actually been pasted in over text around which the artist originally drew.

48. David Freedberg, *The Power of Images* (Chicago: University of Chicago Press, 1989), 365; H. Diane Russell, *Eva/Ave: Woman in Renaissance and Baroque Prints* (Washington, DC: National Gallery of Art, 1990). A state with Antonio Salamanca's address added to the scene with Vulcan and the two lovers exists (in effect this would be the second of four states): see Zoltan Karpati, *The Alchemy of Beauty: Parmigianino Drawings and Prints*, exh. cat. (Budapest: Museum of Fine Arts, 2009), 68, cat. 22.

49. Popham, *Catalogue of the Drawings,* cat. O.C. 54.

50. Registration number 1905,1110.20. Another drawing in the British Museum collection, 1895,0915.751, shows Vulcan leading the other gods toward the scene of Venus and Mars caught in his net. Popham, *Catalogue of the Drawings,* cat. 217; 254.

51. Markings on the verso leave little doubt that the two parts of the current drawing were in fact originally a single sheet. I am grateful to Pat Simons for her insights regarding these and other Parmigianino drawings.

52. On Mantegna, see Jacobsen, "Meaning," 628, and Creighton Gilbert, ed., *Italian Art, 1400–1600: Sources and Documents in the History of Art* (Englewood Cliffs, NJ: Prentice-Hall, 1980), 10–11. On Dürer and Raimondi see Lisa Pon, *Raphael, Dürer, and Marcantonio Raimondi* (New Haven, CT: Yale University Press, 2004). Artisans were viewed as especially envious, and before Lévi-Strauss (Vives, *Passions*, 81). Hesiod, *Works and Days*, lines 11–24: "And potter is angry with potter, and craftsman with craftsman, and beggar is jealous of beggar, and minstrel of minstrel" (Hesiod, *The Homeric Hymns and Homerica*, trans. Hugh G. Evelyn-White [Cambridge, MA: Harvard University Press, 1914], 4–5).

53. Gian Ludovico Masetti Zannini, "Rivalità e lavoro di incisori nelle botteghe Lafréry-Duchet e de la Vacherie," in *Les Fondations nationales dans la Rome pontificale* (Rome: Ecole française de Rome, 1981), 547–66. The case was "Romana suffocationis: Contra Michaelem Angelum incisorem et complices," Archivio di Stato di Roma, Tribunale Criminale del Governatore, *Costituti*, v. 255 (1577).

54. "Io puoi sono andato girando con la fantasia e mi son imaginato che questo sia stato un assasinamentso fatto alla bottega, la qual è invidiata da alcuni de l'arte" (Zannini, "Rivalità," 562). *Invidia* is mentioned repeatedly in other testimony, 563.

55. "Messer Geronimo fosse stato ammazzato per invidia per star lui in botega di messer Claudio, perché quella botega è invidiata dagl'altri perché veramente quella botega è la più grande di Roma et ha le più belle opere di Roma di quest'arte" (ibid., 560).

56. "Anzi era messer Antonio che contrafaceva tutte le stampe nuove de noi altri botegari, perché era un huomo così fatto et havea più il modo di noi altri" (ibid., 555). This episode is also discussed in my introduction to *The Virtual Tourist in Renaissance Rome: Printing and Collecting the Speculum Romanae Magnificentiae* (Chicago: University of Chicago Library, 2008), 15.

57. On Reverdy see Estelle Leutrat, *Les débuts de la gravure sur cuivre en France: Lyon, 1520–1565* (Geneva: Droz, 2007).

58. An evident seam that crosses Venus and Vulcan's legs is the result of the joining of two pieces of canvas. Damage at the canvas's edges also shows that the picture was originally slightly bigger. See Erasmus Weddigen, "The Works of Tintoretto: Sewn, Designed, Patched, and Cut—The Uncertainty of Canvas Measurements," in Miguel Falomir, ed., *Jacopo Tintoretto: Actas del Congreso Internacional* (Madrid: Museo Nacional del Prado, 2009), 157. For more bibliography see the catalogue entry by Robert Echols in Miguel Falomir, *Tintoretto* (Madrid: Museo nacional del Prado, 2007), 200–203. Echols cites Weddigen's 1996 complicated argument that identifies "the theme of the painting as Apollo's betrayal of the adulterous pair: the light of the morning sun entering the window is concentrated by the vase of water and directed to the mirror above the bed, which reflects it to another mirror in the front of the pictorial space (its reflection barely visible in the mirror above the bed . . .), which directs it to Vulcan's forge, alerting him to what is going on in the bedchamber." See Erasmus Weddigen, "Venus Vulkan und Mars: Die Inquisition der Informatik," *Jacomo Tentor F.: Myzelien zur Tintoretto-Forschung* (Munich: Scaneg, 2000).

59. For discussion of an erotic print that displays the devil's behind in a mirror, see Koerner, *Moment*, 351–52, and Kathy Biddick, "The Devil's Anal Eye: Inquisitorial Optics and Ethnographic Authority," in *The Shock of Medievalism* (Durham, NC: Duke University Press, 1998), 105–34, esp. 120–21. John Clarke discusses humorous popular representations of the story in Roman terracotta mold-made medallions from the Rhone Valley in

Looking at Laughter: Humor, Power, and Transgression in Roman Visual Culture, 100 B.C.–A.D. 250 (Berkeley: University of California Press, 2007), 169–72.

60. Michel de Montaigne, "Sur des vers de Virgile," in *Les Essais*, ed. Pierre Villey (Paris: Félix Alcan, 1923), 3:75–150, 118.

EPILOGUE

1. Marguerite de Navarre, *Heptameron*, ed. Renja Salminen (Geneva: Droz, 1991). As with gay men in early twentieth-century urban America, "privacy could only be had in public" (George Chauncey, *Gay New York: Gender, Urban Culture, and the Makings of the Gay Male World, 1890–1940* [New York: Basic Books, 1994]).

2. Castiglione, *Book of the Courtier*, 108.

3. See Alberti, *On the Art of Building*, and Richard Goldthwaite, "The Florentine Palace as Domestic Architecture," *American Historical Review* 77 (1972): 983. Palazzi housed multiple families at least into the Quattrocento, with "constant penetration of street life into the palace" and of different households into each other's rooms. Living arrangements tended to be transient, and "working space . . . was frequently shared among neighbors," as Sandra Cavallo argues in "The Artisan's *Casa*," in Marta Ajmar, ed., *At Home in Renaissance Italy* (London: Victoria and Albert Museum, 2006), 70, 71. See n41 in chapter 6, above.

4. Charles Burroughs (*The Italian Renaissance Palace Façade: Structures of Authority, Surfaces of Sense* [New York: Cambridge University Press, 2002], 44) cites Boccaccio's description of female confinement: women, "forced to follow the whim of their fathers, mothers, brothers, and husbands . . . spend most of their time cooped up within the narrow confines of their rooms." Boccaccio, *Decameron*, trans. David Wallace (Cambridge: Cambridge University Press, 1991), 46. Adrian Randolph notes the confinement and visibility of Mary in Crivelli's *Annunciation*: "Les seuils de l'expérience: L'*Annonciation* de Crivelli et le genre de la peinture," *Perspective* 4 (2007): 649–78; "Renaissance Genderscapes," in Joan Hartman and Adele Seeff, eds., *Attending to Early Modern Women: Structures and Subjectivities* (Newark: University of Delaware Press / London: Associated University Presses, 2006), 21–49.

5. Juan Luis Vives, *Education of a Christian Woman: A Sixteenth-Century Manual*, trans. Charles Fantazzi (Chicago: University of Chicago Press, 2000).

6. Ibid., 127–28.

7. Ibid., 126.

8. Ibid., 111.

9. Paleotti, *Discorso*, 161v: "Se altro lo [the painter] pregarà che voglia rubbare il ritratto, come si dice alla macchia, stando il pittore in una chiesa ò altrove con un libretto in mano, mostrando di leggere, & in effetto lavorando il ritratto, chi non dubitarà di fine cattivo & scelerato, e tanto piu pernicioso, quanto che il mezzo diseruirsi della chiesa a questo fine, è tanto odioso a Dio?"

10. Vives, *Education*, 131–32.

11. Vives, *Passions*, 86.

12. See, e.g., Lance K. Donaldson-Evans, *Love's Fatal Glance: A Study of Eye Imagery in the Poets of the École Lyonnaise* (University, MS: Romance Monographs, 1980), 37.

13. Dizain 186 of Maurice Scève's *Delie*, ed. and trans. I. D. McFarlane (Cambridge:

Cambridge University Press, 1966): "Mais quand ton oeil droit au mien se rencontre, Je suis contrainct de ma teste cliner." (See Donaldson-Evans, *Love's Fatal Glance*, 112.)

14. Hélisenne de Crenne, *Les angoysses douloureuses qui procèdent d'amours* (Paris, 1538), 102–3.

15. Ibid., 106, 108, 111, 115. On the gaze in this text, see also Colette Winn, "La symbolique du regard dans *Les angoysses douloureuses qui procèdent d'amours*," *Orbis Litterarum* 40, no. 3 (1985): 207–21. Winn argues that the author consciously manipulates the gaze as courtly topos, making it the central motif of the text. This gaze is one that seeks recognition in the other's gaze—not only the lover's but also the reader's.

16. Crenne, *Angoysses*, 110.

17. Burton, *Anatomy of Melancholy*, 674.

18. Ibid., 682.

19. Randolph, "Les seuils," 653, 668. "Renaissance Genderscapes," 22, 29.

20. Crenne, *Angoysses*, 105.

21. Ibid., 163–64.

22. Ibid., 188.

23. The two women are adorned not in their bodies but in their hair; the Latin *compta* refers to a "combed" or "braided" woman. The women's hair is reminiscent of styles fashionable in art (as in Michelangelo's *teste divine*), if not in life.

24. See Karen Jacobson, ed., *The French Renaissance in Prints from the Bibliothèque Nationale de France* (Los Angeles: Grunwald Center for the Graphic Arts, 1994).

25. Crenne, *Angoysses*, 102–3.

Bibliography

Abbreviations used:

AB	*Art Bulletin*
AH	*Art History*
BNF	Bibliothèque Nationale de France
JWCI	*Journal of the Warburg and Courtauld Institutes*
PL	Patrologia Latina

Ackerman, James S. "'Ars Sine Scientia Nihil Est': Gothic Theory of Architecture at the Cathedral of Milan." *AB* 31, no. 2 (June 1949): 84–111.

Adler, Daniel. "Painterly Politics: Wölfflin, Formalism, and German Academic Culture, 1885–1915." *AH* 27, no. 3 (June 2004): 431–56.

Adman, A. Fredriksson. "Heymericus de Campo, *Dyalogus super Reuelacionibus beate Birgitte*: A Critical Edition with an Introduction." PhD diss., University of Uppsala, 2003.

Alberti, Leon Battista. *De pictura praestantissima, et numquam satis laudata arte libri tres absolutissimi . . .* Basel: Westheimer, 1540.

———. *De re aedificatoria opus elegantissimum et quam maxime utile.* Florence: Nicolaus Laurentii, 1485.

———. *Gli elementi di pittura.* Ed. Girolamo Mancini. Cortona, Italy: Bimbi, 1864.

———. *Ludi matematici.* Milan: Guanda, 1980.

———. *On Painting.* Trans. John Spencer. New Haven, CT: Yale University Press, 1966.

———. *On Painting.* Trans. Cecil Grayson. Orig. 1972; London: Penguin, 1991.

———. *On the Art of Building in Ten Books (De re aedificatoria)*. Trans. Joseph Rykwert, Neil Leach, and Robert Tavernor. Cambridge, MA: MIT Press, 1988.

———. *La peinture*. Ed. and trans. Thomas Golsenne and Bertrand Prévost. Paris: Seuil, 2004.

———. *The Use and Abuse of Books (De commodis litterarum atque incommodis)*. Trans. Renée Neu Watkins. Prospect Heights, IL: Waveland, 1999.

Albus, Anita. *The Art of Arts: Rediscovering Painting*. New York: Alfred A. Knopf, 2000.

Allen, Michael J. B. "Marsilio Ficino: Daemonic Mathematics and the Hypotenuse of the Spirit." In Anthony Grafton and Nancy Siraisi, eds., *Natural Particulars: Nature and the Disciplines in Renaissance Europe*, 121–37. Cambridge, MA: MIT Press, 1999.

———. *Nuptial Arithmetic: Marsilio Ficino's Commentary on the Fatal Number in Book VIII of Plato's "Republic."* Berkeley: University of California Press, 1994.

Ames-Lewis, Francis. "Domenico Veneziano and the Medici." *Jahrbuch der Berliner Museen* 21 (1979): 67–90.

Andersen, Kirsti. *The Geometry of an Art: The History of the Mathematical Theory of Perspective from Alberti to Monge*. New York: Springer, 2007.

Anonymous. *Vitis mystica*. PL 184. Paris: Migne, 1854.

Antoninus, bishop of Florence. *Summa theologica*. Verona: Augustinum Carattonium, 1740.

Aquinas, Thomas. *Scriptum super libros Sententiarum Magistri Petri Lombardi*, vol. 1. Ed. Pierre Mandonnet and Maria Fabianus Moos. Paris: P. Lethielleux, 1929–47.

Arasse, Daniel. "Annonciation/énonciation." *VS, Quaderni di studi semiotici* 37 (January-April 1984): 3–17.

———. *L'annonciation italienne: Une histoire de perspective*. Paris: Hazan, 1999.

———. "La mémoire de Raphaël." In Wessel Reinink and Jeroen Stumpel, eds., *Memory and Oblivion: Proceedings of the XXIXth International Congress of the History of Art*, 787–93. Dordrecht: Kluwer, 1999.

———. " 'Petit pinceau deviendra grand': Parmigianino et la scène de Vulcain." In *Le Sujet dans le tableau*. Paris: Flammarion, 1997.

———. "Le regard de l'escargot." In *On n'y voit rien*. Paris: Denoël, 2000.

———. "The *Venus of Urbino*, or The Archetype of a Glance." In Rona Goffen, ed., *Titian's "Venus of Urbino."* Cambridge: Cambridge University Press, 1997.

Aristotle. *Metaphysics*. Ed. and trans. Hippocrates G. Apostle. Grinnell, IA: Peripatetic, 1979.

Arnheim, Rudolph. *Toward a Psychology of Art: Collected Essays*. Berkeley: University of California Press, 1966.

Augustine. *Letters 100–155*, vol. 2/2. Trans. Roland Teske, ed. Boniface Ramsey. Hyde Park, NY: New City, 2003.

———. *On the Trinity*. In *A Select Library of the Nicene and Post-Nicene Fathers of the Christian Church*, ed. Philip Schaff. Orig. 1887; Grand Rapids, MI: Eerdmans, 1980.

Bacon, Roger. *Opus Maius*. Ed. John Henry Bridges. Oxford: Clarendon, 1897.

Baldasso, Renzo. "Illustrating the Book of Nature in the Renaissance." PhD diss., Columbia University, 2007.

———. "La stampa dell'editio princeps degli Elementi di Euclide (Venezia, Erhard Ratdolt, 1482." In Lisa Pon and Craig Kallendorf, eds., *The Books of Venice / Il libro vene-*

ziano, 61–100. Venice: La Musa Talìa and Biblioteca Nazionale Marciana; New Castle, DE: Oak Knoll, 2009.

Baxandall, Michael. *Painting and Experience in Fifteenth-Century Italy: A Primer in the Social History of Pictorial Style.* Orig. 1972; New York: Oxford University Press, 1988.

Beck, James H. *Jacopo della Quercia. New York*: Columbia University Press, 1991.

Belting, Hans. *Likeness and Presence: A History of the Image before the Era of Art.* Chicago: University of Chicago Press, 1994.

Benjamin, Walter. *Illuminations*, trans. Harry Zohn. New York: Schocken Books, 1968.

Berger, Pamela C. *The Goddess Obscured.* Boston: Beacon, 1985.

Bessarion, Basilios. *In calumniatorem Platonis libri IV*, vol. 2 of Ludwig Moller, ed., *Kardinal Bessarion als Theologe, Humanist, und Staatsmann: Funde und Forschungen*. Aalen, Germany: Scientia-Verlag, 1967. (Orig. *Quellen und Forschungen aus dem Gebiete der Geschichte*, vol. 22. Paderborn: F. Schöningh, 1927.)

Biddick, Kathy. "The Devil's Anal Eye: Inquisitorial Optics and Ethnographic Authority." In *The Shock of Medievalism*, 105–34. Durham, NC: Duke University Press, 1998.

Black, Antony. *Council and Commune: The Conciliar Movement and the Council of Basle*. Shepherdstown, WV: Patmos, 1979.

———. "Heimericus de Campo: The Council and History." *Annuarium Historiae Conciliorum* 2, no. 1 (1970): 78–86.

———. *Monarchy and Community: Political Ideas in the Later Conciliar Controversy, 1430–1450.* Cambridge: Cambridge University Press, 1970.

———. "The Realist Ecclesiology of Heimerich van de Velde." In Black, *Church, State, and Community*. Aldershot, UK: Ashgate, 2003.

Blanc, Ludwig Gottfried, and Giunio Carbone. *Vocabolario dantesco.* Florence: Barbera, Bianci, 1852.

Bober, Harry. "The Zodiacal Miniatures of the *Très Riches Heures* of the Duke of Berry: Its Sources and Meaning." *JWCI* 11 (1948): 1–34.

Boccaccio. *Decameron.* Trans. David Wallace. Cambridge: Cambridge University Press, 1991.

Boethius. *Arithmetica, duobus discreta libris: Adjecto commentario* . . . Paris: Simon de Colines, 1521.

———. *Boethian Number Theory: A Translation of the "De institutione arithmetica."* Ed. and trans. Michael Masi. Amsterdam: Rodopi, 1983.

———. *Commentaria in Porphyrium.* PL 64. Paris: Migne, 1847.

———. *De consolatione philosophiae.* PL 63. Paris: Migne, 1847.

———. *De institutione arithmetica libri duo* . . . Ed. Gottfried Friedlein. Leipzig: B. G. Teubner, 1867.

Boiardo, Matteo Maria. *Tarocchi.* Ed. Simona Foà. Rome: Salerno, 1993.

Bolzani, G. P. Valeriano. *Hieroglyphica.* Orig. Lyon: Paul Frelon, 1602; New York: Garland, 1976.

Bouleau, Charles. *The Painter's Secret Geometry: A Study of Composition in Art.* New York: Harcourt, Brace and World, 1963.

Bovelles, Charles de. *In hoc opere Caroli Bovilli Samorabrini contenta: Liber cordis, Liber proprie rationis, Liber substantialium propositionum, Liber naturalium sophismatum, Liber cubicarum mensularum*. [Paris]: Josse Bade, 1523.

———. *Liure singulier et vtile touchant l'art et practique de geometrie*. Paris: Simon de Colines, 1542.

———. *Le livre du sage.* Ed. and trans. Pierre Magnard. Orig. 1511; Paris: J. Vrin, 1982.

———. *Physicorum elementorum libri decem.* [Paris]: Josse Bade, 1512.

———. *Proverbes et dicts sententieux avec l'interpretation d'iceux.* Paris: Guillaume Le Noir, 1557.

———. *Que hoc volumine contine[n]tur: Liber de intellectu, Liber de sensu, Liber de nichilo, Ars oppositorum, Liber de generatione, Liber de sapiente, Liber de duodecim numeris, Epistole complures, Insup[er] mathematicu[m] opus quadripartitu[m], De numeris perfecti, De mathematicis rosis, De geometricis corporibus, De geometricis supplementis*. Paris: Henri Estienne, 1510[=1511].

Bradwardine, Thomas. *Geometria speculativa.* Ed. and trans. George Molland. Stuttgart: Franz Steiner, 1989.

Braunfels, Wolfgang. *Die Heilige Dreifaltigkeit.* Düsseldorf: L. Schwann, 1954.

Bredekamp, Horst. "Gotterdammerung des Neoplatonismus." *Kritische Berichte* 14, no. 4 (1986): 39–48.

Brown, Alison. "Pierfrancesco de' Medici, 1430–1476: A Radical Alternative to Elder Medicean Supremacy?" *JWCI* 42 (1979): 81–103.

Burgin, Victor. "Perverse Space." In *In/Different Spaces: Place and Memory in Visual Culture*, 57–76. Berkeley: University of California Press, 1996.

Burke, Jill. *Changing Patrons: Social Identity and the Visual Arts in Renaissance Florence.* University Park: Penn State University Press, 2004.

Burkert, Walter. *Lore and Science in Ancient Pythagoreanism.* Trans. Edwin L. Minar Jr.. Cambridge, MA: Harvard University Press, 1972.

Burroughs, Charles. *The Italian Renaissance Palace Façade: Structures of Authority, Surfaces of Sense*. New York: Cambridge University Press, 2002.

Burtius, Nicolaus. *Musices opusculum Nicolai Burtij parmensis: musices professoris: ac iuris pontificij studiosissimi: musices opusculum incipit: cum defensione Guidonis aretini: aduersus quendam hyspanum veritatis preuaricatorem.* Bologna: Forni, 1969 (facs. of Bologna: Ugo Ruggeri, 1487).

Burton, Robert. *The Anatomy of Melancholy.* Ed. Floyd Dell and Paul Jordan-Smith. New York: Tudor, 1927.

Camerota, Filippo. *La prospettiva del Rinascimento.* Milan: Mondadori, 2006.

Camille, Michael. *The Gothic Idol: Ideology and Image-Making in Medieval Art*. Cambridge: Cambridge University Press, 1989.

Campbell, Stephen. *The Cabinet of Eros: Renaissance Mythological Painting and the Studio of Isabella d'Este*. New Haven, CT: Yale University Press, 2004.

Canaday, John. *What Is Art? An Introduction to Painting, Sculpture, and Architecture*. New York: Alfred A. Knopf, 1980.

Carpo, Mario. *Architecture in the Age of Printing: Orality, Writing, Typography, and Printed Images in the History of Architectural Theory*. Cambridge: MIT Press, 2001.

Carroll, Margaret D. *Painting and Politics in Northern Europe: Van Eyck, Bruegel, Rubens, and Their Contemporaries.* University Park: Penn State University Press, 2008.

Cartaro, Vincenzo. *Immagini.* Venice: Ziletti, 1571.

Cassirer, Ernest. *The Individual and the Cosmos in Renaissance Philosophy*. New York: Harper and Row, 1963.

Castiglione, Baldassarre. *The Book of the Courtier*. Trans. George Anthony Bull. Baltimore: Penguin Books, 1967.

Cavallo, Sandra. "The Artisan's *Casa*." In Marta Ajmar, ed., *At Home in Renaissance Italy*, 66–75. London: Victoria and Albert Museum, 2006.

Celenza, Christopher S. "Pythagoras in the Renaissance: The Case of Marsilio Ficino." *Renaissance Quarterly* 52, no. 3 (Autumn 1999): 667–711.

Cennini, Cennino D'Andrea. *The Craftsman's Handbook: The Italian "Il libro dell' arte."* Trans. Daniel V. Thompson Jr. New York: Dover / Yale University Prerss, 1933.

Champier, Claude. *In librum Aristotelis de Memoria & Recordatione commentarii*. Paris: Sebastianum Niuellium, 1556.

Champier, Symphorien. *Periarchon*. Lyon: Trechsel, 1533.

Chauncey, George. *Gay New York: Gender, Urban Culture, and the Makings of the Gay Male World, 1890–1940*. New York: Basic Books, 1994.

Clarke, John. *Looking at Laughter: Humor, Power, and Transgression in Roman Visual Culture, 100 B.C.–A.D. 250*. Berkeley: University of California Press, 2007.

Cole, Bruce. *Titian and Venetian Painting, 1450–1590*. Boulder, CO: Icon Editions, 1999.

Cole, Michael. "Perpetual Exorcism in Sistine Rome." In Michael Cole and Rebecca Zorach, eds., *The Idol in the Age of Art: Objects, Devotions, and the Early Modern World*, 57–76. Farnham, UK: Ashgate, 2009.

Colomer, Eusebio. "Nikolaus von Kues und Heimeric van den Velde." *Mitteilungen und Forschungsbeiträge der Cusanus-gesellschaft* 4 (1964): 198–213.

———. *Nikolaus von Kues und Raimund Llull aus Handscriften der Kueser Bibliothek*. Quellen und Studien zur Geschichte der Philosophie 2. Berlin: Walter de Gruyter, 1961.

Conley, Tom. *The Self-Made Map: Cartographic Writing in Early Modern France*. Minneapolis: University of Minnesota Press, 1996.

Cosgrove, Denis. "The Geometry of Landscape: Practical and Speculative Arts in Sixteenth-Century Venetian Land Territories." In Denis Cosgrove and Stephen Daniels, eds., *The Iconography of Landscape: Essays on the Symbolic Representation, Design and Use of Past Environments*, 254–75. Cambridge: Cambridge University Press, 1989.

Costa, Paolo, and Francesco Cardinali. *Dizionario della lingua italiana*. Bologna: Fratelli Masi, 1821.

Counet, J.-M. "Mathematics and the Divine in Nicholas of Cusa." In T. Koetsier and L. Bergmans, eds., *Mathematics and the Divine: A Historical Study*, 273–90. Amsterdam: Elsevier, 2005.

Crenne, Hélisenne de. *Les angoysses douloureuses qui procèdent d'amours*. Trans. Paule Demats. Orig. 1538; Paris: Les Belles Lettres, 1968.

Cropper, Elizabeth. "On Beautiful Women, Parmigianino, Petrarchismo, and the Vernacular Style." *AB* 58, no. 3 (September 1976): 374–95.

Crowe, Joseph Archer, and Giovanni Battista Cavalcaselle. *A New History of Painting in Italy, from the II to the XVI Century*. London: Dent, 1909.

Curran, Brian. *The Egyptian Renaissance: The Afterlife of Ancient Egypt in Early Modern Italy*. Chicago: University of Chicago Press, 2007.

Curtius, Quintus. *Historiae Alexandri Magni*. Ed. Carlo M. Lucarini. Berlin: Walter de Gruyter, 2009.

Damisch, Hubert. *The Origin of Perspective*. Trans. John Goodman. Cambridge, MA: MIT Press, 1994.

Danielson, Dennis. *The First Copernican: Georg Joachim Rheticus and the Rise of the Copernican Revolution*. New York: Walker, 2006.

Dehio, G., and G. von Bezold. *Die kirchliche Baukunst des Abendlandes*, vol. 2. Stuttgart: Arnold Bergsträsser, 1901.

Didi-Huberman, Georges. *Devant le temps: Histoire de l'art et anachronisme des images*. Paris: Editions de Minuit, 2000.

———. *Fra Angelico: Dissemblance and Figuration*. Trans. Jane Marie Todd. Chicago: University of Chicago Press, 1995.

Donaldson-Evans, Lance K. *Love's Fatal Glance: A Study of Eye Imagery in the Poets of the École Lyonnaise*. University, MS: Romance Monographs, 1980.

Douët-d'Arcq, L., ed. *Chronique d'Enguerrand de Monstrelet*. Paris: Madame Veuve Jules Renouard, 1861.

Dronke, Peter. *A History of Twelfth-Century Western Philosophy*. Cambridge: Cambridge University Press, 1998.

Duffy, Eamon. *Marking the Hours: English People and Their Prayers, 1240–1570*. New Haven, CT: Yale University Press, 2006.

Dunlop, Anne. *Painted Palaces: The Rise of Secular Art in Early Renaissance Italy*. University Park: Penn State University Press, 2009.

Dupré, Sven. "Printing Practical Mathematics: Oronce Fine's *De speculo ustorio* between Paper and Craft." In Alexander Marr, ed., *The Worlds of Oronce Fine: Mathematics, Instruments and Print in Renaissance France*, 64–82. Donington, UK: Shaun Tyas, 2009.

Edgerton, Samuel. "*Mensurare temporalia facit geometria spiritualis*: Some 15th Century Italian Notions about When and Where the Annunciation Happened." In Irving Lavin and John Plummer, eds., *Studies in Late Medieval and Renaissance Painting in Honor of Millard Meiss*, 115–30. New York: NYU Press, 1977.

———. *The Renaissance Rediscovery of Linear Perspective*. New York: Basic Books, 1975.

Elkins, James. "The Case against Surface Geometry." *AH* 14, no. 2 (June 1991): 143–74.

———. *The Poetics of Perspective*. Ithaca, NY: Cornell University Press, 1994.

Elkins, James, and Robert Williams, eds. *Renaissance Theory*. New York: Routledge, 2008.

Evans, Michael. "The Geometry of the Mind." *Architectural Association Quarterly* 12, no. 4 (1980): 32–55.

———. "An Illustrated Fragment of Peraldus's *Summa* of Vice: Harleian MS 3244." *JWCI* 45 (1982): 14–68.

Falomir, Miguel. *Tintoretto*. Madrid: Museo Nacional del Prado, 2007.

Faraone, Christopher A. *Disappearing Acts: The Text, Image, and Medium of Ancient Greek Curative Amulets*. Forthcoming.

Favre, Antoine. *Les entretiens spirituels*. Ed. Lance K. Donaldson-Evans. Orig. 1602; Paris: Société des textes français modernes, 2002.

Félibien, André. *Conférences de l'Académie royale de peinture et de sculpture*. Paris: Frederic Leonard, 1669.

Ficino, Marsilio. *Commentary on Plato's Symposium on Love*. Trans. Sears Jayne. Dallas: Spring, 2000.

———. *De triplici vita (Three Books on Life)*. Ed. Carol Kaske and John Clark. Binghamton, NY: Medieval and Renaissance Texts and Studies, 1989.

———. *Opera omnia*. Turin: Bottega d'Erasmo, 1962 [facs. of Basel, 1576].

———. *Platonic Theology*. Ed. and trans. Michael J. B. Allen and James Hankins. Cambridge, MA: Harvard University Press, 2006.

Fideler, David R., ed., and Kenneth Sylvan Guthrie, trans. *The Pythagorean Sourcebook and Library*. Grand Rapids, MI: Phanes, 1987.

Filarete (Andrea Averlino). *Trattato di architettura*. Ed. Anna Maria Finoli and Liliana Gras. Milan: Il Polifilo, 1972.

Firenzuola, Agnolo. *On the Beauty of Women*. Trans. Konrad Eisenbichler and Jacqueline Murray. Philadelphia: University of Pennsylvania Press, 1992.

Foister, Susan, and Delphine Cool, eds. *Investigating Jan van Eyck*. Turnhout, Belgium: Brepols, 2000.

Frangenberg, Thomas. "*Auditus visu praestantior:* Comparisons of Hearing and Vision in Charles de Bovelles's *Liber de sensibus*." In Charles Burnett, Michael Fend, and Penelope Gouk, eds., *The Second Sense: Studies in Hearing and Musical Judgment from Antiquity to the Seventeenth Century*, 71–94. London: Warburg Institute, 1991.

Frankl, Paul. "The Secret of the Medieval Masons." *AB* 21, no. 1 (March 1945): 46–60.

Freedberg, David. *The Power of Images*. Chicago: University of Chicago Press, 1989.

Fried, Michael. *Absorption and Theatricality: Painting and Beholder in the Age of Diderot*. Chicago, University of Chicago Press, 1980.

Gadol, Joan Kelly. *Leon Battista Alberti, Universal Man of the Early Renaissance*. Chicago: University of Chicago Press, 1969.

Gage, Frances. "Giulio Mancini and Artist-Amateur Relations in Seventeenth-Century Roman Academies." In Peter M. Lukehart, ed., *The Accademia Seminars: The Accademia di San Luca in Rome, c. 1590–1635*, 247–87. Washington, DC: National Gallery of Art, 2009.

Galis, Diana. "Concealed Wisdom: Renaissance Hieroglyphic and Lorenzo Lotto's Bergamo *Intarsie*." *AB* 72, no. 3 (September 1980): 363–75.

Gallo, Filenio. *Rime di Filenio Gallo*. Ed. Maria Antonietta Grignani. Florence: Leo S. Olschki, 1973.

Gantheret, François. "Du coin de l'oeil." *Nouvelle Revue de Psychanalyse* 44 (Autumn 1991): 107–25.

Garb, Tamar. *Bodies of Modernity: Figure and Flesh in Fin-de-siècle France*. London: Thames and Hudson, 1998.

Garnerius of Rochefort (Garnerius Lingonensis). *Sermones*. PL 205. Paris: Migne, 1855.

Geissen, Angelo, and Wolfram Weiser. *Katalog Alexandrinischer Kaisermünzen der sammlung des Instituts für Altertumskunde der Universität zu Köln, Papyrologica Coloniensa*, vol. 5. Opladen, Germany: Westdeutscher Verlag, 1983.

Gerhoh von Reichersberg (Gerhohus Reicherspergensis). *Syntagma de Henricis IV et V et Gregorio VII*. PL 194. Paris: Migne, 1855.

Gibson, Margaret. *Boethius: His Life, Thought, and Influence*. Oxford: Blackwell, 1981.

Gilbert, Creighton, ed. *Italian Art, 1400–1600: Sources and Documents in the History of Art*. Englewood Cliffs, NJ: Prentice-Hall, 1980.

———. "On Subject and Not-Subject in Italian Renaissance Pictures." *AB* 34, no. 3 (September 1952): 202–16.

Gill, Joseph. *The Council of Florence*. Cambridge: Cambridge University Press, 1959.

Girard, René. *A Theater of Envy: William Shakespeare*. Oxford: Oxford University Press, 1991.

Gnädinger, Franz. "The Case against Surface Geometry: Reply to Elkins." *AH* 15, no. 3 (September 1992): 399–402.

Goffen, Rona. *Giovanni Bellini*. New Haven, CT: Yale University Press, 1989.

———. "Icon and Vision: Giovanni Bellini's Half-Length Madonnas." *AB* 57, no. 4 (December 1975): 487–518.

———. " 'Nostra conversatio in caelis est': Observations on the *sacra conversazione* in the Trecento." *AB* 61, no. 2 (June 1979): 198–222.

———. *Renaissance Rivals: Michelangelo, Leonardo, Raphael, Titian*. New Haven, CT: Yale University Press, 2004.

———. "Sex, Space, and Social History in Titian's *Venus of Urbino*." In Rona Goffen, ed., *Titian's "Venus of Urbino."* Cambridge: Cambridge University Press, 1997.

Goldmann, Lucien. *Marxisme et sciences humaines*. Paris: Gallimard, 1970.

Goldthwaite, Richard. "The Florentine Palace as Domestic Architecture." *American Historical Review* 77 (1972): 867–1012.

Golsenne, Thomas, and Bertrand Prévost. Introduction to Alberti, *La peinture*, ed. and trans. Golsenne and Prévost. Paris: Seuil, 2004.

Gombrich, Ernst. "Botticelli's Mythologies: A Study in the Neoplatonic Symbolism of His Circle." *JWCI* 8 (1945): 7–60.

———. "A Plea for Pluralism." *American Art Journal* 3, no. 1 (Spring 1971): 83–87.

———. *The Sense of Order: A Study in the Psychology of Decorative Art*. Ithaca, NY: Cornell University Press, 1979.

Grafton, Anthony. *Leon Battista Alberti: Master Builder of the Italian Renaissance. New York*: Hill and Wang, 2000.

Greenblatt, Stephen. "Murdering Peasants: Status, Genre, and the Representation of Rebellion." *Representations* 1 (February 1983): 1–29.

Griffiths, J. Gwyn. "Aristotle and Herodotus on Egyptian Geometry." *Classical Review*, n.s. 2, no. 1 (March 1952): 10–11.

Gronau, H. "Zur Geschichte des van Eyckschen Flügelältarchens in der Dresdner Galerie." *Kunstchronik*, n.f. 27 (1916): col. 445–46.

Groves, John. *A Greek and English Dictionary*. Boston: Hilliard, Gray, 1834.

Guido da Pisa. *I fatti d'Enea di frate Guido da Pisa.* Ed. Francesco Foffano. Florence: Sansoni, 1900.

Halfwassen, Jens. *Hegel und der spätantike Neoplatonismus*. Hamburg: Meiner, 1999.

Hamann, Florian. *Das Siegel der Ewigkeit: Universalwissenschaft und Konziliarismus bei Heymericus de Campo.* Münster: Aschendorff, 2006.

———. "Koran und Konziliarismus: Anmerkungen zum Verhältnis von Heymericus de Campo und Nikolaus von Kues." *Vivarium* 43, no. 2 (2005): 276–91.

Hamburger, Jeffrey. *The Rothschild Canticles: Art and Mysticism in Flanders and the Rhineland*. New Haven, CT: Yale University Press, 1990.

Harries, Karsten. *Infinity and Perspective*. Cambridge, MA: MIT Press, 2001.

Hauser, Arnold. *Social History of Art: Renaissance, Mannerism, Baroque.* London: Routledge, 1951.

Heidel, W. A. "The Pythagoreans and Greek Mathematics." *American Journal of Philology* 61, no. 1 (1940): 1–33.

Hendrix, John. *Platonic Architectonics: Platonic Philosophies and the Visual Arts*. New York: Peter Lang, 2004.

Herodotus. *The History*. Trans. G. C. Macaulay, ed. Donald Lateiner. New York: Barnes and Noble, 2004.

Hersey, George L. *Pythagorean Palaces: Magic and Architecture in the Italian Renaissance*. Ithaca, NY: Cornell University Press.

Hesiod. *The Homeric Hymns and Homerica*. Trans. Hugh G. Evelyn-White. Cambridge, MA: Harvard University Press, 1914.

Heymericus de Campo. *Opera selecta*, vol. 1. Ed. and trans. Ruedi Imbach and Pascal Ladner. Freiburg, Germany: Universitätsverlag, 2001.

Hilary of Poitiers (Hilarius Pictaviensis). *Liber de synodis seu fide orientalium*. PL 10. Paris: Migne, 1845.

Hillgarth, J. N. *Ramon Lull and Lullism in Fourteenth-Century France*. Oxford: Clarendon, 1971.

Hoenen, Maarten. "Academics and Intellectual Life in the Low Countries: The University Career of Heymeric de Campo (†1460)." *Recherches de Théologie Ancienne et Médiévale* 61 (1994): 173–209.

Holmes, George. "How the Medici Became the Pope's Bankers." In Nicolai Rubinstein, ed., *Florentine Studies*, 357–80. London, Faber and Faber, 1968.

Holmes, Megan. *Fra Filippo Lippi, the Carmelite Painter*. New Haven, CT: Yale University Press, 1999.

Holt, Elizabeth Gilmore, ed. *A Documentary History of Art*, vol. 1, *The Middle Ages and the Renaissance*. New York: Doubleday, 1957.

Homann, Frederick A. *Practical Geometry [Practica geometria], Attributed to Hugh of St. Victor.* Milwaukee: Marquette University Press, 1991.

Honorius Augustodunensis. *De philosophia mundi*. PL 172. Paris: Migne, 1854.

Hopkins, Jasper. *A Concise Instroduction to the Philosophy of Nicholas of Cusa*. Minneapolis: Arthur J. Banning, 1986.

Hood, William. *Fra Angelico at San Marco*. New Haven, CT: Yale University Press, 1993.

Hugh of St. Victor. *Geometriae practica*. Ed. Roger Baron. Notre Dame, IN: University of Notre Dame Press, 1966.

Huizinga, Johan. *Autumn of the Middle Ages*. Trans. Rodney J. Payton and Ulrich Mammitzsch. Chicago: University of Chicago Press, 1996.

Humfrey, Peter, ed. *The Cambridge Companion to Giovanni Bellini*. Cambridge: Cambridge University Press, 2008.

Hvidberg-Hansen, F. O. *La déesse TNT: Une étude sur la réligion canaanéo-punique*. Copenhagen: Gad, 1982.

Iamblichus (attributed). *The Theology of Arithmetic*. Trans. Robin Waterfield. Grand Rapids, MI: Phanes, 1988.

Imbach, Ruedi. "Das Centheologicon des Heymericus de Campo und die darin enthaltenen Cusanus-Reminiszenzen: Hinweise und Materialien." *Traditio* 39 (1983): 466–77.

Inman, Thomas. *Ancient Faiths Embodied in Ancient Names*. London: S.P., 1868–69.

Ivins, William, ed. *On the Rationalization of Sight, with an Examination of Three Renaissance Texts on Perspective*. New York: Da Capo, 1973.

Jacobsen, Mark. "The Meaning of Mantegna's Battle of Sea Monsters." *AB* 64, no. 4 (December 1982): 623–29.

Jacobson, Karen, ed. *The French Renaissance in Prints from the Bibliothèque Nationale de France*. Los Angeles: Grunwald Center for the Graphic Arts, 1994.

Jeanneret, Michel. *Perpetuum mobile*. Paris: Macula, 1997.

John of Damascus. *Three Treatises on the Divine Images*. Ed. Andrew Louth. Crestwood, NY: St. Vladimir's Seminary Press, 2003.

Jones, W. R. "Lollards and Images: The Defense of Religious Art in Later Medieval England." *Journal of the History of Ideas* 34, no. 1 (January-March 1973): 27–50.

Joost-Gaugier, Christiane. *Measuring Heaven: Pythagoras and His Influence on Thought and Art in Antiquity and the Middle Ages.* Ithaca, NY: Cornell University Press, 2006.

Kandinsky, Wassily. *Complete Writings on Art.* Ed. Kenneth C. Lindsay and Peter Vergo. New York: Da Capo, 1982.

———. Wassily Kandinsky papers, 1921–37. Getty Research Institute. Research Library, accession no. 850910, series I.A. Notes for Semester I, 1925–1931, box 1, folder 16.

Karpati, Zoltan. *The Alchemy of Beauty: Parmigianino Drawings and Prints.* Exh. cat. Budapest: Museum of Fine Arts, 2009.

Kavaler, Ethan Matt. "Renaissance Gothic: Pictures of Geometry and Narratives of Ornament." *AH* 29 (2006): 1–46.

Keller, Harald. *The Renaissance in Italy: Painting, Sculpture, Architecture. New York*: H. N. Abrams, 1969.

Kemp, Martin. "Geometrical Perspective from Brunelleschi to Desargues: A Pictorial Means or an Intellectual End?" *Proceedings of the British Academy* 70 (1984): 89–132.

———. "Science, Non-science, and Nonsense: The interpretation of Brunelleschi's Perspective." *AH* 1, no. 2 (1978): 134–61.

Kent, Francis W. *Household and Lineage in Renaissance Florence: The Family Life of the Capponi, Ginori, and Rucellai.* Princeton, NJ: Princeton University Press, 1977.

Kern, Peter. *Trinität, Maria, Inkarnation.* Berlin: E. Schmidt, 1971.

Klein, Robert. "Pomponius Gauricus on Perspective." In *Form and Meaning*, trans. Madeline Jay and Leon Wieseltier, 102–28. New York: Viking, 1979.

Klibansky, Raymond, Erwin Panofsky, and Fritz Saxl. *Saturn and Melancholy: Studies in the History of Natural Philosophy, Religion, and Art.* London: Nelson, 1964.

Koerner, Joseph. *The Moment of Self-Portraiture in German Renaissance Art.* Chicago: University of Chicago Press, 1993.

Krautheimer, Richard. *Lorenzo Ghiberti.* Princeton, NJ: Princeton University Press, 1970.

Kubovy, Michael. *The Psychology of Perspective and Renaissance Art*. Cambridge: Cambridge University Press, 1986.

Kuhn, Jehane R. "Measured Appearances: Documentation and Design in Early Perspective Drawing." *JWCI* 53 (1990): 114–32.

Kumler, Aden. *Translating Truth: Ambitious Images and Religious Knowledge in Late Medieval France and England.* New Haven and London: Yale University Press, 2011.

Lacan, Jacques. "Of the Gaze as Objet Petit a," in Jacques-Alain Miller, ed., Alan Sheridan. trans., *The Four Fundamental Concepts of Psychoanalysis,* 65-119. W. W. Norton and Company, 1998.

Landino, Cristoforo. *Proemio: Commento di Christophoro Landino fiorentino sopra la comedia di Dante Alighieri poeta fiorentino.* Venice: Bernardino Benali & Matthio di Parma, 1491.

Lawless, Catherine. "Myth, Ritual and Orthodoxy: Cosimo de' Medici and St Peter Martyr." *Cultural and Social History,* vol. 2, no. 3 (Sept. 2005), 273-300.

Lease, Emory B. "The Number Three, Mysterious, Mystic, Magic." *Classical Philology* 14, no. 1 (January 1919): 56–73.

Lecoq, Anne-Marie. *François Ier imaginaire: Symbolique et politique à l'aube de la Renaissance française*. Paris: Macula, 1987.

Lefèvre d'Étaples, Jacques. *De magia naturali.* Biblioteca Apostolica Vaticana MS Vat. Regin. Lat. 1115.

———, ed. *Introductio . . . in arithmecam divi Severini Boetii . . .* N.p., 1507.

Leutrat, Estelle. *Les débuts de la gravure sur cuivre en France: Lyon, 1520–1565*. Geneva: Droz, 2007.

Lille, Alain de. *De articulis Catholicae fidei*. PL 210. Paris: Migne, 1855.

———. "Sermo de Sphaera intelligibili." In Marie-Thérèse d'Alverny, ed., *Textes inédits avec une introduction sur sa vie et ses oeuvres*. Paris: Vrin, 1965.

Lindberg, David C. *Theories of Vision from al-Kindi to Kepler.* Chicago: University of Chicago Press, 1976.

Llull, Ramon. *Raimundi Lulli Opera Latina* 17. Ed. M. Pereira and T. Pindl-Büchel. Turnhout, Belgium: Brepols, 1989.

Loh, Maria. *Titian Remade: Repetition and the Transformation of Early Modern Italian Art.* Los Angeles: Getty Research Institute, 2007.

Lomazzo, Giovanni Paolo. *Trattato dell'arte de la pittura, scultura et architettura.* Milan: Paolo Gottardo Pontio, 1584.

Lynch, T. "The Geometric Body in Dürer's Engraving *Melencolia I.*" *JWCI* 45 (1982): 226–32.

Marguerite de Navarre. *Heptameron.* Ed. Renja Salminen. Geneva: Droz, 1991.

Marino, Eugenio. *Il "Diluvio" di Paolo Uccello in S. Maria Novella ed il Concilio di Firenze (1439–1443): Saggio d'iconoteologia storica.* Pistoia, Italy: Centro riviste della Provincia Romana, 1992.

Marr, Alexander, ed. *The Worlds of Oronce Fine: Mathematics, Instruments and Print in Renaissance France.* Donington, UK: Shaun Tyas, 2009.

Martines, Lauro. *Power and Imagination: City-States in Renaissance Italy*. New York: Alfred A. Knopf, 1979.

Massey, Lyle, ed. *The Treatise on Perspective: Published and Unpublished*. Washington, DC: National Gallery of Art, 2003.

Mayor, Hyatt. *Prints and People: A Social History of Printed Pictures.* New York: Metropolitan Museum of Art, 1971.

Mazzio, Carla. "The Three Dimensional Self: Mathematics, Melancholia, Drama." In David Glimp and Michelle R. Warren, eds., *The Arts of Calculation: Quantifying Thought in Early Modern Europe*, 39–65. New York: Palgrave, 2004.

McCarty, Willard. "The Shape of the Mirror: Metaphorical Catoptrics in Classical Literature." *Arethusa* 22, no. 2 (Fall 1989): 161–95.

McCracken, Charles. "Knowledge of the Soul." In Daniel Garber and Michael Ayers, eds., *The Cambridge History of Seventeenth-Century Philosophy*, 1:796–832. New York: Cambridge University Press, 1998.

McDermott, Peter L. "Nicholas of Cusa: Continuity and Conciliation at the Council of Basel." *Church History* 67, no. 2 (June 1998): 254–73.

McGinn, Bernard. "Cabalists and Christians: Reflections on Cabala in Medieval and Renaissance Thought." In Richard H. Popkin and Gordon M. Weiner, eds., *Jewish Christians and Christian Jews*. Dordrecht: Kluwer Academic, 1994.

———. "Theologians as Trinitarian Iconographers." In Jeffrey F. Hamburger and Anne-

Marie Bouché, eds., *The Mind's Eye: Art and Theological Argument in the Middle Ages*, 186–207. Princeton, NJ: Princeton University Press, 2006.

Meier, Christel. "Figura ad oculum demonstrativa: Visuelle Symbolik und verbale Diskursivitat nach Heymericus de Campo." In Nine Miedema and Rudolf Suntrup, eds., *Literatur, Geschichte, Literaturgeschichte*, 755–81. Frankfurt: Peter Lang, 2003.

Meiss, Millard. "An Early Altarpiece from the Cathedral of Florence." *Metropolitan Museum of Art Bulletin* n.s. 12, no. 10 (June 1954): 302–17.

———. *Painting in Florence and Siena after the Black Death.* Princeton, NJ: Princeton University Press, 1951.

Merriell, D. Juvenal. *To the Image of the Trinity: A Study of the Development of Aquinas' Teaching.* Toronto: Pontifical Institute of Medieval Studies, 1990.

Milburn, Erika. " 'D'Invidia e d'Amor figlia sì ria': Jealousy and the Italian Renaissance Lyric." *Modern Language Review* 97, no. 3 (July 2002): 577–91.

Molho, Anthony. "The Brancacci Chapel: Studies in Its Iconography and History." *JWCI* 40 (1977): 50–98, 322.

Mone, Franz Joseph, ed. *Lateinische Hymnen Mittelalters.* Freiburg im Breisgau, Germany: Herder'sche Verlagshandlung, 1854.

Montaigne, Michel de. *Les Essais*. Ed. Pierre Villey. Paris: Félix Alcan, 1923.

More, Thomas. *Utopia.* Trans. Clarence Miller. New Haven, CT: Yale University Press, 2001.

Morgan, Nigel. "The Coronation of the Virgin by the Trinity and Other Texts and Images of the Glorification of Mary in Fifteenth-Century England." In Nicholas Rogers, ed., *England in the Fifteenth Century.* Stamford, Lincolnshire, UK: P. Watkins, 1994.

Münster, Sebastian. *Cosmographiae Universalis* . . . Basel: Heinrich Petri, 1550.

———. *Rudimenta mathematica*. Basel: Petri, 1551.

Murdoch, John E. *Album of Science: Antiquity and the Middle Ages*. New York: Charles Scribner's Sons, 1984.

Nagel, Alexander, and Christopher S. Wood. *Anachronic Renaissance*. Brooklyn, NY: Zone, 2010.

Naville, Edouard. "The Origin of Egyptian Civilization." *Journal of the Royal Anthropological Institute of Great Britain and Ireland* 37 (January-June 1907): 201–14.

Neidhardt, Uta, and Christoph Schölzel. "Jan van Eyck's *Dresden Triptych.*" In Susan Foister and Delphine Cool, eds., *Investigating Jan van Eyck*, 25–41. Turnhout, Belgium: Brepols, 2000.

Newman, Barbara. *God and the Goddesses: Vision, Poetry, and Belief in the Middle Ages.* Philadelphia: University of Pennsylvania Press, 2003.

Nicholas of Cusa. *Cribratio Alkorani*. In *Nicholas of Cusa's "De pace fidei" and "Cribratio Alkorani."* Trans. Jasper Hopkins. Minneapolis: Arthur J. Banning, 1994.

———. *De docta ignorantia* (*On Learned Ignorance*). Trans. Jasper Hopkins. Minneapolis: Arthur J. Banning, 1981.

———. *Metaphysical Speculations: Six Latin Texts*, vols. 1–2. Ed. and trans. Jasper Hopkins. Minneapolis: Arthur J. Banning, 1998–2000.

———. *Nicholas of Cusa on Wisdom and Knowledge.* Ed. and trans. Jasper Hopkins. Minneapolis: Arthur J. Banning, 1996.

Nicomachus of Gerasa. *Introduction to Arithmetic.* Trans. Martin d'Ooge. New York: Macmillan, 1926.

Nieuwenhove, Rik van. *Jan van Ruusbroec, Mystical Theologian of the Trinity*. Notre Dame, IN: University of Notre Dame Press, 2003.

Nodes, Daniel. "Homeric Allegory in Egidio of Viterbo's *Reflections on the Human Soul*." *Recherches de Théologie et Philosophie Médiévales* 65, no. 2 (1998): 320–32.

Norman, Diana. " 'In the beginning was the Word': An Altarpiece by Ambrogio Lorenzetti for the Augustinian Hermits of Massa Marittima." *Zeitschrift für Kunstgeschichte* 58, no. 4 (1995): 478–503.

Offner, Richard. "The Barberini Panels and Their Painter." In *Medieval Studies in Honor of A. Kingsley Porter*, 1:205–53. Cambridge, MA: Harvard University Press, 1939.

Olah, Iva. "Tarocchi and the Depiction of Christian Princely Virtues in the Schifanoia Frescoes." M.A. thesis, Carleton University, 2005.

O'Malley, J. W. "Egidio da Viterbo and Renaissance Rome." In *Egidio da Viterbo, O.S.A e il suo tempo: Atti del V Convegno dell'Istituto Storico Agostiniano, Roma*, 67–84. Rome: Analecta Augustiniana, 1983.

Oresme, Nicole. *Nicole Oresme and the Medieval Geometry of Qualities and Motions: A Treatise on the Uniformity and Difformity of Intensities Known as "Tractatus de configurationibus qualitatum et motuum."* Ed. and trans. Marshall Clagett. Madison: University of Wisconsin Press, 1968.

Orlandi, Stefano. *La Biblioteca di S. Maria Novella in Firenze dal sec. XIV al sec. XIX*. Florence: Ed. Il Rosario, 1952.

Othlonus, Monachus S. Emmerami. *Dialogus de tribus quaestionibus*. PL 146. Paris: Migne, 1853.

Pacioli, Luca. *De divina proportione*. Orig. Venice: Paganino Paganini, 1509; Milan: Abrosiana, 1956.

———. *Summa de arithmetica, geometria, proportione e proportionalità*. Toscolano, Italy: Paganino Paganini, 1523.

Paleotti, Gabriele. *Discorso intorno alle immagini sacre e profane*. Bologna: Forni, 1990 (facs. of Bologna, 1582).

Palissy, Bernard. *Œuvres*. Ed. Fillon. Niort, France: L. Clouzot, 1888.

———. *Recette veritable* [*par laquelle tous les hommes de la France pourront apprendre a multiplier et augmenter leurs tresors*]. Ed. Frank Lestringant and Christian Barataud. Paris: Macula, 1996 [1563].

Panofsky, Erwin. *Hercules am Scheideweg*. Studien der Bibliothek Warburg 18. Leipzig: Teubner, 1930.

———. *Meaning in the Visual Arts*. Chicago: University of Chicago Press, 1993.

———. *Renaissance and Renascences in Western Art*. New York: Harper and Row, 1960.

Paoletti, John T., and Gary M. Radke. *Art in Renaissance Italy*. Upper Saddle River, NJ: Pearson Prentice Hall, 2005.

Paroncchi, Alessandro. *Studi su la dolce prospettiva*. Milan: A. Martello, 1964.

Patrizi, Francesco. *Nova de universis philosophia libris quinquaginta comprehensa*. Venice: Roberto Meietti, 1593.

Patschovsky, Alexander. *Die Bildwelt der Diagramme Joachims von Fiore: Zur Medialität religiös-politischer Programme im Mittelalter*. Ostfildern, Germany: Thorbecke, 2003.

Pèlerin, Jean. *De artificiali perspectiva*. Toul: Petri Jacobi, 1505; 2nd ed. 1509.

Perisauli, Faustino. *De Triumpho stultitiae*. Ed. A. Viviani and G. Fabbri. Florence: Il Fauno Editore, 1963.

Pico della Mirandola, Giovanni. *Conclusiones sive Theses DCCCC.* Ed. Bohdan Kieszowski. Travaux d'Humanisme et Renaissance 131. Geneva: Droz, 1973.

———. *On the Imagination.* Ed. and trans. H. Caplan. New Haven, CT: Yale University Press, 1930.

Piero della Francesca. *De prospectiva pingendi.* Ed. G. Nicco-Fasola. Florence: Le Lettere, 2005.

Pierozzi, Letizia, and Jean-Marc Mandosio. "L'interprétation alchimique de deux travaux d'Hercule dans le *De Magia naturali* par Lefèvre d'Étaples." In *Chrysopoeia*, 5:191–264. Paris: Société d'etude de l'histoire de l'alchimie, 1992–96.

Plato. *Collected Dialogues of Plato, including the Letters.* Ed. Edith Hamilton. New York: Pantheon, 1961.

Plutarch. *Lives*, vol. 5. Trans. Bernadotte Perrin. Cambridge, MA: Harvard University Press, 1961.

Plutarch. *Moralia*, vol. 5, trans. Frank Cole Babbitt, and vol. 8, trans. Paul Clement. Cambridge, MA: Harvard University Press, 1928–69.

Poggi, G. *Il Duomo di Firenze*. Orig. Berlin: Bruno Cassirer, 1909; Florence: Edizioni Medicea, 1988.

Pon, Lisa. *Raphael, Dürer, and Marcantonio Raimondi.* New Haven, CT: Yale University Press, 2004.

Popham, Arthur E. *Catalogue of the Drawings of Parmigianino*. New Haven, CT: Yale University Press, 1971.

Popper, Karl R. *The Open Society and Its Enemies.* Princeton, NJ: Princeton University Press, 1963.

Postel, Guillaume. *Des admirables secrets des nombres platoniciens.* Ed. Jean-Pierre Brach. Paris: Vrin, 2001.

Proclus. *Commentaire sur la République*, vol. 2. Trans. A. J. Festugière. Paris: Vrin, 1970.

———. *A Commentary on the First Book of Euclid's "Elements."* Trans. Glenn Morrow. Princeton, NJ: Princeton University Press, 1970.

———. *In Platonis Rem Publicam Commentarii*, vol. 2. Leipzig: B. G. Teubner, 1901.

Psellos, Michael. "De daemonibus." In *Index eorvm, qvae hoc in libro habentvr: Iavmlichus de mysteriis . . .*, 50v.Venice: Aldus Manutius and Andrea Soceri, 1516.

Putscher, Marielene. "Das *Buch der Heiligen Dreifaltigkeit* und seine Bilder in Handscriften des 15: Jahrhunderts." In Christoph Meinl, ed., *Die Alchemie in der europäischen Kultur- und Wissenschaftsgeschichte*, 151–78. Wolfenbütteler Forschungen 32. Wiesbaden: Otto Harrassowitz, 1986.

Puttfarken, Thomas. *The Discovery of Pictorial Composition: Theories of Visual Order in Painting, 1400–1800*. New Haven, CT: Yale University Press, 2000.

Quiviger, François. "Benedetto Varchi and the Visual Arts." *JWCI* 50 (1987): 219–24.

Randolph, Adrian. "Renaissance Genderscapes." In Joan Hartman and Adele Seeff, eds., *Attending to Early Modern Women: Structures and Subjectivities*, 21–49. Newark: University of Delaware Press / London: Associated University Presses, 2006.

———. "Les seuils de l'expérience: L'*Annonciation* de Crivelli et le genre de la peinture." *Perspective* 4 (2007): 649–78.

Raw, Barbara Catherine. *Trinity and Incarnation in Anglo-Saxon Art and Thought*. Cambridge: Cambridge University Press, 1997.

Reinhardt, Klaus. "Werke des Heymericus de Campo (†1460) im Codex Cusanus." *Traditio* 50 (1995): 295–310.

Renaudet, Augustin. *Préréforme et humanisme à Paris pendant les premières guerres d'Italie, 1494–1517*. Paris: Librairie d'Argences, 1953.

Reuchlin, Johannes. *On the Art of the Kabbalah (De arte cabbalistica)*. Trans. Martin and Sarah Goodman. Lincoln: University of Nebraska Press, 1983.

Rice, Eugene F., Jr. "The *De magia naturali* of Jacques Lefèvre d'Étaples." In Edward P. Mahoney, ed., *Philosophy and Humanism: Renaissance Essays in Honor of Paul Oskar Kristeller*, 19–29. New York: Columbia University Press, 1976.

Richardson, Carol. *Locating Renaissance Art.* London: Yale University Press, 2007.

Rocke, Michael. *Forbidden Friendships: Homosexuality and Male Culture in Renaissance Florence*. Oxford: Oxford University Press, 1996.

Rosenberg, Charles M. *The Este Monuments and Urban Development in Renaissance Ferrara*. Cambridge: Cambridge University Press, 1997.

———. "The Iconography of the Sala degli Stucchi in the Palazzo Schifanoia in Ferrara." *AB* 61 (1979): 377–83.

Rowland, Ingrid D. "Render unto Caesar the Things Which Are Caesar's: Humanism and the Arts in the Patronage of Agostino." *Renaissance Quarterly* 39, no. 4 (Winter 1986): 673–730.

Rupert of Deutz. *De Trinitate et operibus eius.* PL 167. Paris: Migne, 1854.

Russell, H. Diane. *Eva/Ave: Woman in Renaissance and Baroque Prints*. Washington, DC: National Gallery of Art, 1990.

Sanders, P. M. "Charles de Bovelles's Treatise on the Regular Polyhedra (Paris, 1511)." *Annals of Science* 41 (1984): 513–66.

Sanderson, Warren. "Geometry on a Carolingian Wall." In Nancy Wu, ed., *Ad Quadratum: The Practical Application of Geometry in Medieval Architecture*, 11–24. Aldershot, UK: Ashgate, 2002.

Sanpaolesi, Piero. *Brunelleschi.* Milan: Club del libro, 1962.

Savonarola, Girolamo. *Prediche italiane ai Fiorentini.* Ed. Francesco Cognasso. Perugia: La Nuova Italia, 1930.

Scève, Maurice. *Delie.* Ed. and trans. I. D. McFarlane. Cambridge: Cambridge University Press, 1966.

Schefer, Jean Louis. *Paolo Uccello, le déluge.* Paris: P.O.L., 1999.

Schröder, Eberhard. *Dürer Kunst und Geometrie.* Berlin: Akademie-Verlag, 1980.

Schröder, Richard. *Die verborgene Geometrie in Raffaels "Schule von Athen."* Munich: R. Oldenbourg, 1994.

Sereni, Emilio. *History of the Italian Agricultural Landscape.* Trans. R. Burr Litchfield. Princeton, NJ: Princeton University Press, 1997.

Seuse, Heinrich. *Deutsche Schriften.* Ed. K. Bihlmeyer. Stuttgart: W. Kohlhammer, 1907.

Shahn, Ben. *The Shape of Content*. Cambridge, MA: Harvard University Press, 1957.

Shearman, John. "The Collections of the Younger Branch of the Medici." *Burlington Magazine* 117, no. 862 (January 1975).

———. Review of Janet Cox-Rearick, *Pontormo's Drawings. AB* 154, no. 2 (June 1972): 209–12.

Sisko, John E. "Space, Time and Phantasms in Aristotle, *De memoria* 2, 452B7–25." *Classical Quarterly* 47, no. 1 (1997): 167–75.

Smith, Kathryn Ann. *Art, Identity and Devotion in Fourteenth-Century England: Three Women and Their Books of Hours*. London: British Library, 2003.

Sohm, Philip. "Dürer's *Melencolia I*: The Limits of Knowledge." *Studies in the History of Art* 9 (1980): 13–32.

Soufflin, Pierre. "La *Geometria practica* dans les *Ludi rerum mathematicarum*." *Albertiana* 1 (1998): 87–104.

Stechow, Wolfgang. *Northern Renaissance Art, 1400–1600: Sources and Documents.* Evanston, IL: Northwestern University Press, 1999.

Sticca, Sandro. *The Planctus Mariae in the Dramatic Tradition of the Middle Ages.* Athens: University of Georgia Press, 1988.

Stock, Brian. *Augustine the Reader: Meditation, Self-Knowledge, and the Ethics of Interpretation*. Cambridge, MA: Harvard University Press, 1996.

Strickland, Carol, and John Boswell. *The Annotated Mona Lisa: A Crash Course in Art History from Prehistoric to Post-modern.* Kansas City, MO: Andrews and McMeel, 1992.

Stuhlfauth, Georg. *Das Dreieck: Die Geschichte eines religiösen Symbols.* Stuttgart: Kohlhammer, 1937.

Stuloff, Nikolaus. "Mathematische Tradition in Byzanz und ihr Fortleben bei Nikolaus von Kues." *Mitteilungen und Forschungsbeiträge der Cusanus-gesellschaft* 4 (1964): 420–36.

Summers, David. *Michelangelo and the Language of Art*. Princeton, NJ: Princeton University Press, 1981.

Szulakowska, Urszula. *The Alchemy of Light: Geometry and Optics in Late Renaissance Alchemical Illustration.* Leiden: Brill, 2000.

Tacitus. *The Histories.* Trans. Clifford H. Moore. London: W. Heinemann / New York: G. P. Putnam's Sons, 1925–37.

Talvacchia, Bette. *Taking Positions: On the Erotic in Renaissance Culture*. Princeton, NJ: Princeton University Press, 2001.

Tarán, Leonardo. "Proclus and the Old Academy." In *Collected Papers*, 564–624. Boston: Brill, 2001.

Tavernor, Robert. *On Alberti and the Art of Building.* New Haven, CT: Yale University Press, 1998.

Terry, Allie. "Meraviglia on Stage: Dionysian Visual Rhetoric and Cross-Cultural Communication at the Council of Florence." *Journal of Religion and Theatre* 6, no. 2 (Fall 2007): 94–109, available at http://www.rtjournal.org/vol__6/no__2/terry.html (accessed September 28, 2010).

———. "Politics on the Cloister Walls: Fra Angelico and His Humanist Observers at San Marco." PhD diss., University of Chicago, 2005.

Testi, Maria Laura Cristiani. "La Trinità di Masaccio, Unità di Cosmo." *Critica d'Arte*, ser. 4, 50, no. 4 (1985): 109–12.

Theiss, Peter, and Otto-Joachim Grüsser. "Vision and Cognition in the Natural Philosophy of Albert the Great (Albertus Magnus)." *Documenta Ophthalmologica* 86 (1994): 123–51.

Thenaud, Jean. *Cabale metrifiee.* BNF ms. fr. 882.

———. *Troys resolutions et sentences.* Oesterreichische Nationalbibliothek Cod. 2645. 3v.

Thiofrid of Efternach (Thiofridus Efternacensis). *Flores epitaphii sanctorum.* PL 157. Paris: Migne, 1854.

Tolan, John. *Petrus Alfonsi and His Medieval Readers.* Gainesville: University Press of Florida, 1993.

Tory, Geoffroy. *Champ fleury*. Orig. Paris, 1529; Yorkshire, UK: S.R. Publishers, 1970.

Traversari, Ambrogio. *Latinae epistolae . . .* Bologna: Forni, 1968.

Trinkaus, Charles. "Protagoras in the Renaissance: An Exploration." In Edward P. Mahoney, ed., *Philosophy and Humanism: Renaissance Essays in Honor of Paul Oskar Kristeller*, 190–213. New York: Columbia University Press, 1976.

Turner, A. Richard. *The Vision of Landscape in Renaissance Italy.* Princeton, NJ: Princeton University Press, 1966.

Valois, Noël. "Fra Angelico et le Cardinal Jean de Torquemada." Société Nationale des Antiquaires de France, ed., *Centénaire 1804–1904: Recueil de Mémoires,* 461–70. Paris: C. Klincksieck, 1904.

Varchi, Benedetto. *Opere.* Trieste, Italy: Sezione letterario-artistica del Lloyd austriaco, 1858–59.

Vasari. *Le vite dei più eccellenti pittori, scultori e architetti.* Orig. 1550; Turin: Einaudi, 1986.

———. *Le vite dei più eccellenti pittori, scultori e architetti.* Ed. Licia Ragghianti. Orig. 1568; Milan: Rizzoli, 1942.

Vergil, Polydore. *On Discovery (De inventoribus rerum).* Ed. Brian P. Copenhaver. Orig. 1470; Cambridge, MA: Harvard University Press, 2002.

Vespucci, Amerigo, *The Letters of Amerigo Vespucci, and Other Documents Illustrative of His Career.* Ed. and trans. Clements Markham. London : Hakluyt Society, 1894.

Victor, Joseph M. *Charles de Bovelles, 1479–1553: An Intellectual Biography.* Geneva: Droz, 1978.

Vitruvius. *De architectura libri decem.* Trans. Cesare Cesariano. Como, Italy: Gottardo da Ponte, 1521.

———. *Ten Books on Architecture.* Trans. Ingrid D. Rowland. Cambridge: Cambridge University Press, 1999.

Vives, Juan Luis. *Education of a Christian Woman: A Sixteenth-Century Manual.* Ed. and trans. Charles Fantazzi. Chicago: University of Chicago Press, 2000.

———. *The Passions of the Soul.* Ed. and trans. Carlos Noreña. Lewiston, NY: Edwin Mellen, 1990.

Weddigen, Erasmus. "Venus Vulkan und Mars: Die Inquisition der Informatik." In *Jacomo Tentor F.: Myzelien zur Tintoretto-Forschung.* Munich: Scaneg, 2000.

———. "The Works of Tintoretto: Sewn, Designed, Patched, and Cut—The Uncertainty of Canvas Measurements." In Miguel Falomir, ed., *Jacopo Tintoretto: Actas del Congreso Internacional*, 151–64. Madrid: Museo Nacional del Prado, 2009.

Weiss, Peg. *Kandinsky in Munich: The Formative Jugendstil Years*. Princeton, NJ: Princeton University Press, 1979.

Welliver, Warman. "The Symbolic Architecture of Domenico Veneziano and Piero della Francesca." *Art Quarterly* 36 (1973): 1–30.

White, John. *The Birth and Rebirth of Pictorial Space.* London: Faber and Faber, 1957.

Williamson, Beth. *The Madonna of Humility: Development, Dissemination and Reception, c. 1340–1400.* Woodbridge, UK: Boydell, 2009.

Wind, Edgar. *Bellini's Feast of the Gods: A Study in Venetian Humanism*. Cambridge, MA: Harvard University Press, 1948.

———. *The Eloquence of Symbols: Studies in Humanist Art.* Oxford: Clarendon, 1983.

———. *Pagan Mysteries in the Renaissance*. New Haven, CT: Yale University Press, 1958.

Winn, Colette. "La symbolique du regard dans *Les Angoysses douloureuses qui procèdent d'amours*." *Orbis Litterarum* 40, no. 3 (1985): 207–21.

Winzinger, Franz. "Albrecht Dürers Münchener Selbstbildnis." *Zeitschrift für Kunstwissenschaft* 8 (1954): 43–64.

Witelo. *Vitellionis mathematici doctissimi (Peri optikes)*. Nuremberg: Hans Peterlein, 1535.

———. *Witelonis perspectivae Liber secundus et Liber tertius.* Ed. and trans. Sabetai Ungaru. Studia Copernicana 28. Wrocław, Poland: Ossolineum, 1991.

Wohl, Hellmut. *The Paintings of Domenico Veneziano, ca. 1410–1461: A Study in Florentine Art of the Early Renaissance*. New York: New York University Press, 1980.

Young, Philip H. "The Cypriot Aphrodite Cult: Paphos, Rantidi, and Saint Barnabas." *Journal of Near Eastern Studies* 64, no. 1 (2005): 23–44.

Zaitsev, Evgeny A. "The Meaning of Early Medieval Geometry: From Euclid and Surveyors' Manuals to Christian Philosophy." *Isis* 90, no. 3 (September 1999): 522–53.

Zannini, Gian Ludovico Masetti. "Rivalità e lavoro di incisori nelle botteghe Lafréry-Duchet e de la Vacherie." In *Les fondations nationales dans la Rome pontificale*, 547–66. Rome: École française de Rome, 1981.

Zerner, Henri. "L'eau-forte à Fontainebleau: Le rôle de Fantuzzi." *Art de France* 4 (1964): 70–85.

———. *The School of Fontainebleau: Etchings and Engravings.* Ed. Stanley Baron. London: Thames and Hudson, 1969.

Zöckler, Otto. "Religiöse Sinnbilder aus vorchristlicher und christlicher Zeit," pt. 3, "Trinitäts-Symbole." *Der Beweis des Glaubens* n.f. 2 (1881): 289–301.

Zorach, Rebecca. *Blood, Milk, Ink, Gold: Abundance and Excess in the French Renaissance.* Chicago: University of Chicago Press, 2005.

———. "Love, Truth, Orthodoxy, Reticence, or What Edgar Wind Didn't See in Botticelli's *Primavera*." *Critical Inquiry* 34, no. 1 (2007): 190–224.

———. "Meditation, Idolatry, Mathematics: The Printed Image in Europe around 1500." in Michael Cole and Rebecca Zorach, eds., *The Idol in the Age of Art: Objects, Devotions, and the Early Modern World*, 317–42. Farnham, UK: Ashgate, 2009.

———. "Renaissance Theory: A Selective Introduction." In James Elkins and Robert Williams, eds., *Renaissance Theory*, 3–36. New York: Routledge, 2008.

———, ed. *The Virtual Tourist in Renaissance Rome: Printing and Collecting the Speculum Romanae Magnificentiae*. Chicago: University of Chicago Library, 2008.

Zschelletzschky, Herbert. *Die "drei gottlosen Maler" von Nürnberg.* Leipzig: Seemann, 1975.

Index